Geographies of Tourism and Global Change

Series Editors

Dieter K. Müller, Department of Geography, Umeå University, Umeå, Sweden

Jarkko Saarinen, Geography Research Unit, University of Oulu, Oulu, Finland

Carolin Funck, Graduate School of Integrated Arts and Sciences, Higashi-Hiroshima City, Hiroshima, Japan

In a geographical tradition and using an integrated approach this book series addresses these issues by acknowledging the interrelationship of tourism to wider processes within society and environment. This is done at local, regional, national, and global scales demonstrating links between these scales as well as outcomes of global change for individuals, communities, and societies. Local and regional factors will also be considered as mediators of global change in tourism geographies affecting communities and environments. Thus *Geographies of Tourism and Global Change* applies a truly global perspective highlighting development in different parts of the world and acknowledges tourism as a formative cause for societal and environmental change in an increasingly interconnected world.

The scope of the series is broad and preference will be given to crisp and highly impactful work. Authors and Editors of monographs and edited volumes, from across the globe are welcome to submit proposals. The series insists on a thorough and scholarly perspective, in addition authors are encouraged to consider practical relevance and matters of subject specific importance. All titles are thoroughly reviewed prior to acceptance and publication, ensuring a respectable and high quality collection of publications.

More information about this series at http://www.springer.com/series/15123

Thomas E. Jones · Huong T. Bui · Michal Apollo
Editors

Nature-Based Tourism in Asia's Mountainous Protected Areas

A Trans-regional Review of Peaks and Parks

Foreword by David Weaver

Editors
Thomas E. Jones ⓘ
Environment and Development Cluster
Ritsumeikan Asia Pacific University
Beppu, Japan

Huong T. Bui ⓘ
Tourism and Hospitality Cluster
Ritsumeikan Asia Pacific University
Beppu, Japan

Michal Apollo ⓘ
Department of Tourism
and Regional Studies
Pedagogical University of Krakow
Krakow, Poland

ISSN 2366-5610 ISSN 2366-5629 (electronic)
Geographies of Tourism and Global Change
ISBN 978-3-030-76835-5 ISBN 978-3-030-76833-1 (eBook)
https://doi.org/10.1007/978-3-030-76833-1

© The Editor(s) (if applicable) and The Author(s), under exclusive license to Springer Nature Switzerland AG 2021
This work is subject to copyright. All rights are solely and exclusively licensed by the Publisher, whether the whole or part of the material is concerned, specifically the rights of translation, reprinting, reuse of illustrations, recitation, broadcasting, reproduction on microfilms or in any other physical way, and transmission or information storage and retrieval, electronic adaptation, computer software, or by similar or dissimilar methodology now known or hereafter developed.
The use of general descriptive names, registered names, trademarks, service marks, etc. in this publication does not imply, even in the absence of a specific statement, that such names are exempt from the relevant protective laws and regulations and therefore free for general use.
The publisher, the authors and the editors are safe to assume that the advice and information in this book are believed to be true and accurate at the date of publication. Neither the publisher nor the authors or the editors give a warranty, expressed or implied, with respect to the material contained herein or for any errors or omissions that may have been made. The publisher remains neutral with regard to jurisdictional claims in published maps and institutional affiliations.

This Springer imprint is published by the registered company Springer Nature Switzerland AG
The registered company address is: Gewerbestrasse 11, 6330 Cham, Switzerland

All my love and big thanks to CNHA. Also to RDJSH, for 'having us walk the extra mile.'

Foreword

An innovative way to approach the interdisciplinary investigation of mountains across Asia is to imagine them as multiple focal points. *Culturally*, and as exemplified by Japan's Mt. Fuji and the Five Great Mountains of mainland China, they often serve as national and/or religious icons of great beauty and importance. Particularly in South and Southeast Asia, they also typically serve as the homelands of diverse and distinct indigenous minorities. *Geopolitically*, the frequent use of watershed divides to demarcate international boundaries positions Asian mountains as sites of international tension, illustrated in mid-2020 by violent confrontations between Chinese and Indian military personnel along their disputed Himalayan border. Competition for access to fresh water and other natural resources exacerbate these increasing tensions, while some indigenous minorities, as in Myanmar, have pursued long-standing struggles of self-determination within their mountain heartlands. *Environmentally*, mountains are often studied as places of exceptional biodiversity as well as sites of frequent natural disasters such as floods and landslides. The vulnerability of fragile high-altitude ecosystems to climatic disruption, moreover, has put such places under increased surveillance as lofty 'canaries in the coalmine' of anthropogenic global warming. It is because of their rich environmental and cultural qualities—and sometimes to defuse simmering geopolitical tensions—that most Asian mountains are now designated under one or (usually) more national or trans-national IUCN protected area categories that recognise the unique management challenges of each such setting.

As richly illustrated in the following chapters of this book, Asia's mountains are also now evolving as exceptionally popular settings for rapidly expanding national tourism industries, giving rise to additional *economic*, *social* and *psychological* focal points of note. Long the destination stronghold of a few doughty pilgrims and adventurers, the transformation of Asia from a mainly less economically developed to a mainly more economically developed region during the past five decades has unleashed a tsunami of mostly domestic tourists into upland peripheries throughout the continent, enabled by growing populations, rising discretionary incomes, a desire to periodically escape seething and polluted cities, and the relentless expansion of national transportation networks. The resilience of affected human and non-human mountain communities, accordingly, has been sorely tested by local transformations bestowing a complex array of costs as well as benefits. A few of these contemporary tourists, styled variably by academics as 'ecotourists', 'nature-based tourists' or 'alternative tourists', may well be aware of these transformations and try as a result to behave within the evolving contours of 'sustainable' or 'responsible' tourism. However, the majority, in all likelihood, are oblivious to these issues and stay focused on attaining personally satisfying and socially noteworthy experiences, ephemerally replicated these days in an endless cascade of carefully staged social media posts. Whether Asia's peaks provide 'peak experiences', and how these peak experiences affect the peaks that provide them, are therefore questions well worth asking.

In the absence of effective self-regulation among mountain tourists, protected area managers in Asia face the formidable challenge of ensuring that core biocentric mandates are not sabotaged by parallel mandates to accommodate 'complementary' recreational activity. Recreational demand is not only increasing at a relentless pace (temporary disruptions from pandemics and other external forces notwithstanding), but a widespread pattern of reduced government funding is making protected area managers ever more dependent on revenue from tourism, so that such increases must be at least tacitly embraced by those managers despite the environmental risks they entail. Exposure to Western planning and management strategies focused around the attainment of optimal visitor distribution (as for example through zoning and quotas) and visitor behaviour (as for example through mindful interpretation, pre-emptive policing, and selective demarketing) do provide some extremely useful pathways to the sustainable management of increased visitation. However, these pathways must be qualified by the significant cultural differences which differentiate Asian domestic tourists from Western protected area visitors. The latter, for example, often experience culture shock when visiting Chinese protected areas, where blurred boundaries between the 'cultural' and the 'natural' give rise to temple complexes on high peaks, oversized red calligraphy on cliff faces, and enormous crowds of seemingly happy visitors. Clearly, the sustainable management of Asian protected areas requires the amalgamation of conventional Western strategies with endogenous inputs that reflect the distinct cultural, social, historical and political circumstances of each country and region.

A focus only on sustainability, however, is insufficient; Asia's mountain protected areas must concurrently demonstrate *resilience* in the face of escalating external threats. It is helpful in this respect to imagine mountainous areas not only as multiple focal points, but as multiple focal points interacting and coalescing with increased frequency in obvious and less obvious ways. Ecological and landscape change induced by global warming or direct human intrusions, for example, may make mountain destinations less attractive or more dangerous, or it may stimulate so-called 'last chance' tourism among those hoping to take the last selfie in front of a dying glacier. Cross-border and separatist tensions in affected locations may force some governments to restrict or prohibit the entry of tourists, while other governments might encourage the expanded presence of tourists and associated infrastructure to strengthen their claims to disputed territory. The mountainous pleasure periphery is therefore also a geostrategic pleasure periphery in which national governments and their 'big picture' political calculations play as much a role in the future of Asia's mountain protected areas as local communities, park managers, environmentalists, the tourism industry, and other traditional stakeholders. The need to achieve sustainability and resilience within this broader context of diverse and often incompatible interest groups makes the management of such protected areas a classic 'wicked problem' that increasingly implicates the recreational visitor. Thoughtful analyses on tourism in Asia's mountain protected areas, as systematically engaged on a subregional and country by country basis to capture the region's natural, cultural, political, and social diversity, are therefore a welcome and timely contribution to the field.

Dr. David Weaver
Principal Research Fellow
Queensland University of Technology
Brisbane, Australia

Acknowledgements

Many thanks to the listed parties for their help that was gratefully received for the following:

- *Foreword*: Dr. David Weaver, also for his Keynote presentation in the Webinars.
- *Proofreading*: Prof. R. A. Salazar; Christian Tschirhart; Dr. A. Chakraborty.
- *Student Research Assistants*: Teresa Stetter, Raphaelle Delmas, Zaw Myo Win, Kelvianto Shenyoputro, Hoàng Nguyễn Minh and Nadar Sonny Widyagara.
- *Map-making*: Jin Ziming 金子茗 (using several programs including ArcGIS, CAD and PhotoShop, she created the maps for this book).
- *Financial support*: from the United Graduate School of Agricultural Sciences, Kagoshima University (FY 2017 and 2018) for Chap. 2 (Chen et al.).
- *Contributions to Chap. 10 (Wengel et al.)*: thanks to Andrew Kyaw (founder and leader of the Yangon Hiking & Mountaineering Club), Htet Eaindray (Mandalar Degree College) and Thiha Lu Lin (Thiha, the Traveller).
- *APU Research office*: initial grant for Workshop at AP Conference, and Research Finding Dissemination Subsidy, also for support with the 3 Webinars.

Contents

Part I Introduction

1 **Mountainous Protected Areas & Nature-Based Tourism in Asia** 3
Thomas E. Jones, Michal Apollo, and Huong T. Bui

Part II Northeast Asia

2 **Overcoming Barriers to Nature Conservation in China's Protected Area Network: From Forest Tourism to National Parks** 29
Bixia Chen, Yinping Ding, Yuanmei Jiao, Yi Xie, and Thomas E. Jones

3 **Japan's National Parks: Trends in Administration and Nature-Based Tourism** 49
Thomas E. Jones and Akihiro Kobayashi

4 **South Korea's System of Mountainous Protected Areas and Nature-Based Tourism** 71
Malcolm Cooper

5 **Taiwan's National Network of Protected Areas and Nature-Based Tourism** 91
Chieh-Lu Li and Thomas E. Jones

Part III Southeast Asia

6 **Indonesia's Mountainous Protected Areas: National Parks and Nature-Based Tourism** 111
Wahyu Pamungkas and Thomas E. Jones

7 **Collaborative Management of Protected Areas in Timor-Leste: Stakeholder Participation in Community-Based Tourism in Mount Ramelau** 133
Antonio da Silva and Huong T. Bui

8 **Protected Areas and Nature-Based Tourism in the Philippines: Paying to Climb Mount Apo Natural Park** 153
Aurelia Luzviminda V. Gomez and Thomas E. Jones

9 **Governance and Management of Protected Areas in Vietnam: Nature-Based Tourism in Mountain Areas** 173
Huong T. Bui, Long H. Pham, and Thomas E. Jones

10 **Mountainous Protected Areas in Myanmar: Current Conditions and the Outlook for Nature-Based Tourism** 197
Yana Wengel, Nandar Aye, Wut Yee Kyi Pyar, and Jennifer Kreisz

Part IV South Asia

11 **Indo-Himalayan Protected Areas: Peak-Hunters, Pilgrims and Mountain Tourism** 223
Michal Apollo, Viacheslav Andreychouk, Joanna Mostowska, and Karun Rawat

12 **Nepal's Network of Protected Areas and Nature-Based Tourism** 245
Rajiv Dahal

13 **Mountainous Protected Areas in Sri Lanka: The Way Forward from Tea to Tourism?** 269
Renata Rettinger, Dinesha Senarathna, and Ruwan Ranasinghe

Part V Conclusion

14 **Reflections for Trans-Regional Mountain Tourism** 293
Huong T. Bui, Thomas E. Jones, and Michal Apollo

Editors and Contributors

About the Editors

Thomas E. Jones is Associate Professor in the Environment & Development Cluster at Ritsumeikan APU in Kyushu, Japan. His research interests include Nature-Based Tourism, Protected Area Management and Sustainability. Tom completed his Ph.D. at the University of Tokyo and has conducted visitor surveys on Mount Fuji and in the Japan Alps. ORCID: 0000-0002-4097-1886.

Huong T. Bui is Professor of Tourism and Hospitality cluster, the College of Asia Pacific Studies, Ritsumeikan Asia Pacific University (APU), Japan. Her research interests are Heritage Conservation; War and Disaster-related Tourism, Sustainability and Resilience of the Tourism Sector. ORCID: 0000-0003-1016-6197.

Michal Apollo is an Assistant Professor at the Institute of Geography, Department of Tourism and Regional Studies, Pedagogical University of Krakow, Poland, a Fellow of Yale University's Global Justice Program, New Haven, USA, and a Visiting Researcher at Hainan University – Arizona State University Joint International Tourism College, Haikou, China. Michal's areas of expertise are tourism management, consumer behaviours as well as environmental and socio-economical issues. Currently he is working on a concept of sustainable use of environmental and human resources. He is also an enthusiastic, traveller, diver, mountaineer, ultra-runner, photographer, and science populariser. ORCID: 0000-0002-7777-5176.

Contributors

Viacheslav Andreychouk Faculty of Geography and Regional Studies, University of Warsaw, Warsaw, Poland

Michal Apollo Department of Tourism and Regional Studies, Institute of Geography, Pedagogical University of Krakow, Krakow, Poland

Nandar Aye National Kaohsiung University of Hospitality and Tourism, Kaohsiung, Taiwan

Huong T. Bui College of Asia Pacific Studies, Ritsumeikan Asia Pacific University (APU), Beppu, Japan

Bixia Chen Faculty of Agriculture, University of the Ryukyus, Nishihara, Japan

Malcolm Cooper College of Asia Pacific Studies, Ritsumeikan Asia Pacific University, Beppu, Japan

Antonio da Silva Ministry of Tourism, Commerce and Industry, Dilli, Timor-Leste

Rajiv Dahal Department of Travel and Tourism Management, Faculty of Social Sciences and Humanities, Lumbini Buddhist University, Lumbini, Nepal

Yinping Ding School of Tourism and Geographical Science, Yunnan Normal University, Kunming, China

Aurelia Luzviminda V. Gomez School of Management, University of the Philippines Mindanao, Davao, Philippines

Yuanmei Jiao School of Tourism and Geographical Science, Yunnan Normal University, Kunming, China

Thomas E. Jones Environment and Development Cluster, College of Asia Pacific Studies, Ritsumeikan Asia Pacific University (APU), Beppu, Japan

Akihiro Kobayashi Faculty of Economics, Senshu University, Tokyo, Japan

Jennifer Kreisz Hainan University—Arizona State University Joint International Tourism College, Haikou, China

Chieh-Lu Li Department of Tourism, Recreation, and Leisure Studies, National Dong Hwa University, Hualien, Taiwan

Joanna Mostowska Faculty of Geography and Regional Studies, University of Warsaw, Warsaw, Poland

Wahyu Pamungkas Forestry Agency of South Sumatra Province, Palembang, South Sumatra, Indonesia

Long H. Pham Faculty of Tourism Studies, University of Social Sciences and Humanities, Vietnam National University, Hanoi, Vietnam

Wut Yee Kyi Pyar Mother Earth Tourism Specialists Myanmar, Yangon, Myanmar;
LuxDev Aid & Development Agency, Yangon, Myanmar

Ruwan Ranasinghe Department of Tourism Studies, Faculty of Management, Uva Wellassa University of Sri Lanka, Badulla, Sri Lanka

Karun Rawat Department of Tourism, University of Otago, Dunedin, New Zealand

Renata Rettinger Department of Tourism and Regional Studies, Institute of Geography, Pedagogical University of Krakow, Krakow, Poland

Dinesha Senarathna Department of Geography, University of Kelaniya, Kelaniya, Sri Lanka;
New Zealand Tourism Research Institute, Auckland, New Zealand

Yana Wengel Hainan University—Arizona State University Joint International Tourism College, Haikou, China

Yi Xie School of Economics and Management, Beijing Forestry University, Beijing, China

Abbreviations

ASEAN	Association of Southeast Asian Nations
CBD	Convention on Biological Diversity
CITES	The Convention on International Trade in Endangered Species of Wild Fauna and Flora
ICOMOS	International Council on Monuments and Sites
IPRA	Indigenous Peoples Rights Act (1997)
IUCN	International Union for Conservation of Nature
MAB	Man and the Biosphere
MOE	Ministry of the Environment
MPA	Marine Protected Area
NBT	Nature-Based Tourism
NGO	Non-Governmental Organization
NIPAS	National Integrated Protected Areas System Act (1992)
PA	Protected Area
PAMB	Protected Area Management Board
UNEP-WCMC	UN Environment Programme World Conservation Monitoring Centre
UNESCO	United Nations Educational, Scientific and Cultural Organization
UNWTO	World Tourism Organization
WDPA	World Database on Protected Areas
WHS	World Heritage Site

List of Figures

Fig. 1.1	Mt. Fuji rises from Fuji-Yoshida City on the edge of the national park (*Source* Author)	8
Fig. 1.2	International tourist arrivals 1990–2016 (units: millions) (*Source* UNWTO, 2017)	14
Fig. 2.1	Aggregate number of forest parks and national forest parks (NFP) in China 1982–2019 (*Source* China's Forestry Yearbook 1949–1986, 1987–2005; FGA 2006–2019)	34
Fig. 2.2	The trends in forest park visitors and direct revenues from entrance fees (*Source* China's Forestry Yearbook 1998–2005; FGA 2006–2019)	34
Fig. 2.3	The world's highest outdoor sightseeing elevator in Zhangjiajie National Forest Park (*Source* http://www.qcq.cc/XinDongFang/Class127/2750.html. Retrieved on July 20th, 2012)	38
Fig. 3.1	Map of Japan's national parks and Mount Fuji	57
Fig. 3.2	Annual visits to Japan's nature parks 1950–2016. *Source* MOE (2020). Unit: Millions of visits. (Note: data for Quasi national parks since 1957; Prefectural parks since 1965)	59
Fig. 3.3	Core zone at the top, buffer zone around the base of Mt. Fuji. *Credit* First author	63
Fig. 3.4	Willingness to pay a ¥1000 donation at Mt. Fuji (Jones et al., 2016)	65
Fig. 4.1	Topography of South Korean National Parks	72
Fig. 4.2	Trail map of Hallasan National Park. *Source* Author	74
Fig. 4.3	Crater lakes on the summit of Mt. Hallasan. *Source* Author	86
Fig. 5.1	Visitation trends to Taiwan's national parks (2008–2017). *Source* Adapted from Taiwan National Parks (2020)	98
Fig. 5.2	Amis indigenous people in Hualien interpret their traditional fishery method (*Palakaw*). *Source* first author	101
Fig. 5.3	Paiyun Lodge located at 3,402 m altitude, 2.4 km to Yushan's main peak. *Source* Yushan National Park (2020)	103
Fig. 6.1	Map of Indonesia's protected areas	114

Fig. 6.2	Park rangers torch confiscated consumer goods including Sumatran tiger pelts. https://www.dailymail.co.uk/news/article-3604592/Indonesian-officials-seized-stuffed-Sumatran-tigers-elephant-tusks-endangered-animal-skins-set-fire-warning-illegal-poachers.html	120
Fig. 6.3	Tourist trends to Indonesia's national parks (2014–2018)	123
Fig. 7.1	Timor-Leste's Topography (*Source* Authors' compilation)	135
Fig. 8.1	Location of the Philippines' Protected Areas (PAs)	158
Fig. 8.2	Management zones of the MANP (*Source* DENR XI)	161
Fig. 8.3	Main climbing trails and distances to Mt Apo (*Source* DENR XI)	164
Fig. 8.4	Agreement of climber respondents with motivations for climbing Mt Apo ($n = 431$) (*Source* Gomez, 2015)	165
Fig. 9.1	Distribution of protected areas in Vietnam. *Source* Author compilation based on data from USAID (2013)	177
Fig. 10.1	Myanmar's protected areas (*Source* Authors' elaboration based on data from MONREC)	208
Fig. 10.2	Road to the summit of Mt Natma Taung (*Source* Photo credits: Htet Eaindray and Thiha Lu Lin)	214
Fig. 11.1	Division of the Himalayas (Apollo, 2017b)	228
Fig. 11.2	Altitudinal zones of the Himalayas (2016b). Adapted from Andrejczuk	229
Fig. 11.3	Himalayan areas under conservation. *Source* Author's elaboration	230
Fig. 12.1	International tourist arrivals in Nepal. *Source* MoCTCA (2020; various)	248
Fig. 12.2	Nepal's protected area system. *Source* Adapted from DNPWC (2019)	250
Fig. 12.3	DNPWC organizational structure. *Source* Unofficial translation from DNPWC (2020b)	252
Fig. 12.4	Aggregate revenue from divisions and PAs under DNPWC jurisdiction. *Source* DNPWC (2020c) and MoCTCA (2018)	256
Fig. 13.1	Protected areas in Sri Lanka	272
Fig. 13.2	Horton Plains' international and domestic visitation (2008–2019). *Source* SLTDA (2020)	279
Fig. 14.1	Mountain tourists and prayers flags in Poon Hill with Dhaulagiri Peaks in the background, Annapurna range, Western Nepal. *Source* Author	294

List of Tables

Table 1.1	Mountain tourism and its impact on the natural environment	6
Table 1.2	Ratio of terrestrial and marine protected areas in target countries	11
Table 1.3	A comparison of ecotourism and mass nature-based tourism	11
Table 1.4	Comparative overview of select international conservation sites in NE and SE Asia	16
Table 1.5	Common and official names of target countries and case studies (※ alphabetical order)	20
Table 2.1	Modern chronology of forest tourism in China	33
Table 2.2	A seasonal pricing strategy in Huangshan Scenic Area	41
Table 3.1	Overview of Japans' triple-tiered nature park system	51
Table 3.2	Breakdown of Japan's nature park areas (ha) by conservation zone	54
Table 3.3	Estimated inbound visits to Japan's national parks 2015–2019 and FHINP inbound visits as a per cent	61
Table 4.1	Number of South Korean protected areas	74
Table 4.2	Area, location and designation of South Korean national parks	75
Table 4.3	IUCN classification of protected areas categories	78
Table 4.4	Management elements in the assessment of environmental values	80
Table 4.5	IUCN category and visitation to Korean mountainous national parks	81
Table 5.1	Overview of Taiwan's main systems of protected areas	92
Table 5.2	Overview of Taiwan's nine national parks and one national nature park	95
Table 5.3	Overnight stays and permits at Paiyun Lodge in 2019	103
Table 6.1	Subdivision of Indonesian conservation areas and IUCN protected area categories	115
Table 6.2	Human resources in select national parks	116

Table 6.3	Explanation of the zoning system used in Indonesia's national parks	118
Table 6.4	Indonesian national parks with the highest number of tourist visits	123
Table 6.5	Number of visits to Mount Bromo 2011–2016	125
Table 6.6	Entrance fees at Bromo Tengger Semeru National Park in 2020	126
Table 7.1	Protected areas in Timor-Leste	138
Table 8.1	Categories of Philippine PAs and aggregate area as of 2012	155
Table 9.1	Protected areas in Vietnam	175
Table 9.2	Environmental laws and regulations	179
Table 9.3	Institutions involved in tropical forest and biodiversity management	181
Table 9.4	National Parks in Vietnam	184
Table 9.5	Visitor number and revenue from tourism in Hoang Lien National Park (2015–2019)	190
Table 9.6	Mountain trekkers' season, trails and costs	191
Table 10.1	Myanmar's protected areas	200
Table 10.2	Myanmar's mountain ranges	209
Table 11.1	Percentage of the total area under protection in the Indian Himalaya regions	226
Table 11.2	Domestic and foreign visitor trends to the Himalayas (2011–2012) and rate of change	228
Table 11.3	Overview of Stok Kangri case study site	238
Table 12.1	Policy and statutory instruments related to wildlife and PA management	253
Table 12.2	Tourist trends to Nepal's PAs	258
Table 12.3	Place of visit by tourists 2018–2019	259
Table 12.4	SNP mountain climbers and revenue in 2019	262
Table 13.1	List of national parks in Sri Lanka	273
Table 13.2	Selected mountainous protected areas in Sri Lanka	275
Table 13.3	Mountain protected area visitor numbers and income in 2019	279
Table 13.4	Longitudinal trends in mountain protected area visitor numbers	280
Table 13.5	Characteristics of the most popular tourism destinations in the Knuckles massif	285
Table 14.1	Intra-chapter synopsis of case study peaks' current issues and counter strategies	308
Table 14.2	World's major domestic tourism markets 2018	311

Part I
Introduction

Chapter 1
Mountainous Protected Areas & Nature-Based Tourism in Asia

Thomas E. Jones, Michal Apollo, and Huong T. Bui

1.1 Mountain Environments and Tourism

1.1.1 Characteristics of Mountain Environments

Definitions of mountain areas are unavoidably arbitrary (Messerli & Ives, 1997). Usually no qualitative, or even quantitative, distinction is made between mountains and hills (Barry, 2008). Overall, a mountain is a landform that rises prominently above its surroundings, generally exhibiting steep slopes, a relatively confined summit area, and considerable local relief. A generic typology includes volcanic, fold, plateau, fault-block and dome mountains (Goudie, 2004). However, as our understanding of the mechanisms of mountains' formation from plate tectonics has evolved, Ollier (1981) recognized four types of collisions: (1) continent-continental (*Himalayan* type); (2) continent-to-ocean, related to the continent's overhang and subduction of the ocean floor (*Andean* type); (3) the collision of the continent with the ocean floor and the associated advance of oceanic sediments under the continent, followed by the uplift of the edge of the continent and (4) thickening of the earth's crust as a result of a plate collision, possibly with the accompanying gravity flow of rocks close to the surface. Mountain ranges resulting from continent–continent and continent-to-ocean

---

T. E. Jones (✉) · H. T. Bui (✉)
College of Asia Pacific Studies, Ritsumeikan Asia Pacific University (APU), Beppu, Japan
e-mail: 110054tj@apu.ac.jp

H. T. Bui
e-mail: huongbui@apu.ac.jp

M. Apollo
Department of Tourism and Regional Studies, Institute of Geography, Pedagogical University of Krakow, Krakow, Poland
e-mail: michal.apollo@up.krakow.pl

© The Author(s), under exclusive license to Springer Nature Switzerland AG 2021
T. E. Jones et al. (eds.), *Nature-Based Tourism in Asia's Mountainous Protected Areas*,
Geographies of Tourism and Global Change,
https://doi.org/10.1007/978-3-030-76833-1_1

collisions are thought to be the dominant geological formation on the Earth's surface (Dewey & Burke, 1973).

Mountain geosystems are complex, including elements of abiotic, biotic and anthropic natures interconnected with each other in myriad ways. Humboldt (1807) and Darwin (1859) found that environmental sensitivity increased with altitude and this zonation still represents a core concept in research on the mountain environment (Apollo & Andreychouk, 2020a; Apollo et al., 2020). Overall, mountains' higher elevations produce colder climates than at sea level, by deforming climatic zones to create an aspherical, Koppen's climate H (German: *Hochgebirge*)—mountain climate (see Beniston, 2006). These colder climates strongly affect the ecosystems of mountains: different elevations host different plants and animals. Moreover, endemic species became isolated in altitudinal niches due to inhospitable conditions in the adjacent zones that constrained their movement or dispersal (Barry, 2008).

Meybeck et al. (2001) estimated that about 25% of the land surface is occupied by mountains that are home to 26% of the world's population. Since time immemorial, man has penetrated further and higher into the mountains for hunting and later seeking areas convenient for agriculture (Zurick & Pacheco, 2006). Humans learned to use local raw materials (especially wood), build houses, and cultivate farmlands and pastures. Examples of such activities are abundant in high mountain regions (Barry, 2008). Excessive—related to the extremely dynamic population growth (e.g. Apollo, 2017b)—exploitation of limited resources, as well as poorly-planned development activities lead to degradation of mountain environment and ecosystem services. For example, headwater catchments protect vital supplies of fresh water via precipitation or glacial storage and release systems.

As human populations grew, areas of land use expanded over time, mainly at the expense of forests. In this way, the foothills and lower tiers of the system were anthropogenically transformed (Apollo & Andreychouk, 2020a, 2020b). Due to the upward shift in the range of crops, agricultural land use has also affected high-mountain zones. The indirect influence of man on the plant world is also based on the transformation of the soil, which is inextricably related with vegetation. Soil cultivation directly affects the modeling of the relief and significantly enhances erosive processes (Apollo et al., 2018; Hurni & Nuntapong, 1983). Human influence on the composition of vegetation affected not only forests and arable land, but also meadows. Mountainous environments share some common features, including dynamic and extraordinary sensitivity. The latter is due to the poorness of biotic geosystem which is associated with a harsh climate, unfriendly topography, etc.

Mountains are usually characterized as inaccessible, fragile, diversified and marginal areas (Messerli & Ives, 1997), and therefore over the ages most mountain ecosystems remained relatively isolated from the outside world, as reflected in marginalized or lower income communities. However, rapid improvements in technology, together with access infrastructure have thrust such communities into the spotlight of modernization, typified by the rapid rise of tourism. The variability of climatic conditions, high activity of geodynamic processes and generally poor development of vegetation render the balance in the high-altitude geosystem easily disturbed. Mountains are thus becoming more vulnerable due to accelerating pressures from climate change (Auer et al., 2007). They are also under extra pressure from heightened footfall. For example, Chapter 12 documents the rise in international visitors to Nepal's Sagarmatha National Park from just 20 tourists in 1963, to 20,000 in 1998 and 57,289 in fiscal year 2018. As the number of climbers summiting Mt Everest soared, mountaineers such as Shackley (1993) warned that such peaks had become "giant cash cows." In tandem with climate change and commercialization, mountain tourism disasters are occurring more frequently as in the case of the 2015 Mt Everest avalanches that left 24 dead, or the 2015 Sabah earthquake that left 137 Kinabalu climbers stranded near Low's Peak. Mountain tourism depends on stable climatic conditions that limit "when specific tourism activities can occur (e.g. season length with snow cover or open water), tourism demand (e.g. proportion of people willing to swim or camp under certain conditions), and the quality of a tourism experience (utility) (e.g. hiking in warm, sunny conditions versus a cold rain or extreme heat)" (Scott et al., 2007).

1.1.2 Impacts of Mountain Tourism

Mountainous destinations account for 15–20% of global tourism, ranked second only to sun-sea-and-sand vacations on islands and beaches (Richins et al., 2016). Demand has grown along with access infrastructure as cable cars climb higher and roads reach further up the slopes. Meanwhile on and off-season visits increase as extra tourists seek to avoid the extreme heat of summer (Cavallaro et al., 2017). Even the high-altitude zones, including the inaccessible level, have been exposed for half a century or more to adverse impacts related to mountaineering, mountain trekking, rafting and other types of adventure tourism (Musa et al., 2015). Overall, nature-based tourism (NBT) activities have been increasing around the world since the 1960s (Cordell & Super, 2000; Jin-Hyung et al., 2001; Pröbstl-Haider et al., 2015), and this trend is expected to continue (Apollo, 2017a; Ryan, 2003). Mountains, with their remote and majestic beauty, are among the most popular destinations for NBT (Mieczkowski, 1995). Each year, millions of hikers, trekkers, and climbers swarm to mountains (Apollo, 2017a; Beedie & Hudson, 2003) such as the Seven Summits (Huddart & Stott, 2020) as well as other well-known spots like the Annapurna Circuit (Apollo et al., 2020; Joshi & Dahal, 2019) or Mt. Fuji (Jones et al., 2018). The threat from mass tourism is exacerbated by the tendency of tourists to congregate in mountain honeypots—specific areas, channels and times—which can coincide with biodiversity hotspots (Kruczek et al., 2018). In sum, the increasing volume of tourists presents a serious threat to both the quality of the natural environment (Table 1.1) and the unique cultural identity of local communities.

Tourism also brings cultural revolution via impacts on philosophies, economies and politics, since the commercialization of mountain NBT transforms residents' way of life, culture and customs (Apollo, 2015; Apollo et al., 2020; Musa et al., 2004, 2015). Tourism can stimulate changes in socio-cultural, environmental and economic

Table 1.1 Mountain tourism and its impact on the natural environment

	Trail impacts	Trampling and damage to vegetation	Disturbance or attracting of wildlife	Invasive species of plants	Littering of the mountain environment	Human waste pollution	Noise and light pollution
Apollo and Andreychouk (2020a)	☑		☑		☑	☑	
Apollo and Andreychouk (2020b)		☑					
Apollo (2017c)						☑	
Barros and Pickering (2014)				☑			
Barros et al. (2015)	☑	☑	☑	☑		☑	
Cilimburg et al. (2000)						☑	
Cole (1993)	☑	☑	☑	☑			
Cullen (1986)					☑		
Fidelus (2016)	☑	☑					
Gander and Ingold (1997)			☑				
Hempton and Grossmann (2009)							☑
Kaseva (2009)					☑	☑	
Knight and Gutzwiller (1995)			☑				
Marion and Olive (2006)	☑	☑					☑
Monti and Mackintosh (1979)	☑	☑					
Roe et al. (1997)					☑		
Stevenson et al. (2020)						☑	
Ściężor et al. (2012)							☑
Wall and Wright (1977)		☑	☑	☑			

(continued)

Table 1.1 (continued)

	Trail impacts	Trampling and damage to vegetation	Disturbance or attracting of wildlife	Invasive species of plants	Littering of the mountain environment	Human waste pollution	Noise and light pollution
Weaver and Dale (1978)	☑	☑					
Weaver et al. (2001)				☑			
White et al. (1999)			☑				
Zwijacz-Kozica et al. (2013)			☑				

Source Author original

dimensions in places where such activities come into close contact with local communities (Ap, 1992; Apollo, 2015; Godde et al., 1999; Lama & Sattar, 2004). Mountainous NBT can also play a positive role in promoting an overall improvement in the locals' quality of life through economic development and environmental conservation (Nepal, 2002; Apollo, 2015). Yet the tourism mechanisms that generates such radical transformations must be taken into consideration when developing conservation plans, without which the mantra of 'sustainable development' remains an unobtainable goal (Apollo, 2015; Joshi & Dahal, 2019).

Overall, due to the various levels of economic development in mountainous countries, there is currently little possibility of introducing a comprehensive, rational and balanced approach to the natural environment (Sachs, 2015). However, there are ongoing attempts to set development along a more development trajectory, including the selective designation of PAs characterized by relatively undisturbed natural environments and rare, iconic flora and fauna. This harks back to the American ideal in the late 19th and early twentieth century, when the drive to 'go west' inadvertently culminated in the designation of some of the earliest, largest PAs. As the 'wild' West was gradually opened up, philosophers like John Muir and pragmatic policymakers like Gifford Pinchot pushed for alternative ways to protect the last pockets of 'undeveloped' land. Meanwhile a concurrent dichotomy driven by a similar mix of 'frontier' development spirit and competitive conservation propelled investors ever upward into the mountain areas (Mose & Weixlbaumer, 2007). The Banff Springs Hotel was constructed in 1888 by the Canadian Pacific Railway, at an altitude of 1414 m. In the U.S., the Ahwahnee Lodge was built in 1927 on the floor of Yosemite Valley at an altitude of 1215 m. This mix of railroads, luxury hotels and other tourist infrastructure cemented the tangible 'taming of the highlands' for NBT, portrayed as a moral crusade to civilize the mountain wilderness rife with 'wild animals and evil spirits' (Nash, 2001). Much like the patriotic undertones that pitched American national parks against the castles and cathedrals of Europe, a similar pattern can be detected in today's PAs across Asia. From the 1934 Imperial Hotel in the

Fig. 1.1 Mt. Fuji rises from Fuji-Yoshida City on the edge of the national park (*Source* Author)

Japan Alps, the contemporary equivalents have carried the patriotic competition to civil engineering extremes as typified by the world's 'longest' cable car at Fansipan, Vietnam (the 3-rope non-stop cable car carries tourists a distance of 6293 m up the Muong Hoa Valley to a station near the summit—Chapter 9), or the 'highest' glass elevator in Zhangjiajie, China (326 m high—Chapter 2). However, despite the extant use of PA labels, the global perception remains wrapped in the North American narrative and rarely extends to include iconic Asian mountains such as Mt Everest, Mt Fuji or Mt Jade—all designated 'national parks' but not always recognized as such (Fig. 1.1).

1.2 Protected Areas & Nature-Based Tourism in Asia

1.2.1 The Roots of Protected Areas (PAs) in Asia

Asia is a place of contrasts, an ancient patchwork of cultures that is now the global pacesetter for economic growth. Regionally diverse interactions with mountains have long recognized the value of forests and the need to conserve them. Historical restrictions on hunting and forest exploitation were selectively justified by Confucian thinking and China had set up "offices to oversee the sustainable use of forest resources" by the sixth-second centuries BCE (Miller, 2017). In Japan, references to hunting restrictions date back to the seventh century AD when the Taika Reforms established a separate land category for 'bird hunting and preservation' (Sheppard, 2001). 'Sacred groves,' 'hidden valleys' and 'holy peaks' form another cross-regional nexus with ancient roots. In Mongolia, for example, the custom of protecting certain forested hills dates back to the thirteenth century. The first reserve, the Boghdkhan Mountain Strictly Protected Area, was officially established in the late 1700s, by some estimates the first legally protected natural area in the world (Sheppard, 2001). Throughout history, many other areas and species have been protected across Asia for their cultural and religious significance. For example, Buddhist 'Beyuls' are sacred, hidden valleys found in many parts of the Himalayan region which also host significant biodiversity (Mu et al., 2019). In many cases, such hidden valleys, hunting reserves and sacred groves across Asia became the bedrock for today's PAs.

1.2.2 IUCN Categories of PAs

The term 'protected area' (PA) is a conservation label that corresponds to any "clearly defined geographical space, recognized, dedicated and managed, through legal or other effective means, to achieve the long-term conservation of nature with associated ecosystem services and cultural values (Dudley, 2008)." PA terminology incorporates a mix of land use classifications as diverse as 'national park,' 'nature reserve,' 'wildlife management area' and 'wilderness area.' PAs comprise the core of national conservation strategies and international treaties including the Convention on Biological Diversity (CBD). As biodiversity comes increasingly under threat in an era of the '6th mass extinction,' PAs help safeguard biodiversity, provide such ecosystem services as clean water and air and mitigate climate change. PAs also have a role protecting vulnerable communities, cultural heritage and sacred sites, with the International Union for Conservation of Nature (IUCN—natural heritage) working together with the International Council on Monuments and Sites (ICOMOS—cultural heritage) as gatekeepers to UNESCO's World Heritage Program. Mountains, in particular, symbolized a transformation wrought by a conservation and regional development dichotomy that proved justification for turning 'wasteland into world heritage' (Hall, 1992).

The global trailblazers and early PA templates were large mountainous national parks in North America such as Yellowstone and Banff set up from the end of the nineteenth century onward. Designation drivers included an emerging desire to 'set aside' primeval wilderness in an untouched form that belied the lengthy history of involvement with indigenous peoples (such as First Nations) or certain 'less desirable' species (e.g. the wolf) that were hunted, removed or deliberately eradicated from the parks (Emel, 1995). As PAs were subsequently designated all around the world, a spectrum of management goals evolved to cope with the diverse criteria that has also expanded over time. Classification of the different types of PAs was no simple task, but the IUCN have developed 6 categories (see Table 4.3 in Chapter 4) to cope with the inherent national and regional diversity. These include a spatial mix of large-scale PAs over 10 km^2 (Mose & Weixlbaumer, 2007) versus small-scale ones such as the cat. 'IV' 1.64 km^2 Bukit Timah Nature Reserve near the centre of the city-state of Singapore (Dudley, 2008). At one extreme of the IUCN spectrum, a few sites such as the Swiss National Park are labelled category 'Ia,' with limited tourist access and a management agenda that prioritizes scientific research. Category 'II' corresponds most closely to the original 'Yellowstone model' whereby large areas are 'set-aside' with few permanent populations of people but significant tourism resources including spectacular geoheritage (e.g. waterfalls, hot springs and geysers) and iconic flora and fauna. Category 'V' PAs encompass traditional, inhabited landscapes and seascapes made, modified or maintained by human influences such as farming, forestry and fishing. The IUCN system also reflects the fact that PA management objectives are not static but have changed over time, shifting from sightseeing to biodiversity, and making more use of MPAs as outlined in the next section.

1.2.3 Aichi Target 11: Bigger and Better PAs?

Following the publication of Brundtland's landmark report on sustainable development in 1987, and the 1992 Earth Summit in Rio, the need for robust environmental governance has grown increasingly apparent (Jordan, 2008). In 2012, the IVth World Congress on National Parks & Protected Areas pre-empted the UN CBD's aim to designate over 12% of the earth's terrestrial surface as PAs by 2000 (Mose & Weixlbaumer, 2007). The CBD Framework included Signature Programme No. 2 to "unlock the potential of protected areas, including indigenous and community conserved areas, to conserve biodiversity while contributing towards sustainable development." This in turn paved the way for the Aichi Biodiversity Targets, wherein Target 11 sought to expand protected areas to 17% of terrestrial and inland water areas, and 10% of coastal and marine areas by 2020.

By 2014, there were 10,900 PAs covering almost 14% of terrestrial Asia (Juffe-Bignoli et al., 2014). Table 1.2 presents the ratio of terrestrial and marines PAs in selected countries covered in this volume, with Northeast Asian countries meeting or exceeding the terrestrial 2020 target of 17%. However, many PAs do not match up sufficiently with biodiversity hotspots, or are poorly managed. New and more effective modes of PA governance are thus being sought around the world today, but there is a lack of research related to non-English speaking countries, especially in Asia which has experienced some of the fastest-growth rates for tourism. The focus of this edited volume is to compare mountainous PAs holistically across Northeast, Southeast and South Asia, three regions that are actively seeking to promote NBT to capitalize on tourism resources including impressive landscapes and biodiversity while retaining conservation goals (Table 1.3).

Table 1.2 Ratio of terrestrial and marine protected areas in target countries

	Terrestrial (%)	Marine (%)	Areas of importance for biodiversity (%)
Northeast Asia			
China	16	5	9
Japan	29	8	68
South Korea	17	2	38
Taiwan	20	1	30
Southeast Asia			
Indonesia	12	3	26
Timor Leste	16	1	37
The Philippines	15	1	41
Viet Nam	8	1	39
Myanmar	7	0	25
South Asia			
India	6	0	24
Nepal	24	0	55
Sri Lanka	30	0	44

Source WDPA (2020)

Table 1.3 A comparison of ecotourism and mass nature-based tourism

	Ecotourism 1.0	Pragmatic mass NBT	Mass NBT
Destination	Unspoiled, wild natural destinations	Semi-wild natural or authentic cultural destinations	'3S' settings or PAs (front country only)
Operational logistics	Small scale guided groups	Higher volume offset by spatial & temporal mitigation	High volume during peak season
Access & entry	Visitor permits and limits of use	Cost recovery mechanisms gain revenue for conservation	Connectivity with mass transport
Visitor Education	Actively seek altruistic learning opportunities	Persuasive communication delivers targeted messages	Hedonistic and sight-seeing
Visitor Profile	Dominated by white, male western elites	Multicultural visitors with different cultural backgrounds, preferences, values, expectations	

Source Adapted from Weaver (2001, 2014)

1.3 Nature-Based Tourism in Asian PAs

1.3.1 Nature-Based Tourism

This next section reviews various definitions of 'nature-based' (NBT) or 'ecotourism,' before listing up some of the limitations of earlier models and offering implications for 'enlightened mass tourism' (Weaver, 2014). Various definitions are applied to tourism in PAs, including 'nature-based,' 'sustainable' or 'ecotourism.' The latter emerged as a buzzword in the 1980s and 1990s, with earlier models perceiving the concept as "primarily concerned with the direct enjoyment of some relatively undisturbed phenomenon of nature" (Valentine, 1992, p. 108). Particularly puritanic subsects emphasized "travel to fragile, pristine and usually protected areas that strives to be low impact and (usually) small scale," adding that ecotourism also aims for education, "funds for conservation; directly benefits the economic development and political empowerment of local communities; and fosters respect for different cultures and for human rights" (Honey, 1999, p. 25). Black (1996, p. 4 with italics added) defined ecotourism as: "an experience with a focus on the natural and *cultural* environment, ecologically sustainable activities with a predominant educative and interpretative programme…that contributes to local community groups and projects and to the conservation of the surrounding environment." Such definitions lead to controversy over whether or not ecotourism included cultural aspects not present in the original definition (Boo, 1990).

Meanwhile, NBT was seen by some to be analogous with 'ecotourism.' Weaver (2001, p. 16) defined NBT as "any type of tourism that relies on attractions directly related to the natural environment," without the explicit educational or sustainable aspirations mentioned above. However, umbrella definitions of NBT transcended the addition of *culture* to include '3S' destinations such as island, beach and rainforest attractions, or even consumptive activities such as fishing and hunting. This positioning put NBT on a collision course with 'pure' ecotourism models portrayed as the antithesis of mass tourism that often operated in '3S' (seas, sand, sun) settings with minimal regulations seeking maximum profit (Weaver, 2001). Pure ecotourism, which Weaver later dubbed 'Ecotourism 1.0', could be distinguished from the mass tourism model via (i) education; (ii) the 'triple bottom line' of sustainability; (iii) natural destinations. This 'alternative' to mass tourism differed further in its operating model that called for small groups or limited volume; wild or semi-wilderness settings; and provision of few—if any—visitor services that inculcated a '*Leave No Trace*' ethic to minimize environment impacts. However, while accusations of 'greenwashing' inevitably plagued the implementation of such lofty objectives, the concurrent claims of 'elitism' proved as hard to refute, for ecotourism 1.0 was also a predominantly Western concept, typified by tourists from richer, 'Global North' source countries travelling to poorer 'Global South' destinations. The gulf was starkly demonstrated in differences between mountain destinations in 'developed' and 'developing' countries (Nepal, 2002).

Rather than revisit the ecotourism debate, this book presents Asian PA examples of Weaver's (2014) 'enlightened' or 'pragmatic' mass NBT. Visitors "display 'soft' ecotourism traits such as lower personal commitment to environmental activism, appreciating nature as just one facet of a multipurpose trip, and preference for less strenuous activities confined to well-serviced site-hardened zones" (Weaver, 2013, p. 378). Nonetheless, if mass tourists can be persuaded to venture out of their comfort zones, often just a short stay in a front country setting such as a visitor centre or photogenic scenic spot, there is potential for ecotourism to cross-over to the mainstream market. In fact, trekking is a mostly mountainous hybrid that ascribes to many of the ecotourism ideals including the cultural component via "visits to local villages" and also incorporating "distance hiking...[and] adventure experiences" (Weaver, 2001). Trekking can also be considered an interim version of mountain climbing, that is a more extreme form of adventure tourism associated with 'adrenaline-junkies' and risk management. As well as extra on-site physical exertion, climbers traditionally invested significantly more time and money in specialist training that could require membership in a club or association, or an 'alpine apprenticeship.' Like other types of adventure tourism such as white-water rafting, and scuba or sky-diving, the climber market has merged with aspects of mass tourism to make even high-altitude mountains more accessible for peak-hunters that have the funds, but not always the requisite skills, to attempt iconic summits such as Everest. This is the gentrification process which Beedie and Hudson (2003) described as a "dilution of the essential ingredients of "being a mountaineer" as a result of a democratization process facilitated by the arrival of some urban characteristics in wild mountain regions."

Controversy over the ethics and logistics of PA funding and cost-recovery is a reminder that the lack of consensus over NBT has ramifications beyond the academic world. For NBT is regularly "described as one of the fastest growing sectors of the world's largest industry, and a very important justification for conservation" (Balmford et al., 2009). As early as 1997, the UNWTO estimated that all nature-related forms of tourism accounted for approximately 20% of total international travel. When domestic trips are included, the NBT sub-segment accounts for considerably more of the overall market. However, because it is characterized by trips to parks and PAs, it involves consumption of public as well as private goods, requiring effective and long-sighted policy-making. NBT is thus fundamentally intertwined with the governance and administration of PAs.

Asia's PAs boast pioneers such as Sri Lanka and Japan, with designated PAs that pre-date those in Europe. Asia also hosts higher mountains (Sagarmatha) and greater footfalls (Fujisan) than anywhere else on the planet. Notwithstanding, there have been few empirical attempts to analyse Asia's inventory from an NBT perspective and little research conducted on international and domestic trends. NBT development in Asia is "often seen as large-scale and economically driven, based primarily on leisure, and characterized by resorts utilizing nature as little more than a pleasant background appealing to a wide array of holiday-makers" (Henderson 2011, cited in Frost et al., 2014). Explanations for this hover between conjecture and stereotype, such as the claim that Asian cultures tend to take a more pragmatic and less romanticized view of nature (Kellert, 1995). The debate also has ramifications for the 'wicked problem' of sustainable development, so this book follows a supply and demand dichotomy that focuses on the perspectives of park managers and tourism operators as they strive to achieve pragmatic mass NBT. The chapters recognize that NBT increasingly reflects diverse drivers of domestic and international market demand, including social trends such as population, gender and multiculturalism (Elmahdy et al., 2017). Therefore, overarching tourism trends are summarized in the next section.

1.3.2 Regional Tourism Trends

Asia plays a pivotal role in global tourism, both as a destination region and a source market. The rapid rise of the international tourism sector resulted in growth outstripping that of global trade since 2012, and by 2017 the sector accounted for 7% of all global exports in goods and services (UNWTO, 2017). Asia (including the Pacific) is the world's second most visited region after Europe, with three of the top destinations in China, Japan and Thailand (UNWTO/GTEC, 2019). Leisure, recreation, and holidays is an increasingly popular motivator, accounting for over half of all trips (53%) compared to Visiting-Friends-and-Relatives (27%: VFR, including health and religion); and business and professional (13% Ibid., p. 5; Fig. 1.2).

Asia also posted the fastest pre-pandemic regional growth in terms of international arrivals in 2018, with average growth rate of 7%, accounting for a 25% global share of visitor numbers and a 30% share of international tourism receipts (UNWTO, 2019). Double digit growth in South Asia (19%) was driven by arrivals in India, Iran, Nepal, and Sri Lanka. In South-East Asia, most destinations posted strong growth (7%), particularly Vietnam. Outbound travel from China and India fuelled growth

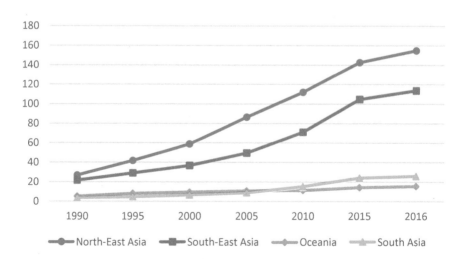

Fig. 1.2 International tourist arrivals 1990–2016 (units: millions) (*Source* UNWTO, 2017)

in many destinations in the sub-region. Growth in North-East Asia, the largest sub-region in Asia, was steady at 6%, much of which continued to be fuelled by China. Intraregional tourism dominates Asia and the Pacific, with arrivals that originated from within Asia the majority in Northeast Asia (88%), Southeast Asia (81%) and South Asia (33%) (UNWTO/GTEC, 2019).

Asia was the first region to suffer the impact of COVID-19, experiencing a record drop of 72% in arrivals during the first half of 2020. North-East Asia (−83%) recorded the largest decrease, followed by South-East Asia with a 64% decline, while Oceania and South Asia recorded a drop of 58% and 55% respectively. By June-July 2020, international travel had ground to a halt reflecting travel restrictions amid efforts to contain Coronavirus outbreaks. However, Asia has the world's largest domestic markets due to the vast populations and geographical size. India reported almost 1.9 billion domestic tourist trips (overnight trips) in 2018, while an estimated 5.5 billion domestic visitor trips in 2018 were recorded in China which includes both overnight and same-day visitors. In Japan, 317 million domestic overnight hotel stays accounted for 89% of bed occupancy (UNWTO, 2020). Despite the recent increase in inbounds, Japanese tourism is still dominated by the domestic market that spent USD 201 billion in 2017, almost six times the receipts from inbound tourism (UNWTO, 2020). The huge volume and high spending propensity of the domestic market in Asia are expected to be the engine of recovery for tourism industry of the region, especially as international travel restrictions are starting to be relaxed as of October 2020.

As tourism in Asia has developed in the past decade according to such strong growth figures, its well-established prominence as a nature-based tourism 'destination' is diversifying to include more aspects of an 'origin' region (Frost et al., 2014). Yet despite rapid growth in international tourist arrivals in Asia's emerging economies, the role of the region's PAs remains unknown or under-reported, creating a gap in monitoring and marketing the changing dynamics of mountainous NBT. An inventory approach (Table 1.4) testifies to the rich natural and cultural capital that is increasingly accessible across Asia. However, many of the prestigious PA labels are linked to international organizations such as UNESCO, with few examples of regional certification equivalents such as ASEAN heritage parks. This raises the question of externally-imposed values, especially as the UNESCO World Heritage program's underlying philosophy has been criticised as a top-down, Eurocentric approach (Winter, 2014). UNESCO ideologies such as 'outstanding universal values' do not always align neatly with host communities' own idea of the significance of heritage (Jones et al., 2020). Thus, in order for park rangers and local communities to feel a continuing sense of ownership and involvement, further reconciliation is required between PA planners at the regional, national and destination level.

Table 1.4 Comparative overview of select international conservation sites in NE and SE Asia

Country	WHS	MAB	Ramsar sites	ASEAN Heritage Parks	GGN
China	16	29	48	0	37
Indonesia	4	7	7	6	4
Japan	4	5	50	0	9
Malaysia	2	1	7	3	1
Philippines	3	2	7	8	
Rep. Korea	1	5	22	0	3
Vietnam	3	8	8	6	2
Source	UNEP-WCMC (2018)			ASEAN (2016)	UNESCO (2017)

WHS = World Heritage Sites (natural and mixed only); ASEAN = ASEAN Heritage Parks; MAB = Biosphere Reserves; GGN = Global Geopark Network; ∗Taiwan not included on this list due to jurisdictional barriers to UN listings

1.4 Content of This Book

The outline of the book is as follows. First, this chapter uses a literature review to identify major trends in mountain tourism and PAs to set the stage for a transnational comparison, contextualizing criteria with which to investigate (i) conservation governance; and (ii) sustainable tourism. This framework is then applied to empirically explore 12 country case studies, including diverse mountainous PA destinations across Northeast, Southeast and South Asia regions. Some strong intra-regional synergies emerged and are discussed in Chapter 14, with some implications for trans-regional NBT.

These target countries and PA destinations selected are at diverse stages of the development spectrum. The subsequent set of comparative PA case studies identify best practise and pitfalls. Relationships between supply-side PA destinations and stakeholders are analysed in the context of demand trends shaped by attitudes of local residents and tourists. A convenience sample of mountainous case studies was selected to incorporate a diverse mix of Asian peaks. These include volcanoes (Bromo and Hallasan); iconic ranges (the Himalaya and Huangshan) and free-standing peaks (Apo and Fuji).

Throughout this manuscript, the common English names are used (Eagles et al., 2001). Official names (Table 1.4) are abbreviated thereafter, e.g. People's Republic of China (hereafter referred to as 'China,' including Hong Kong). It is acknowledged that this list is not complete, however, due to data deficiencies, lack of reference sources or limits to the authors' own network of researchers, several countries had to be be omitted (e.g. Cambodia, Laos PDR, Macau, Mongolia, North Korea, Singapore, Thailand etc.). Furthermore, it is also recognized that the Asia region contains numerous contested territories and jurisdictions. The potential to promote trans-boundary 'peace parks,' and establish more of them, is recognized in Chapter 14.

1.4.1 Northeast Asia

In Chapter 2, Chen, Ding, Jiao, Xie and Jones summarize the status quo of China's PAs with a focus on the 3,505 forest parks. 'Forest tourism' accounts for 30% of all domestic trips but regional development is prioritized over conservation. Forest parks and PAs have been increasingly 'opened up' for tourism purposes, resulting in widespread environmental impacts in spite of significant revenue streams generated by entrance fees. To recalibrate resource management toward conservation, China introduced new national parks in 2015, and the case of Pudacuo, one of the inaugural batch, is investigated alongside Huangshan Scenic Area.

In Chapter 3, Jones and Kobayashi focus on Japan which in 1931 became "the first politically independent non-Western country to establish national parks," as distinguished from equivalents designated "at the behest of European colonial authorities" (Frost et al., 2014). The number of national parks, and visits to them, increased rapidly in line with post-WWII economic growth. Domestic visitation peaked in 1991 at 415 million before gradually declining. The number of inbound visits was estimated to have increased to 6.7 million visits in 2019, with almost half of them recorded at Fuji-Hakone-Izu National Park, centred on Mt. Fuji—Japan's tallest peak at 3776 m. As the Fuji case study shows, unlocking the potential of Japan's 'multi-purpose' parks requires cross-cutting partnerships that emphasize co-management to resolve conservation and development trade-offs.

Cooper (Chapter 4) finds considerable common ground in Korean national parks that comprise mostly mountainous PAs. The parks correspond to IUCN categories II and V, using a similar 'multi-purpose' agenda to restrict forms of development in the interests of conservation. 22 National Parks cover nearly 7% of the country and are controlled by the Korean National Park Service, except for Hallasan in Jeju, which was established in 1987 and has been managed separately since 1998. As an IUCN category II 'national park,' and simultaneously one of the main drawcards on an island that receives more than 15 million annual visitors, Hallasan seeks pragmatic mass NBT solutions for the domestic and increasingly international market segments.

In Chapter 5, Li explores Taiwan's 9 'national' and 1 'national nature' parks that cover 8.6% of the landmass, mostly in mountainous regions. The three most popular parks in 2015 were Kenting, Taroko and Yangminshan, accounting for some two-thirds of all visitation. NBT trends (2008–2017) revealed visitation fluctuations that reflect international and domestic demand trends. Mountain hiking remains pre-eminent among visitor activities, but new niches such as rock climbing, mountain biking and canyoneering are also on the rise. Popular mountain hikes such as Yushan implement a permit system drawn by lottery, and national park user fees have been charged since 2016.

1.4.2 Southeast Asia

Pamungkas and Jones (Chapter 6) explain how Indonesia's state forests are divided into conservation; protected; and production categories. Within the former are 54 national parks that account for around 9% of the archipelago's landmass. However, the Indonesian parks reflect a dilemma of the Aichi Biodiversity Targets, for even as the area under protection has expanded rapidly, the effectiveness has been called into question. Ineffective conservation is exacerbated by the influx of tourists at popular destinations such as Mt. Bromo (2329 m), an active volcano in the Bromo Tengger Semeru National Park. Unregulated NBT has significant environmental impacts. Management tools such as cost recovery mechanisms and climber permits are investigated as potential counter-strategies.

In Chapter 7, Da Silva and Bui offer an overview of the system for governance and management of PAs in Timor-Leste. Discussion on nature-based tourism in mountain protected areas leads us to Mt. Ramelau, one of the most popular destinations for tourism and pilgrimage, where issues concerning stakeholder's collaboration are highlighted. A recent increase in visitation provides positive future prospects for tourism, but has also brought a variety of negative impacts, leading to temporary site closure for revision of visitor management. This article highlights the importance of engaging stakeholders in NBT management.

The Philippines (Chapter 8) is a biodiversity hotspot in Southeast Asia. The establishment and expansion of protected areas has been one of the major counterstrategies to conserve the country's vulnerable biodiversity and the National Integrated Protected Areas Systems (NIPAS) Act of 1992, provides the legal framework for PA's designation and management. Nature-based tourism has increased in PAs such as Mt. Apo, a Nature Park designated under NIPAS. Fees are currently collected from climbers but only represent a fraction of the Total Economic Value. Gomez and Jones proved that NBT activities could provide additional funds to enhance PA management effectiveness and incentivize a wider range of stakeholders.

The current system of PAs in Vietnam contributes to protection of biodiversity, albeit in an overlapping and complicated manner. Bui, Pham and Jones (Chapter 9) investigate the 33 national parks, of which 22 are designated in mountainous regions. Governance is changing with decentralization of state control and delegation of additional powers to local authorities. However, by analyzing the resource politics in mountain area of Hoang Lien National Park (HLNP), the authors contend that a re-centralization of resource management is evident in a new form of partnership investment in mega tourism projects such as the cable car to the top of Mt. Fansipan (3143 m).

Myanmar (Chapter 10) is the largest country in Southeast Asia, with Mt. Hkakaborazi (5881 m) the highest peak. The broader landscape is characterised by lowland plains and major rivers, with PAs covering 41,156 km^2 (6.1%) of the country's landmass. Most of the PAs are wildlife sanctuaries; there are seven ASEAN Heritage Parks, five Ramsar Sites and four Marine Protected Areas. Wengel, Aye, Yee Kyi Pyar and Kreisz investigate the conservation and the governance of Myanmar's national park management from policy and practice perspectives. We delve the potential for mountainous NBT while reviewing the challenges and opportunities of pioneering mountainous PAs with an example from a popular and accessible peak—Mt. Natma Taung (3053 m).

1.4.3 South Asia

The Himalayas (Chapter 11) are among the world's most popular mountaineering destinations for millions of climbers, trekkers and pilgrims. Apollo, Andreychouk, Mostowska and Rawat outline the state of the natural environment within PAs and the trends in the changes, implications on NBT in the Indian Himalayan PAs and its challenges. There are ongoing efforts to preserve areas characterised by rare flora and fauna, where nature protection is on a par with commercial goals, that is tourism. As a destination attracting over 50 million tourists annually, more and more PAs in the Himalayas have been opened to tourism, with some such as Stok Kangri in Hemis National Park forced to take mitigating action such as temporary trail closures.

Nepal's 20 PAs cover 34,419 sq. km (23%) of total land area that is home to the majority of Nepal's tourism inventory and the hub of NBT activities (Chapter 12). More than 60% of foreign visitors to Nepal visit PAs and pre-pandemic, there has been exceptional growth of domestic visitors too. Dalal finds that the sustainability of Nepal's PAs depends upon community access, inclusion, and control with attitudinal change in the mindset of the park officials suggested as a means to handle communities' grievances more effectively. Also, visitor management and better interpretation for visitors, community and industry alike are vital along with wildlife management to mitigate wildlife-human conflicts.

In Chapter 13, Rettinger, Senarathna and Ranasinghe focus on Sri Lanka's unique mountainous national parks, whose enormous biodiversity has faced significant impacts for centuries from tea plantations that transformed the island's upland environment. The struggle to turn from tea to tourism is epitomized by the 26 national parks that cover 5734 km^2, accounting for 8.74% of the landmass. Most of the mountainous PA are clustered within the central highlands, with the Horton Plains recording the highest number of domestic and international tourists. The opportunities and threats facing Sri Lanka's mountainous NBT are contextualized in one of the island's most famous mountainous PAs, the Knuckles massif.

1.4.4 A Trans-Regional, Trans-Boundary Approach to Asian PAs

This edited book adopts a transboundary perspective that aims to investigate parallel PAs in twelve countries across East, Southeast and South Asia. Using a twin-tier structure, each chapter sets the scene of national PA networks before drilling down to destination-level case studies. In addition to country case studies, the cross-cutting analysis of governance and NBT dynamics helps map out the administrative and legal contours that shape PA trends, identifying institutional strengths and weaknesses across Asia's mountains. From an epistemological standpoint, the list of contributors reflects the 'trans-boundary' subject matter by attempting to combine geological, anthropological, socio-economic approaches. We thus present findings on a diverse emic–etic continuum, rather than recognizing a clash between opposing styles (Zhu & Bargiela-Chiappini, 2013). The interdisciplinary essence extends to active inclusion of numerous native, non-English language sources that are used to contextualize site-specific PA management cases written up where possible by in situ authors, including a mix of local researchers and long-term foreign residents from academic and practitioner backgrounds. A final boundary to cross is therefore the delineator between academic and applied arenas, but even for the former considerable discrepancy exists between PA definitions provided by government agencies, intergovernmental organizations such as IUCN, and peer-reviewed academic sources.

Diverse models of governance are expanded from a workshop held at Ritsumeikan Asia–Pacific University in November 2017, with applied examples from countries presented via diverse case studies as presented in Table 1.5. The Edited book project aims to implement the comparative criteria from the workshop by combining theory with practice. The cross-cutting panel included participants of diverse nationalities, drawn from a cluster of APU scholars working in tandem with researchers from universities in Japan and around the world. The panel's presentations formed a nucleus of the book, providing insights for research fields as diverse as human geography; tourism; landscape architecture; and public policy. The transboundary approach enables holistic discussion of sustainable PA trajectories, including plans to develop a pan-Asian PA system akin to ASEAN Heritage Parks or Natura 2000.

Table 1.5 Common and official names of target countries and case studies (※ alphabetical order)

Common name	Official name	Case study
China	People's Republic of China	Huangshan (1864 m)
India	Republic of India	Stok Kangri (6153 m)
Indonesia	Republic of Indonesia	Mt. Bromo (2329 m)
Japan	Japan	Mt. Fuji (3776 m)
Myanmar	Republic of the Union of Myanmar	Mt. Natma Taung (3053 m)
Nepal	Federal Democratic Republic of Nepal	Sagarmatha (8848 m)
Philippines	Republic of the Philippines	Mt. Apo (2954 m)
Sri Lanka	Democratic Socialist Republic of Sri Lanka	The Knuckles Masif (1906 m)
South Korea	Republic of Korea	Hallasan (1947 m)
Taiwan	Taiwan, Republic of China	Yushan (3952 m)
Timor Leste	Democratic Republic of Timor-Leste	Mt. Ramelau (2986 m)
Vietnam	Socialist Republic of Viet Nam	Fansipan (3147 m)

Our findings reiterate the challenges that mountainous PAs face due to rapid increases in visitation, including tourists, climbers and pilgrims. Due to their topography and socio-cultural conditions, the visitor flow is compressed into narrow spatial or temporal confines, with the majority using the same trails in the same season and even at the same time of day. Many mountain base camps and gateways, both in the PA itself or surrounding buffer zones, have been rendered radically more accessible by recent infrastructure developments such as roads, rail or cable car connections, as in the case of Fansipan (Chapter 9), to link growing urban populations with hitherto inaccessible peaks. Trekking tourism has exacerbated commercialization because "sacred landscape is an intangible concept and vulnerable to external influences such as westernization, tourism, and cultural assimilation" (Apollo, 2015; Rutte, 2011 cited in Mu et al., 2019). To counter the consequences of 'over-tourism' in mountainous PAs, many of the destinations in this book have introduced economic or physical restrictions. For the former, iconic peaks such as Sagarmatha have raised their rates, while others such as Stok Kangri have closed sections of trail. During the current coronavirus crisis, other destinations such as Fujisan have shut down for an entire season due to the risk of infection. Despite the temporary negative impacts for the NBT sector, there is hope that such action could inspire a radical re-think in the direction of pragmatic mass NBT towards a more sustainable relationship with mountainous PAs across Asia.

References

Ap, J. (1992). Residents' perceptions on tourism impacts. *Annals of Tourism Research, 19*(4), 665–690.

Apollo, M. (2015). The clash-social, environmental and economical changes in tourism destination areas caused by tourism the case of Himalayan villages (India and Nepal). *Current Issues of Tourism Research, 5*(1), 6–19.

Apollo, M. (2017a). The true accessibility of mountaineering: The case of the High Himalaya. *Journal of Outdoor Recreation and Tourism, 17*, 29–43.

Apollo, M. (2017b). The population of Himalayan regions—By the numbers: Past, present, and future. In R. Efe & M. Ozturk (Eds.), *Contemporary studies in environment and tourism* (pp. 143–159). Cambridge Scholars Publishing.

Apollo, M. (2017c). The good, the bad and the ugly-three approaches to management of human waste in a high-mountain environment. *International Journal of Environmental Studies, 74*(1), 129–158.

Apollo, M., Andreychouk, V., & Bhattarai, S. S. (2018). Short-term impacts of livestock grazing on vegetation and track formation in a high mountain environment: A case study from the Himalayan Miyar Valley (India). *Sustainability, 10*(4), 951. https://doi.org/10.3390/su10040951.

Apollo, M., & Andreychouk, V. (2020a). Mountaineering and the natural environment in developing countries: An insight to a comprehensive approach. *International Journal of Environmental Studies*. https://doi.org/10.1080/00207233.2019.1704047.

Apollo, M., & Andreychouk, V. (2020b). Trampling intensity and vegetation response according to altitude: An experimental study from the Himalayan Miyar Valley. *Resources, 9*, 98. https://doi.org/10.3390/resources9080098.

Apollo, M., Andreychouk, V., Moolio, P., Wengel, Y., & Myga-Piątek, U. (2020). Does the altitude of habitat influence residents' attitudes to guests? A new dimension in the residents' attitudes to tourism. *Journal of Outdoor Recreation and Tourism, 31*, 100312.

ASEAN Clearing House Mechanism. (2016). ASEAN Heritage Parks. http://chm.aseanbiodiversity.org/index.php?option=com_wrapper&view=wrapper&Itemid=110&current=110.

Auer, I., Böhm, R., Jurkovic, A., Lipa, W., Orlik, A., Potzmann, R., Schöner, W., Ungersböck, M., Matulla, C., Briffa, K., Jones, P., Efthymiadis, D., Brunetti, M., Nanni, T., Maugeri, M., Mercalli, L., Mestre, O., Moisselin, J.-M., Begert, M., ... Nieplova, E. (2007). HISTALP—Historical instrumental climatological surface time series of the Greater Alpine Region. *International Journal of Climatology: A Journal of the Royal Meteorological Society, 27*(1), 17–46.

Balmford, A., Beresford, J., Green, J., Naidoo, R., Walpole, M., & Manica, A. (2009). A global perspective on trends in nature-based tourism. *PLoS Biol, 7*(6), e1000144.

Barros, A., & Pickering, C. M. (2014). Non-native plant invasion in relation to tourism use of Aconcagua Park, Argentina, the highest protected area in the Southern Hemisphere. *Mountain Research and Development, 34*(1), 13–26.

Barros, A., Monz, C., & Pickering, C. (2015). Is tourism damaging ecosystems in the Andes? Current knowledge and an agenda for future research. *Ambio, 44*(2), 82–98.

Barry, R. G. (2008). *Mountain weather and climate*. Cambridge University Press.

Beedie, P., & Hudson, S. (2003). Emergence of mountain-based adventure tourism. *Annals of Tourism Research, 30*(3), 625–643.

Beniston, M. (2006). Mountain weather and climate: A general overview and a focus on climatic change in the Alps. *Hydrobiologia, 562*, 3–16.

Black, R. (1996). Ecotourism: What does it really mean? *Range, 34*, 4–7.

Boo, E. (1990). *Ecotourism: The potentials and pitfalls: Country case studies*. WWF.

Cavallaro, F., Ciari, F., Nocera, S., Prettenthaler, F., & Scuttari, A. (2017). The impacts of climate change on tourist mobility in mountain areas. *Journal of Sustainable Tourism, 25*(8), 1063–1083.

Cilimburg, A., Monz, C., & Kehoe, S. (2000). Wildland recreation and human waste: A review of problems, practices, and concerns. *Environmental Management, 25*(6), 587–598.

Cole, D. N. (1993). Minimizing conflict between recreation and nature conservation. In D. S. Smith & P. C. Hellmund (Eds.), *Ecology of greenways: Design and function of linear conservation areas* (pp. 105–122). University of Minnesota Press.

Cordell, H. K., & Super, G. R. (2000). Trends in Americans' outdoor recreation. In Gartner, W. C. & Lime, D. W. (Eds.), *Trends in outdoor recreation, leisure and tourism* (133–144). Cabi.

Cullen, R. (1986). Himalayan Mountaineering Expedition Garbage. *Environmental Conservation, 13*, 293–297. https://doi.org/10.1017/S0376892900035335.

Darwin, C. (1859). *On the origin of the species by means of natural selection*. Murray.

Dewey, J. F., & Burke, K. (1973). Tibetan, Variscan and Precambrian basement reactivation: Products of continental collision. *The Journal of Geology, 81*, 683–692.

Dudley, N. (Editor) (2008). *Guidelines for applying protected area management categories*. IUCN. x + 86 pp.

Eagles, P. F. J., Bowman, M. E. & Tao, T. C.-H. (2001). *Guidelines for tourism in parks and protected areas of East Asia*. IUCN, x + 99 pp.

Elmahdy, Y. M., Haukeland, J. V. & Fredman, P. (2017). Tourism megatrends: A literature review focused on nature-based tourism.

Emel, J. (1995). Are you man enough, big and bad enough? Ecofeminism and wolf eradication in the USA. *Environment and Planning d: Society and Space, 13*(6), 707–734.

Fidelus, J. (2016). Slope transformations within tourist footpaths in the northern and southern parts of the Western Tatra Mountains (Poland, Slovakia). *Zeitschrift Für Geomorphologie, 60*(3), 139–162.

Frost, W., Laing, J., & Beeton, S. (2014). The future of nature-based tourism in the Asia-Pacific region. *Journal of Travel Research, 53*(6), 721–732.

Gander, H. & Ingold, P. (1997). Reactions of male alpine chamois Rupicapra r. rupicapra to hikers, joggers and mountain bikers. *Biological Conservation, 79*(1), 107–109.

Godde, P. M., Price, M. F., & Zimmermann, F. M. (1999). Tourism and development in mountain regions: Moving forward into the new millennium. In P. M. Godde, M. P. Price, & F. M. Zimmermann (Eds.), *Tourism and development in mountain regions* (pp. 1–25). CABI Publishing.

Goudie, A. (2004). *Encyclopaedia of geomorphology* (Vol. 2). Routledge.

Hall, C. M. (1992). *Wasteland to World Heritage*. Melbourne University Press.

Hempton, G., & Grossmann, J. (2009). *One square inch of silence: One man's search for natural silence in a noisy world*. Simon and Schuster.

Honey, M. (1999). *Ecotourism and sustainable development: Who owns paradise?* Island Press.

Huddart, D. & Stott, T. (2020). *Outdoor recreation: Environmental impacts and management*. Springer Nature.

Humboldt, A. (1807). *Ideen zu einer Geographie der Pflanzen nebst einem Naturgemälde der Tropenländer*. Cotta, Tübingen.

Hurni, H., & Nuntapong, S. (1983). Agro-forestry improvements for shifting cultivation systems: Soil conservation research in northern Thailand. *Mountain Research and Development, 3*(4), 338–345. https://doi.org/10.2307/3673037.

Jin-Hyung, L., Scott, D., & Floyd, M. F. (2001). Structural inequalities in outdoor recreation participation: A multiple hierarchy stratification perspective. *Journal of Leisure Research, 33*(4), 427.

Jones, T. E., Beeton, S., & Cooper, M. (2018). World heritage listing as a catalyst for collaboration: Can Mount Fuji's trail signs point the way for Japan's multi-purpose national parks? *Journal of Ecotourism, 17*(3), 220–238.

Jones, T. E., Bui, H. T., & Ando, K. (2020). Zoning for world heritage sites: Dual dilemmas in development and demographics. *Tourism Geographies*. https://doi.org/10.1080/14616688.2020.1780631.

Jordan, A. (2008). The governance of sustainable development: taking stock and looking forwards. *Environment and planning C: Government and policy, 26*(1), 17–33.

Joshi, S., & Dahal, R. (2019). Relationship between social carrying capacity and tourism carrying capacity: A case of Annapurna Conservation Area, Nepal. *Journal of Tourism and Hospitality Education, 9*, 9–29.

Kaseva, M. E. (2009). Problems of solid waste management on Mount Kilimanjaro: Challenge to tourism. *Waste Management Research, 28*(8), 695–704.

Juffe-Bignoli, D., Bhatt, S, Park, S., Eassom, A., Belle, E. M. S., Murti, R., Buyck, C., Raza Rizvi, A., Rao, M., Lewis, E., MacSharry, B., & Kingston, N. (2014). *Asia Protected Planet 2014*. UNEP-WCMC.

Kellert, S. R. (1995). Concepts of nature East and West. In M. E. Soulé & G. Lease (Eds.), *Reinventing nature? Responses to postmodern deconstruction* (pp. 103–121). Island.

Knight, R. L., & Gutzwiller, K. J. (Eds.). (1995). *Wildlife and recreationists*. Island Press.

Kruczek, Z., Kruczek, M., & Szromek, A. R. (2018). Possibilities of using the tourism area life cycle model to understand and provide sustainable solution for tourism development in the Antarctic Region. *Sustainability, 10*(1), 89. https://doi.org/10.3390/su10010089.

Lama, W. B., & Sattar, N. (2004). Mountain tourism and the conservation of biological and cultural diversity. In M. F. Price, L. Jansky, & A. A. Iatsenia (Eds.), *Key issues for mountain areas* (pp. 111–148). United Nations University Press.

Marion, J. L., & Olive, N. (2006). Assessing and understanding trail degradation: Results from Big South Fork National River and recreational area. US Geological Survey. http://www.pwrc.usgs.gov. Accessed 10 October 2016.

Meybeck, M., Green, P., & Vorosmarty, C. (2001). A new typology for mountains and other relief classes. *Mountain Research & Development, 21*(1), 34–45.

Messerli, B., & Ives, J. D. (1997). *Mountains of the world: A global priority*. Parthenon.

Mieczkowski, Z. (1995). *Environmental issues of tourism and recreation*. University Press of America.

Miller, I. M. (2017). Forestry and the politics of sustainability in early China. *Environmental History, 22*(4), 594–617.

Monti, P. W., & Mackintosh, E. E. (1979). Effect of camping on surface soil properties in the boreal forest region of northwestern Ontario, Canada. *Soil Science Society of America Journal, 43*(5), 1024–1029.

Mose, I., & Weixlbaumer, N. (2007). A new paradigm for protected areas in Europe. *Protected areas and regional development in Europe. Towards a new model for the 21st century*, 3–20.

Mu, Y., Nepal, S. K., & Lai, P. H. (2019). Tourism and sacred landscape in Sagarmatha (Mt. Everest) National Park, Nepal. *Tourism Geographies, 21*(3), 442–459.

Musa, G., Hall, C. M., & Higham, J. E. (2004). Tourism sustainability and health impacts in high altitude adventure, cultural and ecotourism destinations: A case study of Nepal's Sagarmatha National Park. *Journal of Sustainable Tourism, 12*(4), 306–331.

Musa, G., Higham, J., & Thompson-Carr, A. (2015). Mountaineering tourism: Looking to the Horizon. In G. Musa, J. Higham, & A. Thompson-Carr (Eds.), *Mountaineering tourism* (pp. 328–348). Routledge.

Nash, R. (2001). *Wilderness and the American mind*. Yale University Press.

Nepal, S. K. (2002). Mountain ecotourism and sustainable development. *Mountain Research & Development, 22*(2), 104–109.

Ollier, C. D. (1981). *Tectonics and landforms*. Longman.

Pröbstl-Haider, U., Haider, W., Wirth, V., & Beardmore, B. (2015). Will climate change increase the attractiveness of summer destinations in the European Alps? A survey of German tourists. *Journal of Outdoor Recreation and Tourism, 11*, 44–57. https://doi.org/10.1016/j.jort.2015.07.003.

Richins, H., Johnsen, S. & Hull, J. S. (2016). Overview of mountain tourism: Substantive nature, historical context, areas of focus. In *Mountain tourism: Experiences, communities, environments and sustainable futures* (pp. 1–12). CABI.

Roe, R., Leader-Williams, N., & Dalal-Clayton, B. (1997). *Take only photographs, leave only footprints: The environmental impacts of wildlife tourism*. Environmental Planning Group, International Institute for Environment and Development.

Ryan, C. (2003). *Recreational tourism: Demand and impacts* (Vol. 11). Channel View Publications.

Sachs, J. D. (2015). *The age of sustainable development*. Columbia University Press.

Ściężor, T., Kubala, M., & Kaszowski, W. (2012). Light pollution of the mountain areas in Poland. *Archives of Environmental Protection, 38*(4), 59–69.

Scott, D., Jones, B., & Konopek, J. (2007). Implications of climate and environmental change for nature-based tourism in the Canadian Rocky Mountains: A case study of Waterton Lakes National Park. *Tourism Management, 28*(2), 570–579.

Shackley, M. (1993). No room at the top?. *Tourism Management, 14*(6), 483–485.

Sheppard, D. (2001). Twenty-first century strategies for protected areas in East Asia. In *The George Wright Forum* (Vol. 18, No. 2, pp. 40–55). *The George Wright Society*. USA.

Stevenson, L. C., Allen, T., Mendez, D., Sellars, D., & Gould, G. S. (2020). Is open defaecation in outdoor recreation and camping areas a public health issue in Australia? A literature review. *Health Promotion Journal of Australia*. https://doi.org/10.1002/hpja.300.

UNEP-WCMC. (2018). Protected area profile for Asia & Pacific from World Database of Protected Areas. https://protectedplanet.net/region/AS.

UNESCO. (2017). List of UNESCO Global Geoparks. http://www.unesco.org/new/en/natural-sciences/environment/earth-sciences/unesco-global-geoparks/list-of-unesco-global-geoparks/.

Valentine, P. (1992). Nature-based tourism. Belhaven Press.

Wall, G., & Wright, C. (1977). *The environmental impact of outdoor recreation*. University of Waterloo.

Weaver, D. B. (2001). Ecotourism in the context of other tourism types. In *The encyclopaedia of ecotourism* (pp. 73–83).

Weaver, D. B. (2013). Protected area visitor willingness to participate in site enhancement activities. *Journal of Travel Research, 52*(3), 377–391.

Weaver, D. B. (2014). Asymmetrical dialectics of sustainable tourism: Toward enlightened mass tourism. *Journal of Travel Research, 53*(2), 131–140.

Weaver, T., & Dale, D. (1978). Trampling effects of horses, hikers and bikes in meadows and forests. *Journal of Applied Ecology, 15*, 451–457.

Weaver, T., Gustafson, D., & Lichthardt, J. (2001). Exotic plants in early and late seral vegetation of fifteen northern Rocky Mountain environments (HTs). *Western North American Naturalist, 61*(4), 417–427.

White, D., Kendall, K. C., & Picton, H. D. (1999). Potential energetic effects of mountain climbers on foraging grizzly bears. *Wildlife Society Bulletin, 27*(1), 146–151.

Winter, T. (2014). Beyond Eurocentrism? Heritage conservation and the politics of difference. *International Journal of Heritage Studies, 20*(2), 123–137.

World Database on Protected Areas (WDPA). (2020). Global partnership on Aichi Target 11. Retrieved on 16 October 2020 from https://www.protectedplanet.net/en/thematic-areas/global-partnership-on-aichi-target-11.

World Tourism Organization (UNWTO). (2017). *UNWTO tourism highlights* (2017 edition). UNWTO. https://doi.org/10.18111/9789284419029.

World Tourism Organization (UNWTO). (2019). *International tourism highlights* (2019 edition). UNWTO. https://doi.org/10.18111/9789284421152.

World Tourism Organization (UNWTO). (2020). *UNWTO briefing note—Tourism and COVID-19, Issue 3. Understanding domestic tourism and seizing its opportunities.* UNWTO. https://doi.org/10.18111/9789284422111.

Zhu, Y., & Bargiela-Chiappini, F. (2013). Balancing emic and etic: Situated learning and ethnography of communication in cross-cultural management education. *Academy of Management Learning & Education, 12*(3), 380–395.

Zurick, D., & Pacheco, J. (2006). *Illustrated atlas of the Himalaya.* The Unsiversity Press of Kentucky.

Zwijacz-Kozica, T., Selva, N., Barja, I., Silván, G., Martínez-Fernández, L., Illera, J. C., & Jodłowski, M. (2013). Concentration of fecal cortisol metabolites in chamois in relation to tourist pressure in Tatra National Park (South Poland). *Actatheriologica, 58*(2), 215–222.

Thomas E. Jones is Associate Professor in the Environment & Development Cluster at Ritsumeikan APU in Kyushu, Japan. His research interests include Nature-Based Tourism, Protected Area Management and Sustainability. Tom completed his PhD at the University of Tokyo and has conducted visitor surveys on Mount Fuji and in the Japan Alps.

Michal Apollo is an Assistant Professor at the Pedagogical University of Krakow, Institute of Geography, Department of Tourism and Regional Studies, and a Fellow of Yale University's Global Justice Program, New Haven, USA. Michal's areas of expertise are tourism management, consumer behaviours as well as environmental and socio-economical issues. Currently he is working on a concept of sustainable use of environmental and human resources. He is also an enthusiastic, traveller, diver, mountaineer, ultra-runner, photographer, and science populariser.

Huong T. Bui is Professor of Tourism and Hospitality cluster, the College of Asia Pacific Studies, Ritsumeikan Asia Pacific University (APU), Japan. Her research interests are Heritage Conservation; War and Disaster-related Tourism, Sustainability and Resilience of the Tourism Sector.

Part II
Northeast Asia

Chapter 2
Overcoming Barriers to Nature Conservation in China's Protected Area Network: From Forest Tourism to National Parks

Bixia Chen, Yinping Ding, Yuanmei Jiao, Yi Xie, and Thomas E. Jones

2.1 Introduction

Since the establishment of the Dinghushan Nature Reserve in Guangdong Province in 1956, China has designated a total over 10,000 protected areas (PAs), accounting for 18% of China's landmass (Wang, 2016). China's PA network encompasses many different categories of nature reserves, scenic spots, geoparks, forest parks, wetland parks, water parks, grassland nature parks, desert parks, etc. (Wang et al., 2011). The PA system has evolved through three distinct phases. In the first phase, prior to 2000, only national nature reserves, national parks and national forest parks were recognised. During the second phase, from 2001, national geoparks and national water parks were added, and later national urban wetland parks joined the list in 2004. The third phase, from 2005, recognised national wetland parks and national mining parks, followed by national (urban) important parks in 2007.

B. Chen (✉)
Faculty of Agriculture, University of the Ryukyus, Nishihara, Japan
e-mail: chenbx@agr.u-ryukyu.ac.jp

Y. Ding · Y. Jiao
School of Tourism and Geographical Science, Yunnan Normal University, Kunming, China
e-mail: 2726294590@qq.com

Y. Jiao
e-mail: 670062770@qq.com

Y. Xie
School of Economics and Management, Beijing Forestry University, Beijing, China
e-mail: xybjfu@126.com

T. E. Jones
College of Asia Pacific Studies, Ritsumeikan Asia Pacific University (APU), Beppu, Japan
e-mail: 110054tj@apu.ac.jp

© The Author(s), under exclusive license to Springer Nature Switzerland AG 2021
T. E. Jones et al. (eds.), *Nature-Based Tourism in Asia's Mountainous Protected Areas*, Geographies of Tourism and Global Change,
https://doi.org/10.1007/978-3-030-76833-1_2

However, the most comprehensive global database on terrestrial and marine protected areas, the World Database on Protected Areas (WDPA), initiated by the International Union for Conservation of Nature (IUCN) and the United Nations Environment Programme (UNEP), has not yet recognized many of these PAs (Wang et al., 2011). By June 2020, only 122 sites from China have been included in the world's PA dataset, under the IUCN categories 'VI', 'Ia', 'Ib', and 'IV', while many sites are classified under 'not reported' or 'not applicable' (WDPA, 2020 Protected Planet). This disparity reveals a mismatch between Chinese PAs system and global PA benchmarks, resulting in the invention of concepts in conservation and tourism unique to Chinese context.

Nature conservation has increasingly relied on political and economic support from tourism and recreation (Buckley et al., 2017), with nature-based tourism (NBT), in particular, being a significant subsector of the tourism industry worldwide (Buckley et al., 2008; Weaver & Lawton, 2007). Tourism to PAs has increased faster than other types of tourism in China (Li, 2004), with some of the most popular destinations in the country located inside PAs. Although the concept of ecotourism (*shengtai lüyou*) appeared in the Chinese-language academic literature in the early 1990s (Wang, 1993), the Chinese version is a cultural analogue of the globally-used 'ecotourism', with key differences from the Western concept. These include emphasis on promoting human health; a predilection for human artefacts to enhance nature; and no limitations on scale (Buckley et al., 2008). Against the principle of limited human access to the core zone, over 93% of China's PAs have been 'opened' to visitors, with access granted and tourist facilities constructed but entry fees or tour costs charged. The percent of PAs open to visitors increased from 20% in the 1980s, to 31% during the 1990s, and to 42% in the 2000s (Zhong et al., 2015). To date, only 7% of PAs have never been officially opened to the general public.

Among various PA categories, the national forest park (NFP) is a term used in mainland China (Chen et al., 2019), which emerged in early 1982 with the establishment of Zhangjiajie National Forest Park by the State Council in China (Hu et al., 2004; Li, 1994; Liu & Tao, 2003; Wei et al., 2006; Wu & Wu, 1998; Yu, 2001). A forest park was first officially defined in the Forest Park Management Regulation (China. The State Forestry Administration, 1994, revised 2016, Article 6 & 7) as:

> an area of scenic forest landscape with intensive historic and cultural heritage for the purposes of tourism, recreation, scientific, cultural, and educational activities. It was further classified into three ranks of national, provincial, and municipal levels. A national forest park is one with exceptional scenic beauty of forest landscape, high value of cultural landscape, being representative in a region and well-known nation-wide, and endowed with sufficient tourism facilities. To establish a national forest park, Provincial Forestry Department submits an application to the State Forestry Bureau for approval.

As this definition states, the primary objective of the forest park is to provide recreational or educational opportunities to the public, without overt requirements for environmental protection or forest conservation (Huang et al., 2008). Nonetheless, since their introduction in the early 1980s, NFPs have taken significant steps in terms of protecting specific forest resources, promoting forest-based tourism and local communities' development (Ma et al., 2009). Forest tourism has been defined as appreciating the forest biological environment (Wang & Wang, 1998), using forest biological integrity directly and indirectly, including its natural and cultural scenic attraction (Dong, 2002). The activities of forest tourism include picnics, camping, hunting, fishing, boating, paragliding, horse-riding, and forest education. Venues comprise land designated as 'natural forest,' as well as plantations, but exclude forest in urban parks or gardens (Dong, 2002). The local governments and bureaus set targets for managing these PAs for tourism and poverty alleviation. Unfortunately, after three decades of development for tourism, the current management strategy has jeopardized nature conservation, and the central government recognises the need for change. As a result of this transitional mindset, an entirely new network of national parks is in the trial phase, and will start as the first such national park, Sanjiangyuan National Park, has been approved to open in Qinghai Province in 2020. Being independent from the stakeholders who benefit from the tourism industry, university researchers objectively address the shortage of a scientific management plan and the environmental conservation for forest parks (Hu & Zhang, 1998). Many recent researches have mentioned the shortage of investment capital of forest parks (Ma & Lin, 2007; Zhu, 2007). While limited investment has been mostly used to establish recreational facilities for tourists, funding for research related to forest tourism is thus far insufficient (Hu et al., 2004).

In this chapter, we review China's history of PA establishment and tourist development, as well as the problems and disputes related to balancing economic benefits and nature conservation using forest parks as examples. Against the backdrop of the transformation of the national park system the development of forest tourism and associated problems brought by increasing numbers of visitors and tourism facilities are discussed. Dual case studies are introduced from Pudacuo National Park to illustrate the establishment of the new system as well as Huangshan Scenic Area that symbolizes certain impediments to a more holistic nature conservation system in China.

2.2 Methods

The authors reviewed Chinese-language academic literature on forest tourism, forest parks and national parks, as obtained from the CNKI academic research database using keywords such as forest tourism (Chinese: *senlin lüyou*), forest park (*senlin gongyuan*), ecotourism (*shengtai lüyou*), national park (*guojia gongyuan*), protected area, and nature reserve. Forest tourism is an integral part of ecotourism, particularly in China (Huang et al., 2008). China's Forestry Yearbooks were used for statistical data and the official Internet homepage of the state's Forestry and Grassland Administration (hereafter FGA) was accessed for updated news and related legal documents. In addition, research on China's forest parks, forest tourism and ecotourism in the English-language were also collected and analysed, whereupon we found that English works (e.g. Buckley et al., 2008, 2017; Huang et al., 2008;

Zhong et al., 2008, 2015) were relatively fewer, but have been increasing quickly in the last few years (e.g., Duan & Wen, 2017; Wang, 2019; Wu et al., 2018).

For the case study of Pudacuo National Park, authors conducted interviews with national park experts in March 2018 to explore their in-depth insights towards this new national park system and its extremely complicated network disputes, which could significantly hamper the conservation objectives of the national parks system. A second round of interviews with host community and park management administrative staff was conducted in March 2019. The first and second authors conducted the first-round interviews, and two volunteer graduate students from Yunnan Normal University conducted the second-round of interviews. A semi-structured interview approach was applied in both rounds. The interviews were conducted in Chinese for 30–60 minutes, and field notes taken in the original language. The notes were then transcribed in a word file for data management and analysis. Notes were independently cross-checked to verify that the major ideas, concepts, and issues raised by the participants were accurately documented. A content analysis approach has been applied to elicit common themes.

2.3 Forest Parks and Forest Tourism in China

2.3.1 The System of Forest Parks

The ninth national survey on forest resources (2014–2018) reveals that the forest area in China was 220 million ha, accounting for 23% of the territorial landmass in 2018 (National Forestry & Grassland Administration, 2019). This ranks fifth in the world after Russia, Brazil, Canada and the United States in terms of forest area. The majority was classified as natural forest, although the planation area reached 79.5 million ha, the largest in the world (Ibid.). A total of 3392 forest parks and 897 national forest parks (NFP) had been designated by 2018 (Ibid.). The national forestry agency in China started to promote the use of forest land for recreational purposes with the establishment of forest parks from the late 1970s onwards (Li & Chen, 2007). Although the first forest park was established in 1982, the number of parks began to increase significantly after 1991 (Li & Chen, 2007; Wei et al., 2006). Together with their establishment, promotion of forest tourism evolved through a series of landmark events (See Table 2.1).

The number of forest parks and NFPs has increased exponentially since 1991, due to the following four reasons (Li & Chen, 2007). First, after Deng Xiaoping's inspection tour of the south of China, more attention was paid to developing tertiary industries, including tourism. Second, due to the depletion of forest resources, a revaluation of the forest's ecological services was urgently required. Third, ecological and economic profits generated by the forest parks and tourism became more prominent. Fourth, the then Forestry Ministry held discussions on nationwide forest parks and forest tourism. Thus, from 1994 to 2000, China's forest parks continued to increase steadily in number with the issuance of a series of related regulations. This

Table 2.1 Modern chronology of forest tourism in China

Year	Event
1981	A nationwide forest tourism symposium held by the then Ministry of Forestry in Beijing and prepared a "Memorandum on Forest Tourism's Development"
1982	The first national forest park (NFP) was established in Zhangjiajie, Hunan Province
1992	The then Ministry of Forestry held a symposium in Dalian on forest parks and forest tourism and required the state-owned forest farm which possesses scenic landscape, rich biodiversity, and intensive cultural landscape to establish a forest park
1993	Forest tourism branch established within Chinese Forestry Society to discuss forest parks and Ecotourism (*shengtai luyou*) in forests
1994	Forest Park Administration Regulations were issued by the then Ministry of Forestry
1994	*Shengtai luyou* branch established in China Tourism Association
1996	General Design Standards of China's Forest Parks was issued by the then Ministry of Forestry (LYPT5132 – 95)
1999	The State Technical Supervision Bureau issued the China Forest Park landscape resources grade evaluation (GB/T1805-1999)
2002	Forest Park Management Office was established in the State Forestry Administration
2008	General Planning Standards of National Forest Park was edited with the cooperation of the State Forestry Administration and two universities
2008	National Forest Park Standardization Technical Committee was approved by the State National Standardization Technical Committee
2009	Tianqiaogou NFP was the first to be deprived of NFP status, heralding the start of a selection criteria and evaluation/elimination system for forest parks

Source Shengtai Luyou: cross-cultural comparison in ecotourism (Buckley et al., 2008); China's Forestry Yearbook 1998–2005; FGA 2006–2018

sharp increase resulted in triple the number of designated forest parks and NFPs by 2019 compared with 2001 (See Fig. 2.1).

2.3.2 Forest Tourism

Forest tourism has been defined as a form of tourism that appreciates the forest's biological environment (Wang & Wang, 1998), while utilizing biological integrity as an attraction (Dong, 2002). However, the volume of forest tourism in China has expanded dramatically since 2001 (Li & Chen, 2007). According to the Forest Park Management Office of the FGA, forest parks received a total of 1.6 billion visits in 2018, accounting for 30% of all domestic trips in China. In 2018, revenue from entry fees reached 1.5 trillion RMB ($\approx$USD 0.21 trillion), counted as the forest parks' direct income. The trends in visitation and revenue from forest tourism are shown in Fig. 2.2.

The majority of forest park visitors are domestic, with international visitors only accounting for 15% of the total number of visitors in 2016. The length of stay tends

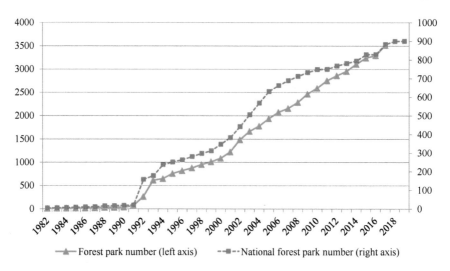

Fig. 2.1 Aggregate number of forest parks and national forest parks (NFP) in China 1982–2019 (*Source* China's Forestry Yearbook 1949–1986, 1987–2005; FGA 2006–2019)

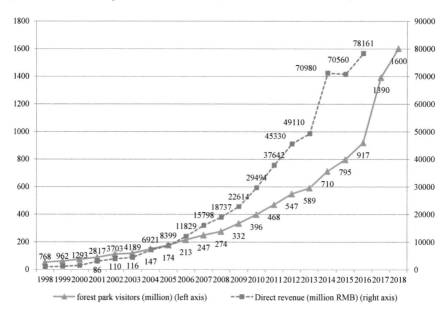

Fig. 2.2 The trends in forest park visitors and direct revenues from entrance fees (*Source* China's Forestry Yearbook 1998–2005; FGA 2006–2019)

to be short–for example, a survey in Xi'an showed that half of those visitors surveyed only stayed one day (Liu et al., 2008). However, there is considerable variation among parks and another survey in Zhangjiajie NFP found that a majority (71.9%) of the tourists stayed for more than 3 days with a mean stay length of 3.6 days (Zhong et al., 2008). In Jiangsu Province, survey results showed that the average length of stay was less than 3 days, only about half of the typical length of stay of North American ecotourists (Buckley et al., 2008).

Forest tourism incorporates forest-based activities, including nature-appreciation, camping, fishing, boating, skiing, hunting, scientific observation and educational activities (Yu & Jin, 2007). Chinese tourists perceived nature-appreciation to be the most important motivation. For instance, surveys covering Zhangjiajie and Qiandaohu NFPs found that the motivation of about 90% of the tourists who visited was to appreciate the scenic landscape (Wei et al., 2006). Reports from other provinces also listed scenic landscape as the major motivation to visit in Fuzhou (Huang et al., 2002) and Xi'an (Liu et al., 2008). Forest tourists were found to be better educated and have above average income levels (Liu et al., 2008; Zhang & Zhao, 2005). In addition, the value of the forest landscape to individuals' physical and mental wellbeing is significantly promoted in China (Buckley et al., 2008). The vast majority of tourists visit the forest for fresh air (Chen et al., 2018; Wu, 2003; Wu & Wu, 1998; Zhong et al., 1998), which is in line with the aims of Chinese park agencies to promote the potential health benefits of visiting parks. It is commonplace for residents of all ages to do exercise inside an urban forest park (Chen et al., 2018). A significant difference exists in forest recreation style between China and Western countries, especially in terms of the role of exercise in promoting human health, that has been attributed to a cultural divergence (Buckley et al., 2008).

Natural landscapes, such as forests or woodlands, are the setting for forest tourism and recreational activities (Font & Tribe, 1999), while culture and history also adds value to China's nature-based tourism (Buckley et al., 2008). Since time immemorial, nature appreciation has been recorded in traditional Chinese paintings (Wen & Dai, 2004). Also, China has a long history of travel to and appreciation of famous mountains and rivers. Ancient Chinese philosophies view humans as part of nature, so millions of Chinese tourists visit sites immortalized by poets and artists (Sofield & Li, 1998). These aesthetic ideals may include human artefacts or calligraphy that has been deliberately built, carved or painted into the landscape as part of the iconic scenery. However, to Western eyes, this often results in visitor infrastructure that intrudes on the natural landscape (Buckley et al., 2008).

2.4 Issues in Management of Forest Parks and Forest Tourism

2.4.1 Overlapping Management of Forest Parks

In China, nature reserves, forest parks, and scenic sites refer to scenic areas with natural or cultural landscapes, albeit under the jurisdiction of different government agencies. Nature reserves are managed by the Environmental Protection Administration of the State Council; forest parks are managed by FPMO, and scenic sites are supervised by the Construction Administration of the State Council. Meanwhile, forest park development plans are usually drawn up by the tourism administration, which tends to prioritize economic development, resulting in excessive travel routes and tourism facilities (Liu & Tao, 2003).

The presence of multiple overlapping actors involved with PA management raises serious challenges for nature conservation. The same area can be redesignated or endowed with multiple PA labels, such as nature reserve, forest park, or scenic site, primarily for the sake of the local government's prestige (Hu & Zhang, 1998). Consequently, multi-sector management leads to conflict between the tourism and conservation agendas (Wang, 1991). For example, Jiuzhaigou Nature Reserve was redesignated as a scenic site in 2019, Dali Cang'er Nature Reserve was additionally labelled as Cang'er Scenic Site, a national scenic site was built within the outer-ring of Wuyishan National Nature Reserve, and Beijing Bihuashan was simultaneously classified as a nature reserve and as a forest park. Moreover, each type of PA (nature reserves, forest parks and scenic sites) has its own respective set of policies and regulations. The lack of law enforcement leads to a failure of effective protection of natural environment and biodiversity of forest parks (Han & Zhuge, 2001; Huang et al., 2008; Zhong et al., 2008).

Additionally, many forest parks emerged out of state-owned forest farms, where most of the staff had been previously engaged in timber production. Consequently, most of the parks face a shortage of investment capital and dedicated tourist management personnel being under the management of forest farm workers who lacked professional training until the mid-1990s, when the specialized forestry universities started programs covering forest tourism (Li, 1994). The forest tourism industry has provided direct employment opportunities for some 500,000 farmers and has benefited about 20 million other people living in areas adjacent to forest parks. About 4,654 villages have realized poverty reduction by surpassing the $1 daily per capita living cost that constitutes the international poverty line (Liu, 2010).

2.4.2 Overuse and Misuse of Forest Tourist Resources

According to the IUCN, a national park in developed countries usually contains the following characteristics: (1) a natural landscape of beauty consisting of ecosystems that are not materially altered; (2) exploitation or occupation is mitigated or eliminated; (3) used for inspirational, educational, cultural and recreational purposes; (4) a minimum size of 1000 ha of protected area; (5) statutory legal protection; (6) a budget and staff sufficient for effective protection; and (7) prohibition on the exploitation of natural resources (Gulez, 1992).

Currently, there is no legislation in China that deals with the protection of forest parks despite their high usage levels. The existing Forest Park Management Regulation places emphasis on tourism use rather than on protection (Huang et al., 2008). The definition of a Chinese forest park does not contain any legal requirements for the protection of forest resources or ecological biodiversity (Huang et al., 2008). Table 2.1 shows the slow development of standardization of forest park management, planning and design into a legal framework. Based on Clause 11 of the current *Forest Park Management Regulations*, commercial development is allowed anywhere inside the park other than in rare natural or cultural spots, or core scenic zones (Deng et al., 2003). For example, Zhangjiajie NFP spent over RMB 100 million (approximately equivalent to USD 14 million) on an outdoor sightseeing lift with a vertical height of 326 m (Fig. 2.3), although its construction was strongly criticized by researchers (Li & Shou, 2003). Likewise, according to China's Man and Biosphere National Committee, among nature reserves where forest tourism was developed, about 44% had been polluted by waste, 11% by noise, 3% air and 2% water. Overall, 22% suffered from damage to the PA, and 11% suffered from deteriorating tourist resources (Liu & Tao, 2003). Consequently, following the ecological deterioration attributed to increasing visitors and nearby hotels and restaurants (Tian, 2004), the death of a large number of *Abies chensiensis,* or Shenshi fir trees (Abies chensiensis, n.d.), an endangered species known as a 'living fossil' in Shirenshan Nature Reserve, was reported in 2004 as the worst ecological damage to occur in Hunan Province.

However, neither the deteriorating natural environment nor the excessive construction inside scenic spots has restrained tourists from swarming to famous nature-based tourism destinations. The 'over-use' phenomenon can be attributed to tourist activities and their needs. Sightseeing is the prevalent activity and tourists, in particular elderly people on packaged tours, who have higher requirements for accessibility, convenience and comfort (Zhong et al., 2008). The world's highest outdoor elevator was built in 2002 to transport the tourists between scenic spots and meet the demand for comfort (Fig. 2.3).

The commercialization of forest parks is universal. Visitor facilities are built in order to enhance opportunities for visitors to enjoy nature. However, interviews found that those built environments conversely resulted in negative impacts on natural landscape. Hotels and restaurants are commonly constructed inside scenic spots, leading to deterioration in air quality and ground water pollution linked to the rapid increase of hotels and other facilities, for example in Zhangjiajie NFP (Shi, 2005; Yang & Zhou, 2005). In addition, the capacity of sewage treatment facilities has lagged behind the rapid expansion in visitors and accommodation infrastructure in and around NFPs (Zhou & Yu, 2004).

Fig. 2.3 The world's highest outdoor sightseeing elevator in Zhangjiajie National Forest Park (*Source* http://www.qcq.cc/XinDongFang/Class127/2750.html. Retrieved on July 20th, 2012)

2.5 From Pudacuo to Huangshan

2.5.1 Case Study of Pudacuo National Park, Yunnan Province

To counter the commercialization of forest parks, a new network of 'national parks' is emerging. Pudacuo was the first park dedicated to achieve the concurrent goals of conservation, ecological and cultural tourism, environmental education, and community benefit. With a mandate to protect world natural and cultural heritage, whilst providing sightseeing opportunities for domestic and international tourists, Pudacuo was designated as China's first national park in August 2006, albeit by Yunnan Province and not by the Chinese central government. The park is in the centre of "Three Parallel Rivers," a UNESCO World Natural Heritage site in Shangri-La County in the northwest of Yunnan Province. It comprises Bita Lake Nature Reserve, that is a Ramsar Wetland site, and the 'Three Parallel Rivers' world natural heritage site, which includes Redhill Area's Shudu Lake. The park is 22 kms away from Shangri-La County, with a total area of approximately 300 square kilometres. Northwest Yunnan Province contains zones of alpine temperate coniferous forest vegetation, the highest point of which is the northern peak of Mi li Tang Mountain, 4159 m above sea level. The lowest altitude is Gold Ditch, to the east of Bita Lake, 3200 m above sea level. Pudacuo Park hosts diverse habitats, including marshland, lakes, and springs, with native forests that contain rare flora and fauna such as black-necked cranes. Pudacuo National Park opened to visitors in 2007. Since opening, the number of tourists and tourism-related income grew quickly, doubling by 2014. The number of tourists was 1.1 million in 2018, but the first quarter of 2019 saw an increase of about 20% (Gong & Ye, 2017) with improved access due to a new high-speed railway connection with Lijiang, another UNESCO world heritage site approximately 200 km away from Pudacuo.

Selected as one of the first ten pilot sites of China's national park management system, Pudacuo National Park Authority was established within the Forestry Protection Office in August, 2018, under the Diqing Tibetan Autonomous Prefectural Government, who administers the unified governance and development planning of the park, and Diqing Prefectural Tourism Enterprise, who manage the park. The company also takes responsibility for environmental conservation, e.g. forest management and water quality monitoring. However, institutional friction has come out of having a private tourist company in charge of management, especially as Pudacuo failed the first round of the national park evaluation process, together with two other sites, Mount Wuyi and the Great Wall. Interviewee Y. W. (male, 60 years old), one of the chief planning committee members of Pudacuo National Park in Yunnan Province, revealed:

> I was unhappy with the dysfunctional Pudacuo National Park Administrative Office. I had been dedicated to the planning of a strong management office directly overseen by the provincial government. However, the park management office prioritizes the economic benefits from tourist development, but neglects to enhance ecological and social aspects of the park (March 25, 2018)

Our interviews with members of the community living inside the park also mentioned negative impacts of tourism industry on the ecological environment. A villager (male, 40 years old) said:

> Ecotourism programs caused irreparable damage to *Ptychobarbus chungtienensis*. It is a representative rare and endemic fish in the plateau. It is also a unique and protected species in Bitahai Nature Reserve and Pudacuo National Park. There were plenty of *Ptychobarbus* in

the past, however, the number decreased sharply after the tourist company started ecotourism programs on the lake and motor cruises burned petrol. Later, some villagers requested that the company use a clean fuel for the motor (March 8, 2019).

Nonetheless, most villagers living inside the park are involved in the tourist industry to some degree. An interviewee (female, 27-year-old) told us that her husband and other male villagers provide pickup and transfer services for tourists who sometimes eat meals at their houses. They also sell souvenirs such as dried yak meat. Another interviewee (a male 50-year-old with one daughter and one son) treated tourists friendlily, regularly inviting them into his house to experience Tibetan culture. He also sold dried yak meat and oranges to the tourists.

Park planner Y.W. also expressed his concerns about two other major problems related to the management of Pudacuo National Park: the low education level of residents inside the national park and lack of sufficient eco-compensation. Of the 33 households of Tibetan folk people in four hamlets living inside the national park, the majority have not received any education. Land use transformation from grazing grassland to protected areas has deprived these residents of their traditional livelihood. Moreover, the low educational level of these residents hinders them from participating in tourism-related industries, except for some simple labour jobs. Another incentive was 'eco-compensation' paid by the tourism enterprise to local communities. The amount of eco-compensation was calculated each year based on the recipient's place of residence and land area. However, disputes have frequently broken out over the figure as local residents tended to request increased amounts of eco-compensation.

Another interviewee (male, 40-year-old) revealed the varying amount of eco-compensation. In the first 5-year contract with the tourism enterprise, each household member received RMB 2000 per year, and thereafter the amount increased by RMB 5000 per year. However, the tourism enterprise stopped paying any eco-compensation after October 2016. In June 2016, the ecotourism program at Bita Lake and horse-riding activities in the meadows were halted, while entry fees for visitors declined from the original RMB 238 to RMB 100. The interviewee felt the eco-compensation amount to be closely related to the performance of the tourism industry, especially the entry fee and certain lakeside ecotourism programs as well as the horse-riding activities. He stated that because of inadequate education, the community feels helpless in dealing with the loss of eco-compensation, and that he did not know how to negotiate with the company.

In principle, Pudacuo National Park was established to conserve nature as well as improve the standard of living for local people. The livelihoods of many community members have been largely transformed from dependence on primary industry to dependence on the tourism industry. Local people benefited from the direct payment from the eco-compensation scheme as well as employment from tourist industry. However, the flawed nature management institutions have hampered the effective protection of endangered species from the threats of heavy tourism use of the national park.

2.5.2 Case Study of Huangshan Scenic Area: Monitoring Climbers Inside the WHS

Huangshan (the Chinese name for the Yellow Mountain), is located about 410 km southwest of Shanghai in Anhui province. It was listed as a UNESCO World Heritage site in 1990, one of only 39 World Heritage Sites globally classified in the 'mixed heritage' category (WHC, 2017). Located in the humid subtropical monsoon zone, Huangshan covers an area of 15,400 ha with a buffer zone of 14,200 ha. With its characteristic granite peaks and twisted pine trees emerging from a vast sea of clouds, Huangshan's landscape has long been held in affection by Chinese pilgrims, poets, and philosophers (Singh, 1992). Tourism took off in the early 1980s, and Huangshan was among the first National Key Scenic Parks (*guo jia zhong dian feng jing ming sheng qu*) to be designated by the party (Xu et al., 2016). There were 1.6 million visitors in 1990 at the time of the UNESCO listing and the number had exceeded 3 million by 2013. Neighbouring villages, Xidi and Hongcun, listed as cultural sites, also experienced a rapid increase in the number of international visitors (including those from Taiwan, Hong Kong and Macao) from 2,205,000 in 2000 to 3,256,000 in 2004 (Yan & Morrison, 2008). Huangshan Scenic Area (HSA) is a public–private partnership under the jurisdiction of the local government (HSA Administrative Committee, headed by the mayor) and listed on the stock market as Huangshan Tourism Development (HTD) Co., Ltd. This creates an unusual mix of civil servants simultaneously executing senior management roles within a limited company.

Two main cost recovery fees are charged at Huangshan, an entry fee and ticket for the cableway (Table 2.2). The entry fee increased from RMB 0.5 (1980) to RMB 230 during peak seasons (Xu et al., 2016), but has remained constant since 2008. Both fees vary according to the season: during peak times (from March 1 to November 30), the adult entry fee is RMB 230, but in the winter off-season, the entry fee is reduced to RMB 150. This seasonal pricing strategy aims to encourage more visitors to come in the winter slack season, while demarketing during the peak season. Pricing is set by HTD Co. Ltd via control over site management: cable way operations, hotels, travel agencies and food & beverage franchises.

Table 2.2 A seasonal pricing strategy in Huangshan Scenic Area

	On season (November–March)	Off season (December–February)
Ticket price	RMB 230 ($34)	RMB 150 ($22)
Cloud valley, Pacific cableway ticket (one way)	RMB 80 ($12)	RMB 65 ($10)
Yuping cable ticket (one way)	RMB 90 ($13)	RMB 75 ($11)
West Sea Grand Canyon cable ticket (one way)	RMB 100 ($15)	RMB 80 ($12)

※ USD 1.00 = RMB 6.79207 (calculated on 1 June, 2017).
Source Authors' compilation

Carrying capacity was calculated in the HSA 2007–2025 master plan, with the plan for instant visitor capacity estimated to fall within the range of 15,000 to 22,000 visitors. By 2025, instant capacity will fall to 9900, with a daily capacity of 12,000 visitors. The annual visitor capacity was set at 2.6 million according to the 2006 benchmark, including 2.4 million domestic and 201,000 international visitors (HMG, 2006.). However, the actual number of visits has consistently exceeded the stated carrying capacity. For example, according to HTD's 2015 Annual report, 3.1828 million visitors entered Huangshan, a year on year increase of 7.1%. Revenue reached 1.665 billion RMB, an increase of 11.73% year on year. Net profit attributed to shareholders of listed companies reached RMB 296 million, an increase of 41.36%. (HTD Co., Ltd, 2016). The site has expanded to incorporate 4 different cableways located in different parts of the Huangshan covering distances of up to 3.7 km.

Huangshan's public–private partnership paved the way for establishment of a market-based cost recovery model. HTD Co. wields extensive control over HSA via site management which includes daily operations, infrastructure construction, and environmental sanitation, as well as cableway operations, hotels, travel agencies, and food & beverage franchises. However, the system places considerable responsibility on the efficiency and ethics of appointed civil servants. Meanwhile, the significant revenue stream created by HSA for local government coffers could conversely create conservation quandaries and sideline some segments of residents (Zinda, 2012). The realities of 'overtourism' placed pressure on environmental and social carrying capacity, as shown by the example of 23.6% overvisitation in 2015. In search of sustainability, HTD Co. Ltd. has pioneered a range of solutions linked to water and waste management, while freezing entry fees and using price-discount mechanisms to encourage off-peak visitation. Nonetheless, significant conservation challenges remain to protect Huangshan.

2.6 Conclusion

This chapter discusses the nexus of tourism development and nature conservation related to the protected area (PA) system in China over the past four decades. Among various categories of PAs, the forest park network was among the first to be established and has a comparatively larger number of sites, so it is used an example to analyse the dramatic trends observed in development and nature conservation issues. China has developed many diverse PA labels that seek to marry conservation goals with the promotion of economic development within and around the PAs (Li et al., 2008; Su et al., 2007; Zhou & Grumbine, 2011). In fact, this emphasis on economic development distinguishes China's PA network from Western equivalents (Miller-Rushing et al., 2017). The central government wants to make conservation the top priority by establishing a new system of national parks. However, the provincial governments which oversee the PAs tend to consider a PA designation as a poverty alleviation strategy and prioritize tourism development. As this chapter has shown, forest tourism provides revenue for some poor local communities and provides funding for the construction of infrastructure in the PAs.

Deteriorating natural resources inside many of the PAs due to tourist development and overlapping networks of management authorities are some of the factors that have forced the central government to rebuild the nature conservation system and make it more effective. Since the 18th National Congress of the Communist Party held in 2012, the central government has mentioned 'ecological civilization construction' at an unprecedented level. The new national parks system represents an important actualization of this concept.

The new national parks system attempts to protect endangered species by eliminating tourist activities inside the PAs' core zones while achieving natural resource conservation goals. However, the fledgling national parks have a long way to go before they can function effectively to protect nature and resolve the existing dilemma. China's national park system adopts a top-down approach to planning and management, leaving very limited room to local communities' rights and sustainable livelihoods. The conservation paradigm worldwide has shifted to a participatory and inclusive approach (Wang, 2019). This top-down approach will eventually risk hampering conservation goals without local community's support by depriving them of their interests (Ibid). Following a U.S. model, China has decided to stress the natural environment and has built a strong governance framework to protect threatened and endangered species. However, excluding human presence within protected areas is impractical and may be counterproductive to conservation work (Guo, 2016, as cited in Wang, 2019). For humans and wild species and forests have a long history of coexistence in China and the PAs that the government is trying to protect are anthropogenic rather than the pristine wilderness of the American continent (Olwig, 2002). One recent study suggests integrating biodiversity, ecosystem services, and human activities to achieve sustainable development goals for biodiversity conservation (Xu et al., 2017).

The past decades have witnessed rapid growth in tourist numbers and related revenues in forest parks, as in other types of PAs. In fact, 'forest park' is a PA category not widely used outside China (Chen et al., 2019), where 'forest tourism' encompasses ecotourism activities conducted in forest-based settings. However, Chinese ecotourism activities tend to be passive when compared to Western equivalents, with an overwhelming majority of such activities involving the appreciation of natural or cultural attractions (Chen et al., 2018; Wei et al., 2006). Park agencies in China have thus promoted the healthy function of the fresh forest air, and the cultural heritage that also comprises an important attraction for Chinese ecotourists (Buckley et al., 2008; Chen et al., 2018).

However, the fast-increasing tourist numbers have induced overuse of natural resources, in combination with other problems related to the lack of competent management personnel and pertinent monitoring. The large number of tourists who seek out nature and fresh air consequently put too much pressure on PA conservation. Overexploitation of tourism, a chaotic governance system, conflicts with local communities living in and around park, and a shortage of professional park managers partly due to the sharp expansion of PAs have all threatened the sustainable management of natural resources, thus the central government started to redesign a centrally-controlled nature park system (Li et al., 2016).

The first case study of Pudacuo National Park illustrates that transformation to a new national park system, instructed by the provincial government, achieved some preliminary gains in protecting endangered species through halting ecotourism activities inside the PA's core zone. However, this new PA scheme has not assured the automatic establishment of an effective conservation system. A stricter management approach has further deprived local communities' interests by benefiting from important natural resources and may result in the loss of support by local communities. It is recommended that a sustainable tourism industry be established to benefit the local community without comprising environmental problems. Such a sustainable approach should consider not only the residents, who may focus on the short-term benefits, taking consideration of their low education level (Wang et al., 2012) but this approach should also include increased participatory roles for NGOs and the tourism industry in planning and development to ensure diverse views and to better equip PAs to foresee possible issues in future operations of the new national parks.

The second case study of Huangshan Scenic Area details the experience of a seasonal pricing strategy and setting both a daily and annual carrying capacity. Moreover, HTD Co. Ltd. has pioneered a range of sustainable solutions linked to water and waste management. However, as with the first case study, the public–private partnership management system has impaired its effectiveness to resolve the 'overuse' or 'over-tourism' issue. HSA has created enormous revenue for the local government and huge profits for HTD Co. Ltd. This kind of partnership management system represents the crux of conservation quandaries.

References

Buckley, R., Cater, C., Zhong, L., & Chen, T. (2008). Shengtai Luyou: Cross-cultural comparison in ecotourism. *Annals of Tourism Research, 35*(1), 945–968.
Buckley, R., Zhong, L., & Ma, X. (2017). Visitors to protected areas in China. *Biological Conservation, 209*, 83–88.
Chen, B., Qi, X., & Qiu, Z. (2018). Recreational use of urban forest parks: A case study in Fuzhou National Forest Park China. *Journal of Forest Research, 23*(3), 183–189.
Chen, Z., Fu, W., Konijnendijk, C., Pan, H., Huang, S., Zhu, Z., Qiao, Y., Wang, N., & Dong, J. (2019). National forest parks in China: Origin, evolution, and sustainable development. *Forests, 10*(4), 323. https://doi.org/10.3390/f10040323
China Forestry Yearbook 1989–1986, 1987–2005. China Forestry Press. Beijing (in Chinese).
China. The State Forestry Administration. (1994). Forest park management regulation. http://www.forestry.gov.cn/ (in Chinese).
Deng, J., Bauer, T., & Huang, Y. (2003). Ecotourism, protected areas, and globalization: Issues and prospects in China. *ASEAN Journal on Hospitality and Tourism, 2*(1), 17–32.
Dong, Z. (2002). *Forest tourism in China*. Petroleum Industry Press (in Chinese).
Duan, W., & Wen, Y. (2017). Impacts of protected areas on local livelihoods: Evidence of giant panda biosphere reserves in Sichuan Province, China. *Land Use Policy, 68*, 168–178.
Font, X., & Tribe, J. (1999). *Forest tourism and forest recreation*. CABI Publishing.

Gong, H. & Ye, W. (2017). Theoretical explore and practice of ecotourism in Pudacuo National Park. In Ye, W., Zhang, Y., & Li, H. (Eds.), *China's Ecotourism Development Report* (pp. 435–442) (in Chinese).

Gulez, S. (1992). A method of evaluating areas for national park status. *Environmental Management, 16*(6), 811–818.

Han, Z., & Zhuge, R. (2001). Ecotourism in China's nature reserves: Opportunities and challenges. *Journal of Sustainable Tourism, 9*(3), 228–242.

Hu, J., Zhang, P., & Mei, Y. (2004). On sustainable development of forest tourism in China. *Journal of Zhejiang Forestry College, 21*(2), 194–197 (in Chinese).

Hu, Y., & Zhang, Q. (1998). Discussion on the theoretical problems of forest parks-also relationship among nature reserves, scenery spots and famous sites, and forest parks. *Journal of Beijing Forestry University, 20*(3), 49–57 (in Chinese).

Huang, S., Lan, S., Lian, Q., Fu, Y., & You, Y. (2002). Investigation and analysis about tourist preference of national forest park in Fuzhou area. *Problems of Forestry Economics, 22*(5), 308–311 (in Chinese).

Huang, Y., Deng, J., Li, J., & Zhong, Y. (2008). Visitors' attitudes towards China's national forest park policy, roles and functions, and appropriate use. *Journal of Sustainable Tourism, 16*(1), 63–84.

Huangshan Municipal Government – HMG. (2006). Overall Planning of Huangshan Scenic Spot 2007–2025. Statistics Bureau of Huangshan, Huangshan Statistics Yearbook 2005. p. 338 (in Chinese).

Huangshan Tourism Development (HTD) Co., Ltd. (2016). Annual report 2015. http://app.finance.ifeng.com/data/stock/ggzw.php?id=15639337&symbol=600054.

Li, C., & Shou, P. (2003). The restart of the construction of Zhangjiajie sightseeing lift encounters the embarrassment of "up or down." *People.* http://www.people.com.cn/GB/huanbao/1074/2029784.html.

Li, J., Wang, W., Axmacher, J. C., Zhang, Y., & Zhu, Y. (2016). Streamling China's protected areas. *Science, 351*(6278), 1160. https://doi.org/10.1126/science.351.6278.1160-a.

Li, M., Wu, B., & Cai, L. (2008). Tourism development of World Heritage sties in China: A geographic perspective. *Tourism Management, 29*, 308–319.

Li, S. (1994). Evaluation of present situation and trend prediction of China's forest parks. *Journal of Central South Forestry University, 14*(2), 163–167 (in Chinese).

Li, S., & Chen, X. (2007). Study on the developing of track of China's forest parks and forest tourism. *Tourism Tribune, 22*(5), 66–72 (in Chinese).

Li, W. (2004). Environmental management indicators for ecotourism in China's nature reserves: A case study in Tianmushan Nature Reserve. *Tourism Management, 25*, 559–564.

Liu, Y. (2010). Forestry Bureau: Forest tourism injects vitality and vitality into modern forestry construction. Retrieved on June 19, 2020 from http://www.gov.cn/jrzg/2010-05/27/content_1615118.html.

Liu, Y., & Tao, Y. (2003). The analysis of the development obstacles for Chinese forest tourism. *Problems of Forestry Economics, 23*(1), 49–52 (in Chinese).

Liu, Y., Wu, S., & Li, M. (2008). Econometrics study on forest travel consumption behaviour of Chinese citizen: An example of citizen of Xi'an. *Journal of Arid Land Resources and Environment, 22*(11), 145–149 (in Chinese).

Ma, X., & Lin, M. (2007). Forest tourism in Guangdong. *Tropical Geography, 26*(3), 285–289 (in Chinese).

Ma, X., Ryan, C., & Bao, J. (2009). Chinese national parks: Differences, resource use and tourism product portfolios. *Tourism Management, 30*, 21–30.

Miller-Rushing, A. J., Primack, R. B., Ma, K., & Zhou, Z. (2017). A Chinese approach to protected areas: A case study comparison with the United States. *Biological Conservation, 210*, 1010–1112.

National Forestry and Grassland Administration. (2019). Forest Resources in China. http://www.china-ceecforestry.org/wp-content/uploads/2019/08/Forest-Resources-in-China%E2%80%94%E2%80%94The-9th-National-Forest-Inventory.pdf.

Olwig, K. (2002). *Landscape, nature, and the body politics: From Britain's renaissance to Amercia's new world*. University of the Wisconsin Press.

Shi, Q. (2005). *Environmental impact assessments of forest parks*. China Science Press.

Singh, S. (1992). Tourism Development of Huangshan Scenic Area, Letter published in WHC (UNESCO World Heritage Commission) (2017). Properties inscribed on the World Heritage List, China. http://whc.unesco.org/en/statesparties/cn.

Sofield, T., & Li, F. (1998). Tourism development and cultural policies in China. *Annals of Tourism Research, 25*, 362–392.

Su, D., Wall, G., & Eagles, P. F. J. (2007). Emerging governance approaches for tourism in the protected areas in China. *Environmental Management, 39*, 749–759.

Tian, Y. (2004). The death of a large number of living fossils due to the over utilization of tourism in Henan. http://www.people.com.cn/GB/14838/14839/22019/3035061.html (in Chinese).

UNEP-WCMC, & IUCN. (2020). Protected planet: The World Database on Protected Areas (WDPA). UNEP-WCMC and IUCN. Retrieved August 01, 2020 from www.protectedplanet.net.

Wang, G., Innes, J., Wu, S., Krzyzanowski, J., Yin, Y., Dai, S., & Liu, S. (2012). National park development in china: Conservation or commercialization? *Ambio, 41*(3), 247–261.

Wang, J. Z. (2019). National parks in China: Parks for people or for the nation? *Land Use Policy, 81*, 825–833.

Wang, L., Chen, A., & Gao, Z. (2011). An exploration into a diversified world of national park systems: China's prospects within a global context. *Journal of Geographical Sciences, 21*(5), 882–896.

Wang, X., & Wang, J. (1998). Forest park and ecotourism. *Tourism Tribune, 2*, 58–62 (in Chinese).

Wang, X. (1991). Review report on the course of protected areas in China. Document prepared for the Palaearctic Asia Regional Review, presented at the Congress on World Parks and Protected Areas at Caracas, Veneduela, February 1992.

Wang, X. (1993). Significance and methods for developing ecotourism in protected areas. *Journal of Plant Resources and Environment, 2*, 49–54 (in Chinese).

Wang, Y. (2016). National parks, system overhaul/new system, new mentality/finding peace, China Report, 20–33.

Weaver, D., & Lawton, L. (2007). Twenty years on: The state of contemporary ecotourism research. *Tourism Management, 28*, 1168–1179.

Wei, Z., Li, J., & Wang, Z. (2006). The summary of the forest tourism development in China. *Problems of Forestry Economics, 26*(2), 142–145 (in Chinese).

Wen, H., & Dai, M. (2004). On the historical and cultural characteristics of forest tourism in China. *Qiu Suo, 12*, 159–161 (in Chinese).

World Heritage Center (WHC). (2017). World Heritage List. Retrieved on June 16, 2020 from https://whc.unesco.org/en/list/?search=&type=mixed&order=country.

Wu, C., & Wu, Z. (1998). Developing prospect of forest recreation in China. *Journal of Central South Forestry University, 18*(3), 96–100 (in Chinese).

Wu, J., Hu, Y., Liu, T., & He, Q. (2018). Value capture in protected areas from the perspective of common-pool resource governance: A case study of Jiuzhai Valley National Park China. *Land Use Policy, 79*, 452–462.

Wu, Z. (2003). Further exploitation of healthcare tourism resources in forestry recreation areas. *Journal of Beijing Forestry University, 25*, 63–67 (in Chinese).

Xu, H., Zhu, D., & Bao, J. (2016). Sustainability and nature based mass tourism: Lessons from China's approach to the Huangshan Scenic Park. *Journal of Sustainable Tourism, 24*(2), 182–202.

Xu, W., Xiao, Y., Zhang, J., Yang, W., Zhang, L., Hull, V., Wang, Z., Zheng, H., Liu, J., Polasky, S., Jiang, L., Xiao, Y., Shi, X., Rao, E., Lu, F., Wang, X., Daily, G. C., & Ouyang, Z. (2017). Strengthening protected areas for biodiversity and ecosystem services in China. *PNAS, 114*(7), 1601–1606.

Yan, C., & Morrison, A. (2008). The influence of visitors' awareness of world heritage listings: A case study of Huangshan, Xidi and Hongcun in Southern Anhui, China. *Journal of Heritage Tourism, 2*(3), 184–195.

Yang, M., & Zhou, G. (2005). Study on effective measures of environmental protection for sustainable development of tourism industry in the world natural heritage Wulingyuan. *Ecological Economy of China, 1*(2), 84–88 (in Chinese).

Yu, H. (2001). A strategic study on the development of China's forest tourism industry in the 21st century. *Tourism Tribune, 16*(5), 67–69 (in Chinese).

Yu, K., & Jin, Y. (2007). A literature review of forest tourism theory research in home and abroad. *Problems of Forestry Economics, 27*(4), 380–384 (in Chinese).

Zhang, H., & Zhao, Z. (2005). A survey of ecotourism market structure based on ehaviour and attitude: Case study of Taibai Mountain National Forest Park. *Tourism Tribune, 20*, 34–38 (in Chinese).

Zhong, L., Buckley, R., Wardle, C., & Wang, L. (2015). Environmental and visitor management in a thousand protected areas in China. *Biological Conservation, 181*, 219–225. https://doi.org/10.1016/j.biocon.2014.11.007.

Zhong, L., Deng, J., & Xiang, B. (2008). Tourism development and the tourism area life-cycle model: A case study of Zhangjiajie National Forest Park, China. *Tourism Management, 29*, 841–856.

Zhong, L., Wu, C., & Xiao, W. (1998). Aeroanion researches in evaluation of forest recreation resources. *Chinese Journal of Ecology, 17*(6), 56–60 (in Chinese).

Zhou, D. Q., & Grumbine, R. E. (2011). National Parks in China: experiments with protecting nature and human livelihoods in Yunnan province, People's Republic of China (PRC). *Biological Conservation, 144*, 1314–1321.

Zhou, N., & Yu, K. (2004). The urbanization of national park and its countermeasures. *Urban Planning Forum, 1*, 57–61 (in Chinese).

Zhu, Y. (2007). Discussion on the sustainable development of forest eco-tourism in Henan Province. *Journal of Xinyang Normal University (Philos & Sci Edit), 27*(6), 56–58 (in Chinese).

Zinda, J. A. (2012). Hazards of collaboration: Local state co-optation of a new protected-area model in Southwest China. *Society & Natural Resources: An International Journal, 25*(4), 384–399.

Bixia Chen is Associate Professor in the Faculty of Agriculture, University of the Ryukyus in Okinawa, Japan. Her research interests include culturally preserved forests, forest park management, forest tourism, and rural tourism. Chen obtained her PhD at Kagoshima University and has conducted field surveys relevant to the spatial distribution and ecosystem services of Garcina subelliptica (Fukugi) homestead windbreak on Ryukyu Archipelago.

Yinping Ding is a doctoral candidate at Geographical Science School of Yunnan Normal University (YNNU) in Kunming, China. Her research interests include Aesthetic Ecosystem Service, Landscape Ecology, Eco-Tourism, Heritage conservation and Sustainability.

Yuanmei Jiao is Professor in the Faculty of Geographical Science at Yunnan Normal University (YNNU) in Kunming, China. Her research interests include Landscape Ecology, Eco-Tourism, Heritage conservation and Sustainability. Jiao completed her PhD at the Chinese Academy of Science and Post-Doctor fellow at the University of Tokyo.

Yi Xie is Professor at School of Economics and Management, Beijing Forestry University, China. His research interests are forest management institutional efficiency evaluation, farm household forestry in China. Xie completed his PhD at Beijing Forestry University.

Thomas E. Jones is Associate Professor in the Environment & Development Cluster at Ritsumeikan APU in Kyushu, Japan. His research interests include Nature-Based Tourism, Protected Area Management and Sustainability. Tom completed his PhD at the University of Tokyo and has conducted visitor surveys on Mount Fuji and in the Japan Alps.

Chapter 3
Japan's National Parks: Trends in Administration and Nature-Based Tourism

Thomas E. Jones and Akihiro Kobayashi

3.1 Introduction

Given its prevailing image as a high-tech Asian economic powerhouse, visitors to Japan are often surprised to find that over two-thirds of the country comprises mountainous woodland. In 2010, Japan ranked in the world's top ten most-forested countries (GFW, 2020), with 80% of the archipelago defined as 'mountains' or 'highlands' (Karan, 2005). However, extensive infrastructure improvements have helped 'conquer' and 'tame' these upland areas for regional development and to facilitate mass tourism (Totman, 2014). One consequence of the combination of extensive mountain forests and convenient access arrangements is that Japan also ranks high in terms of per capita visits to national parks (Pergams & Zaradic, 2008). However, visitation has declined since the 1990s, and after a recent rebound due to the rise in inbound tourism, has suffered from the onset of the coronavirus.

Like many global PA networks outside of the 'new world' nations, Japan's national parks must juggle multiple mandates, seeking to balance traditional land use with nature-based tourism (NBT) and biodiversity conservation (Kato, 2008). The latter is shaped by geophysical extremes that range from Subarctic (northern Hokkaido) to Subtropical zones (southern Okinawa). The eastern Pacific seaboard has a relatively dry climate, whereas certain regions along the North-west coast experience heavy dumps of 'powder snow' and Yakushima island—a temperate rain forest off the south coast of Kyushu- records some of the highest precipitation on the planet (CEPF, 2020). The wildlife spectrum has over 90,000 native species, including iconic

---

T. E. Jones (✉)
College of Asia Pacific Studies, Ritsumeikan Asia Pacific University (APU), Beppu, Japan
e-mail: 110054tj@apu.ac.jp

A. Kobayashi
Faculty of Economics, Senshu University, Tokyo, Japan
e-mail: 33kobayasi@isc.senshu-u.ac.jp

© The Author(s), under exclusive license to Springer Nature Switzerland AG 2021
T. E. Jones et al. (eds.), *Nature-Based Tourism in Asia's Mountainous Protected Areas*, Geographies of Tourism and Global Change,
https://doi.org/10.1007/978-3-030-76833-1_3

examples such as the Brown Bear (*Ursus arctos*), Giant Salamander (*Andrias japonicus*), and the Macaque (*Macaca fuscata*) more commonly known as 'snow monkeys' (IUCN, 2020). Viewing such species is one of the main targets of wildlife tourism (Usui & Funck, 2018), a growing niche within the broader market for nature-based tourism (NBT).

However, the wider context for Japan's rural regions includes demographic decline and the socio-economic hollowing out of the countryside (Matanle & Rausch, 2011). Biodiversity is also under threat. For example, Yakushima is one of 6,852 islands that are home to distinct flora and fauna including many endemic species, some of them Critically Endangered, such as the Okinawa woodpecker (*Dendrocopos noguchii*) or Endangered, such as the Bonin Flying Fox (*Pteropus pselaphon*) (Vincenot, 2017). At the Convention on Biodiversity meeting in 2010, Target 11 sought to expand global Protected Areas (PAs) to 17% of terrestrial and inland water areas, and 10% of coastal and marine areas by 2020 (Woodley et al, 2019). This Aichi Target 11 provides a timely opportunity to take stock of Japan's PAs and review their fundamental objectives. This chapter investigates the longitudinal development of Japan's 'multi-purpose' parks from the twin perspectives of supply and demand, the former exploring land ownership and zoning policy before the latter examines evolving trends for NBT. Within an extensive PA network, the focus falls on the national parks that represent the 'gold standard' of Japan's triple-tiered nature park system and form a rallying point for conservation goals while attracting the bulk of NBT. Section 3.3 uses the case study example of climbers on iconic Mount Fuji's slopes to symbolize the potentials and pitfall of Japan's multi-purpose parks in practice. Finally, Sect. 3.4 discusses the key characteristics and challenges for parks and peaks and the future direction of NBT.

3.2 Japan's National Parks

3.2.1 Historical Roots Prior to World War II

The first countries to promote their version of a 'national park' philosophy were Australia, Canada, South Africa and the United States at the end of the nineteenth century (Kupper, 2009), but the history of Japan's PAs was not far behind. Among the earliest efforts to pass PA legislation was a petition presented to Parliament in 1911 to recognize Nikko as an 'Imperial Park' (MOE, 2018). However, a fierce debate over the designation criteria raged for many years. Later, this evolved into an ideological dispute between academics Honda Seiroku and his former student Uehara Keiji that echoed the rift between American preservationists, led by John Muir, and conservationists, such as Gifford Pinchot, that took a more utilitarian approach to natural resource management (Meyer, 1997). Political and bureaucratic manoeuvring also stalled the early efforts to establish PAs in Japan. It took the prolonged economic depression in the 1920s to finally convince lawmakers to pass the National Parks Act in 1931, as part of a tourism policy geared towards regional development that also aimed to attract international tourists (Kato, 2008).

Thereafter the first 12 parks were designated in quick succession between 1934 and 1936. In 1937, three further parks were gazetted in Taiwan, at the time a Japanese colony. These new national parks at Tatun, Tsugitaka-Taroko, and Niitaka-Arisan, extended the designation criteria based around sacred mountains from the 'motherland' (Kanda, 2012). However, during the increasingly desperate conflict of the Pacific war, PA objectives were radically re-worded to reflect the era's facist policy-making imperatives—even the flagship journal's title 'National Parks' was re-named in 1943 to the more nationalistic 'National Land, Healthy Nation' (Nishimura, 2016). In the aftermath of World War II, the direction of park planning was transformed again following an influential report by Walter Popham, an American landscape architect commissioned by the Allies' General Headquarters. Japan's era of rapid economic growth had begun, and as 'national park' status became a coveted label for regional development the number of designations quickly increased. Ise-Shima was the first post-war listing in 1947 (Mizuuchi, 2012), with a further seven had been added by 1955. In an era of economic re-building, the national parks became place brands, seen as 'solutions that motivate and co-ordinate the various stakeholders for the regional interests' (Rainisto, 2003).

3.2.2 A System of 'Multi-Purpose' Nature Parks

Attempts were made to rein in the more rampant development projects occurring in and around PAs by bolstering the conservation capacity. The original National Parks Act of 1931 was superseded in 1957 by the Nature Parks Law, an umbrella legislation that capped the number of 'national' parks at 20 maximum and introduced the subcategories of 'quasi-national' and 'prefectural' nature parks. A hierarchical system of Special Conservation Zones was also introduced, and despite several subsequent amendments, this zoned, triple-tiered system survives to this day (Table 3.1). As the top tier, the national parks represent the highest level of environmental protection

Table 3.1 Overview of Japans' triple-tiered nature park system

Type of park	Number	Area (sq km)	%[a]	Designated	Maintained
National	34	21,949.31	5.81	central	central
Quasi-national	57	14,451.50	3.82	central	local
Prefectural	311	19,487.30	5.16	local	local
Total	402	55,888.11	14.79		

[a]Area as a per cent of Japan's total land mass, calculated as 377,974.92 sq km in the 2019 national census
Source MOE (2020)

afforded to a broad range of habitats from mountainous forests to lakes, rivers, coastline and wetlands. However, the national parks do not consist solely of untouched nature, instead displaying varying degrees of human impact. These include cultural heritage sites such as temples and shrines, traditional land uses such as farming and forestry, and also sizeable resident populations living around or even inside the park boundaries, as discussed next through the lens of land ownership in the parks.

3.2.3 Ownership of National Park Land

Diverse land ownership mosaics are the key to understanding why so many of Japan's national parks are protected as IUCN Category V 'working landscapes' rather than recognized as Category II 'national parks.' Covering a sizable area that accounts for almost 6% of the total landmass of Japan (Table 3.1), national parks must make space for 'secondary nature,' including traditional forestry or farming landscapes managed by local communities. In fact, 25.6% of Japan's national park land is privately owned, with a further 12.4% of the total owned by local government, including prefectures and municipalities. Most of the parkland (an average of 62% across the country) is designated national forestland owned directly by the state. However, ownership titles belong not to the Ministry of Environment (MOE) but instead fall under the jurisdiction of the Forestry Agency, a branch of the Ministry of Agriculture, Forestry and Fisheries (MAFF).

Notably, the aforementioned national averages conceal large discrepancies between the 34 individual parks. Some, such as Ise-Shima in Mie Prefecture, are almost entirely privately owned (96.1%) due to the extensive Shinto shrine complex of Ise Jingu. Others, such as Daisetsuzan in Hokkaido, comprise mostly state-owned land (94.7% or 99.1% if all non-private land is included). If the latter resembles a US-style 'set-aside' park, the former is closer to European 'co-management' models. This diversity within the national parks network, coupled with a strong legal tradition of protecting private landowners' rights, ensures that achieving conservation goals across parkland with multiple owners is no simple matter. In addition, the MOE's legal mandate under the Nature Parks Act is undermined by a lack of land ownership. The MOE owns only 0.2% of all national parkland (Kato, 2008), compared to the 62% owned by the MAFF. Furthermore, the MOE's park administration division has traditionally been hamstrung by insufficient funding and a lack of human resources (Imura, 2005). Instead, the MOE relies on legal 'sticks' and 'carrots' laid out by the Nature Parks Act to steer stakeholders toward conservation goals assisted by a range of environmental laws such as the Wildlife Protection Act, Existence of Species Act and the Ecotourism Act (2008).

3.2.4 Managing 'Multi-Purpose' Parks with Core and Buffer Zones

In order to manage the 'multi-purpose' parks without large-scale MOE land ownership, the Nature Parks Act permits a range of uses and attempts to funnel them toward twin goals of resource conservation and enjoying nature, delineated by the MOE (2018) as follows:

(1) to restrict development projects and other human activities with a view to protecting the exceptional natural landscapes that are characteristic of Japan; and (2) to foster a joyful experience of nature, including an appreciation of landscapes.

In the spirit of Yellowstone's epitaph 'for the benefit and enjoyment of all,' sightseeing and landscape appreciation were among the original designation goals (Murakushi, 2005). However, numerous other objectives have been added over time to meet the changing needs of PAs. For example, a system of wilderness areas was added in 1972 based on the US model, but according to Kato (2008), the few examples are too small and remote to offer effective conservation. 1972 was also a benchmark for Marine Protected Areas with the triple listing of two tropical archipelagos (Iriomote-Ishigaki in Okinawa and Ogasawara, also known as the Bonin Islands) and the Ashizuri-Uwakai coastline along the southwestern tip of Shikoku. Kushiro-shitsugen was Japan's first wetland to be listed under the Ramsar Convention in 1980, and in 1987, it was also designated as a national park. 'Biodiversity' itself was a more recent addition to official PA objectives. After Japan ratified the Convention on Biological Diversity in 1993, a National Biodiversity Strategy came into force in 1995. Thereafter, references to biodiversity began to feature more prominently in park and PA planning. In order to achieve the overarching goals to balance protection and promotion, each national park has an individual management plan, updated every five years and divided into a Regulatory Plan and a Facility Plan.

To regulate the activities of different landowners, the parks are administered using a system known as *chiikisei*, translated literally as "national park management by zoning and regulation" but interpreted by Hiwasaki (2005) to mean 'multiple-use parks.' Unlike in the U.S. and Canada, for example, where state land has been 'set-aside' for the exclusive purpose of creating parkland, Japan's 'multi-purpose' system instead seeks to superimpose conservation goals via a sliding scale of regulations on land use and development, with planning permission levers pulled to limit human impact on the environment (Kato, 2008). Based on prior land use patterns, and specific conservation targets, the parkland is divided into 'Special areas,' equivalent to core zones that prioritize conservation goals, and 'Ordinary areas' akin to buffer zones (Table 3.2.). At the top of the hierarchy, Special Protection Zones are subject to the strictest regulations, with MOE permission needed for any kind of cutting or collecting of wood, plants, or rocks. Activities such as wild camping, picking plants or flowers, or building campfires are also illegal in these zones. However, it is worth noting that these Special Protection zones amount to only 13% of all national parkland (Table 3.2) and 5% of quasi national parkland—often in the most mountainous or inaccessible regions. In Special Zones (sub-divided into Class I to

Table 3.2 Breakdown of Japan's nature park areas (ha) by conservation zone

Type of park	Special Protection Zones	Special Zones			Ordinary Zones	Total
		Class 1	Class 2	Class 3		
National park	292,399	290,792	517,133	518,749	575,859	2,194,931
(percentage of total)	13.3%	13.2%	23.6%	23.6%	26.2%	100.0%
Quasi-national park	65,202	176,050	385,189	707,186	111,523	111,523
(percentage of total)	4.5%	12.2%	26.7%	48.9%	7.7%	100.0%
Prefectural nature park	–	69,093	178,897	447,186	1,253,555	1,948,730
(percentage of total)	0.0%	3.5%	9.2%	22.9%	64.3%	100.0%
Total	357,601	535,935	1,081,219	1,673,121	1,940,937	5,588,811
(percentage of total)	6.4%	9.6%	19.3%	29.9%	34.7%	100.0%

Source Adapted from MOE (2020)

III), commercial activities such as construction, forestry, and vehicular access (by snow mobiles and so on) face restrictions (Kato, 2008). However, in the Ordinary (buffer) zones, development projects face less rigorous restraints—even mining and large-scale construction projects merely requiring notification to the MOE rather than their explicit permission to proceed. Compared to stricter conservation areas, these 'ordinary' zones comprise the largest single category that accounts for 26% of all parkland (see Table 3.2), which has ramifications for the PA's effectiveness as discussed later in the case study of Mt Fuji.

3.2.5 National Park Administration

The next section investigates park administration to see why MOE struggles at times to achieve its zoning-based management goals for a number of institutional reasons. Besides its insignificant landownership, MOE is a relative newcomer to Kasumigaseki (the Japanese equivalent to Washington or Westminster) since it was established only in 1971 as the Environment Agency (Imura, 2005). It is thus consigned to play a coordinating role in park planning, yet its institutional objectives do not always chime with those of other key stakeholders. One of the longest-running rivalries has been with the MAFF, a senior agency, established in 1881, whose National Forest covers 62% of national parkland and around 20% of the entire landmass of Japan (Tanaka, 2014). MAFF dominates MOE in terms not only of land ownership but also human and fiscal resources. Yet the managerial objectives of the Forestry

Agency within the MAFF have traditionally been distinct from those of the MOE, tending to prioritize timber extraction over conservation or tourism (Kato, 2008). This conflict of interests undermines the MOE's ability to implement effective park administration, as demonstrated by research in Hokkaido which used GIS to overlay a national forest management plan with that of Shikotsu-Toya National Park. The results confirmed that half of the park boundaries were indistinguishable from those of the national forest, while 36% matched municipal boundaries (Aikoh & Tomidokoro, 2010). Such evidence implies that the national park plan, which in theory should provide varying levels of regulation to achieve biodiversity conservation, was in reality zoned around a priori forestry and governance interests that are often inconsistent with the park values espoused by the MOE.

The MAFF's historical legacy of allowing timber extraction from national forest inside the national parks has occasionally resulted in outright conflict. One example was the Shiretoko logging controversy, which erupted from 1981 to 1987 following the publication of a plan to clear-cut swathes of Shiretoko National Park, one of Japan's last-surviving old growth forests on the north-eastern tip of Hokkaido (Natori, 1997). The MAFF's logging faced unprecedented resistance from a conservation alliance underpinned by the Shiretoko National Trust movement, but ultimately centred on the MOE. This was a rare example of the MOE straying from its conflict-averse neutral stance to petition strongly in favour of biodiversity conservation, citing the intact 'sea-to-summit' ecology of a site that would go on to be listed as a natural World Heritage Site (Mitsuda & Geisler, 1992). The MAFF reluctantly reduced the target area size and style of timber extraction. The Shiretoko incident thus helped accelerate the diversification of Japanese forest policy away from the post-war norm of monoculture plantations, hastening the MAFF's re-orientation from production toward 'alternative' goals including biodiversity and recreation. However, the policy shift also reflected an ongoing economic decline in the domestic forestry industry, echoed by a wholesale down-sizing in the scale of MAFF budget and number of bureaucrats (Tanaka, 2014). Despite the restructuring, a controversial merger between the two agencies was never realized and to date the MAFF still owns the majority of parkland and dominates the MOE in both budget and manpower, with an approximate 4:1 ratio for permanent employees (e-Stat, 2020). An unofficial stalemate has thus been reached in the national parks between a depleted but embattled MAFF and the MOE, the legal park manager that continues to 'reign not rule.'

3.2.6 Human Resource Dimension of Park Administration

This intra-agency rivalry also reflects a longer running debate to find the most suitable manager for Japan's parks. The pre-war roots of administration lie in the Department of Interior's Sanitary Bureau, but in 1938, jurisdiction was transferred along with the Health Division to the newly formed Ministry of Health and Welfare. However, human resources were insufficient in the pre-war era, and visitor services almost nonexistent (Murakushi, 2005). Having already been downsized, the skeleton staff was stripped back to one single ranger and briefly disbanded altogether in 1945 (Tanaka, 2014). Even in the post-war period, regional development agenda were commonplace with conservation personnel few and far between. However, following a series of major industrial pollution outbreaks and court cases, the Environment Agency was established in 1971, assuming responsibility for Nature Parks. It was upgraded to Ministry status in 2001 and the current MOE administers the parks via seven Regional Environmental Affairs Offices with 1,901 permanent staff in 2019 (e-Stat, 2020). These include Rangers—permanent employees that rotate periodically between the Head office in Tokyo and local branch offices responsible for park administration on a daily basis. The nomenclature of these 'rangers' has evolved from *kokuritsu koen kanriin* (national park supervisors) between 1953–1984, to *kanrikan* or (national park officers) until 2000, before becoming *shizen hogokan* (literally conservation officers) in 2000 (Kato, 2008). Their number increased in the 1970s after the Environment Agency was set up, then again in the 1990s, and most recently in 2001 when the MOE was established. Yet the current total of around 300 Rangers across all national parks is trivial considering the territorial size of their administrative domains and the number of visitors such areas attract (Kato, 2008). One extreme example is at Fuji-Hakone-Izu National Park, where only 4 tenured Rangers are responsible for the Mount Fuji district that attracts tens of millions of visitors each year. The bulk of the Rangers' work thus comprises administrative deskwork related to planning permission applications although they are theoretically mandated to carry out a wide range of tasks including conservation of native flora and fauna, updating park plans, patrolling and conducting surveys (Tanaka, 2014).

The core of MOE Rangers have been boosted since 2005 by the addition of Active Rangers, a complementary system that plays a more hands-on role carrying out patrols in the field and coordinating volunteers. As of 2007 the MOE Rangers and 80 Active Rangers were supported by a voluntary or non-contract workforce of Nature Park Leaders, Park Volunteers and Green Workers who assist with nature regeneration and extermination of invasive species programs. Other stakeholders include the Natural Parks Foundation that plays a role in certain parks via management of car parks, toilets and other vital tourist facilities (Fig. 3.1).

3.3 Nature-Based Tourism

3.3.1 Nature Based Tourism as a Lens for Viewing Visitation

Nature Based Tourism (NBT) is an umbrella term that describes "any type of tourism that relies of attractions directly related to the natural environment" (Weaver, 2001: 16). A conceptual definition is summarized in Chap. 1. Briefly, some see it as a harbinger for 'ecotourism' that is "primarily concerned with the direct enjoyment of some relatively undisturbed phenomenon of nature" (Valentine, 1992, p. 108), but

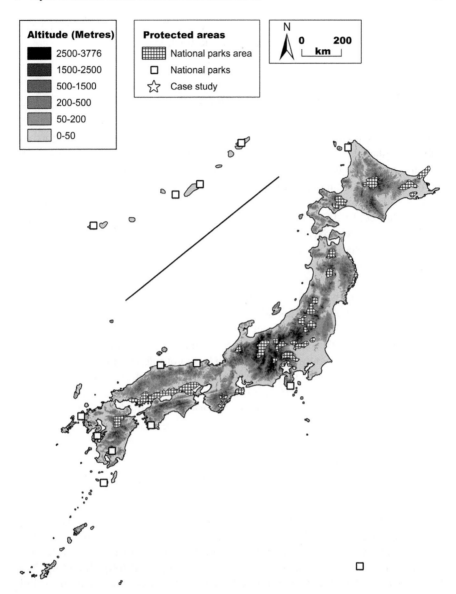

Fig. 3.1 Map of Japan's national parks and Mount Fuji

we interpret it more broadly as PA visitation. To date, there has been little empirical research that investigates Asian NBT in a holistic manner covering supply and demand drivers simultaneously, hence this chapter's focus on Japan's national parks. This next section begins by introducing the data sets used for monitoring domestic and international market segments, then provides contextual evidence from the case of Mt. Fuji, one of Japan's pre-eminent NBT destinations.

3.3.2 Monitoring Nature Based Tourism

Although stock-taking of ecological inventories and environmental indicators are well-established in PAs, visitor monitoring surveys are a more recent addition to the park manager's toolbox (Cessford & Muhar, 2003). PA managers use various counting devices, including automated trail counters with passive infrared sensors, as a cost-efficient means to track visitors (Hadwen et al., 2007; Kahler & Arnberger, 2008; Muhar et al., 2002). Collecting and reporting a central database can provide insights into long-term NBT trends and visitation data has been compiled in Japan's national parks since 1950 based on four sources: (i) individual national parks; (ii) core facilities zones; (iii) visitor centres; and (iv) long-distance trails. Sources (i) and (ii) are generated from sample days and tourism surveys, while (iii) and (iv) derive from infrared counters. Data is compiled and released in print and electronic form with a reporting lag of two years (MOE, 2020). As in the UK examples investigated by Cope et al. (2000), national level data is amalgamated from a variety of local government sources. This multi-agency approach has shortcomings related to reliability and consistency among sources. It also lacks consideration for repeat visitors, or those motivated by highly specific or non-park related factors. However, although the data relies on an eclectic mix of sources, including traffic counters and carparks, the estimates do provide a longitudinally consistent benchmark of visitation trends (Fig. 3.2).

3.3.3 Domestic Demand for Nature Based Tourism

The number of designated national parks increased gradually to 17 in 1950 and 19 by 1960, before leaping to 27 by 1974. As the parks grew in number, annual visitation also rose rapidly to exceed 50 million in 1950. Between 1960 and 1963, estimated visits increased from 90 to 145 million, and by 1971 surpassed 300 million (Fig. 3.2). The 1964 Tokyo Olympic Games pushed planners to surpass a notional upper limit of 20 parks. The new parks were part of a broader surge in infrastructure and amenity projects to be rendered viable by the Olympics; "multiple train and subway lines were completed, as was a large highway building project crisscrossing the metropolitan area" accompanied by new bullet-trains and toll roads (Droubie, 2008). The improved access arrangements encouraged new waves of recreationists, and growing numbers of private car owners sent park visitation rates soaring to record highs in the 1960s.

This was an era in which GDP also grew steadily at an annualized average of 11% from 1955 to 1973. Rapid economic growth and urbanization brought construction of new bullet-trains, highways and other access infrastructure in tandem with

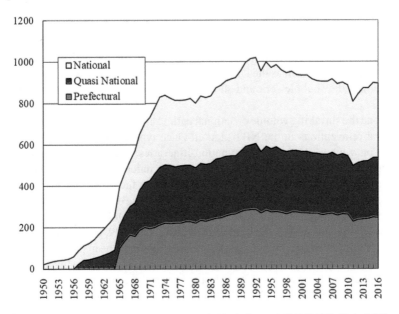

Fig. 3.2 Annual visits to Japan's nature parks 1950–2016. *Source* MOE (2020). Unit: Millions of visits. (Note: data for Quasi national parks since 1957; Prefectural parks since 1965)

increasing disposable income, leisure time and car ownership (Oyadomari, 1989). The correlation between park visits and macro-economic trends was demonstrated by the oil crisis in the 1973, leading to the first year-on-year decline in 1975. However, visitation quickly rebounded and increased steadily until the end of the 1980s real estate 'bubble' economy, when the number of annual visitors peaked in 1991 at 415 million, or 3.4 annual visits per capita—one of the highest in the world (Pergams & Zaradic, 2008). Thereafter came an era of extended shrinkage wherein aggregate visits fell 25% to 309 million in 2011, the year of the triple disaster, before recovering to 359 million in 2016. The cause of the decline has not been established but reflects various socio-economic and demographic drivers that 'flattened the curve' following the extraordinary post-war rise in visitation and the increase in the number of national parks.

Little research has attempted to link the decline in Japan's population with NBT trends. Boosted by the 'baby-boomer' generation, Japan's populace grew from 82 million (1950) to a peak of 128 million (2010), but thereafter population has begun to decline. In particular, the rural regions in which many PAs are located face severe demographic challenges due to shrinking, ageing populations (Matanle, 2007). Large-scale revitalization projects have been implemented experimentally, epitomized by the 1987 Resort Law plan to convert nearly 40,000 km^2 (11% of Japan's total landmass) into purpose-built rural resorts (Havens, 2011). Yet the financial and environmental shortcomings of these 'white elephants' quickly became apparent and by 1992, 39% of the 77 designated infrastructure hubs and 83.8% of the planned

2,046 special facilities were incomplete or abandoned (Oura, 2008). Following the bursting of the real estate bubble, the post-1991 stagnation of Japan's 'lost decades' coincided with a prolonged downturn in visits to national parks. Even today, basic economic indicators such as the land price in rural regions remains a fraction of their value during the 'bubble' era and show no sign of recovery (Matanle & Rausch, 2011).

Beyond the shrinking volume of national park visits, domestic demand has adapted to radical reinventions in the NBT market. Once typified by bus-loads of package tourists on whistle-stop, multi-destination itineraries, contemporary trends favour smaller groups that travel at their own pace. The independent travel style of such smaller groups has reduced demand for package tour facilities, causing the closure of large-scale hotels, restaurants and souvenir shops. The domestic decline was exacerbated by a corresponding increase in overseas travel. Outbound Japanese tourists doubled from 5 million in 1986 to exceed 10 million in 1989, peaking at 18.5 million in 2012 (JTBF, 2014). At home, seasonal destinations have been particularly susceptible to NBT fads. For example, the number of skiers and skaters using mountainous Nagano Prefecture's winter sports industry declined from a peak of 22 million in 1990 to under 8 million in 2006 (Kureha, 2008). More encouragingly, the average stay time at destinations has risen with sporadic demand for special interest tourism, such as 'green tourism' and 'ecotourism' (Yamada, 2011).

3.3.4 Inbound Nature Based Tourism

Trends in international tourism also reflect the appetite for NBT. Although promotion of international tourism was one designation criteria for Japan's original National Parks Act passed in 1931 (Murakushi, 2005), the initial batch of 12 parks, designated from 1934 to 1936, and the subsequent post-war additions developed almost exclusively as destinations for domestic visitors (Jones, 2014). National park visitation data shown in Fig. 3.2 is yet to include many international visitors, but overseas interest in Japan's national parks is now increasing rapidly for the first time (Jones & Ohsawa, 2016). Although international NBT remains an underreported segment, the embryonic market has expanded significantly since a renewed policy-focus on 'inbounds' from 1996 (Soshiroda, 2005). Steady growth in the total number of arrivals into Japan has seen a 'spill-over' effect as inbounds stay longer and venture further 'off the beaten track' to rural destinations located in national parks. After hosting the FIFA Soccer World Cup (2002), the subsequent Visit Japan Campaign launched in 2003 encouraged international arrivals to increase from 3.8 million to 8.3 million in 2008, when the Japan Tourism Agency was established to promote international tourism (Jones, 2014). However, the timing coincided with a global dip in tourism arrivals, and the initial target of 10 million inbound arrivals by 2010 was not achieved until 2013, when the awarding of the 2020 Olympic Games to Tokyo inspired a revised target of 20 million arrivals by 2020. Visitors from nearby Taiwan, South Korea and China comprise the bulk of inbound travellers, with selective deregulation of

Table 3.3 Estimated inbound visits to Japan's national parks 2015–2019 and FHINP inbound visits as a per cent

Year	All national parks	Fuji-Hakone-Izu	%
2015	4,900,000	2,341,000	48
2016	5,460,000	2,577,000	47
2017	6,000,000	2,580,000	43
2018	6,940,000	2,991,000	43
2019	6,670,000	3,093,000	46

Source MOE (2019)

visa requirements also encouraging more visitors from Southeast Asian countries such as Thailand and Malaysia. The number of international arrivals reached 24 million in 2016 and 32 million in 2019. Apart from the national policy emphasis on promotion of inbound tourism in the run-up to the 2020 Olympics, the increase also reflected macro-economic factors such as the weak yen-dollar exchange rate, making Japanese goods and services comparatively cheaper. The timing dove-tailed with the rise of Asia's 'middle class,' with increasing time and disposable income available to explore NBT in Japan's parks.

The MOE began to monitor the number of inbound visits to national parks in 2012, and results show a year-on-year increase of 28% in 2013. Thereafter the total increased to 6.7 million visits in 2019, that still represents under 2% of aggregated park visits. However, the rapid growth of inbounds is likely to be underreported being based on an airport exit survey which asked visitors to report the destinations visited during their trip, so despite the substantial sample size, results face a substantial recall bias exacerbated by 'unfamiliarity with [Japanese] place names' (Funck, 2013). Despite this underreporting, the most visited park for inbounds, Fuji-Hakone-Izu (FHINP), was consistent with overall trends, easily accounting for the largest proportion of visitors. The MOE (2019) reports an increase at FHINP from 1 million in 2013 to over 3 million in 2019—Table 3.3 (right hand column) reveals that FHINP inbounds account for almost half of all inbound park visits across Japan.

3.4 Mount Fuji in Context

3.4.1 The 'Most-Visited' National Park in the World?

Mount Fuji, Japan's tallest peak at 3776 m, is located some 120 km southwest of Tokyo between Yamanashi and Shizuoka Prefectures. Fuji's symmetrical shape that curves up from the Pacific Ocean was formed about 10,000 years ago. The cone is a combination of at least three volcanoes that emerged from multiple eruptions along a triple plate junction (Chakraborty & Jones, 2018). The active stratovolcano is located in the Fuji-Hakone-Izu National Park that follows a volcanic belt stretching south through the mountains of Hakone across the Izu Peninsula.

Due to its iconic shape and cultural legacy, its height above sea level, and proximity to the Tokyo plain, Mount Fuji is a 'must-see' NBT destination. Few who have visited Fuji, particularly during the summer peak period, would doubt that Fuji-Hakone-Izu is among the world's most visited national parks. MOE estimates the annual total at over 110 million annual visits, over one third of total visits to all of Japan's national parks. Congestion is exacerbated by the clustering of holidaymakers in the summer peak, especially during the official climbing season from July to mid-September. Fuji's fifth station on the North side in Yamanashi attracts an estimated 3–4 million annual visitors, of whom some 10% attempt to reach the summit. As Japan's highest peak and one of Asia's premier mountain tourism destinations, Mt. Fuji attracts increasing numbers of climbers, including many internationals tourists who select it as a readily accessible, nontechnical 'trophy' peak. Fuji thus stands out among Japan's mountains for its comparatively large numbers of young, inexperienced, first-time climbers (Jones & Yamamoto, 2016), whose demographics are distinct from other peaks and whose climbing style bears little resemblance to the 'worship-ascent' of yesteryear.

3.4.2 'Worship-Ascent' of Mount Fuji: from Pilgrims to Peak-hunters

The stratovolcano has been worshipped since ancient times in pacification rites, and in the fourteenth century, Shugendo practitioners established a trail that led pilgrims to the summit (Polidor, 2007). In the Edo era (1603–1868), Fuji's summit became the goal for guided groups of pilgrims known as '*ko*'. Trailhead towns around Fuji's foot established clusters of *oshi* houses to cater for the pilgrims' spiritual and logistical needs. Working closely with local shrines and temples, *oshi* presided over the Fuji *ko* style of worship-ascent, and also provided in-house accommodation for the pilgrims. With the help of *sendatsu* guides, the *oshi* organized logistics such as porters and provided spiritual services including blessings, rites and ablutions. Pilgrims paid a package fee which included the devotional diet, purification and banners, as well as interpretation of local customs, thus ensuring a cost recovery mechanism for local communities. Although the Edo era witnessed a major eruption in 1707, by the early 19th Century there were 808 Fuji-*ko* groups in Tokyo alone, pooling contributions from among members to send representatives on a summer pilgrimage (Bernstein, 2008). In their heyday, numbers of Fuji *ko* pilgrims reached twenty thousand per season (Ito, 2009). This history of worship-ascent, combined with Fuji's artistic legacy, formed the core components of inscription on UNESCO's list of cultural World Heritage Sites in 2013.

However, during the post-WWII period of extended economic growth, rising living standards brought increased leisure time and improved access infrastructure, including the construction of many toll roads within national parks (Oyadomari, 1989). At Fuji, the Subaru line toll road on the Yamanashi side was completed in

Fig. 3.3 Core zone at the top, buffer zone around the base of Mt. Fuji. *Credit* First author

time for the Tokyo Olympics in 1964. This 30 km paved road transports cars, buses and bikes up to the Yoshida fifth station trailhead, located at an altitude of 2,300 m above sea level. A dramatic reduction in round-trip climbing time from the fifth station radically altered climbing behaviour by transforming Fuji's summit into an overnight expedition. Today, most climbers reach the summit via high-speed roundtrips from Tokyo, including less than 24 h climb time (Jones & Yamamoto, 2016) (Fig. 3.3).

3.4.3 Monitoring & Monetizing Climbers on Mount Fuji

The convenient access for climbers from urban hubs has also facilitated a rise in footfall. In 1981, approximately 100,000 summer climbers were counted on the Yoshida trail. The figure doubled thereafter, exceeding 200,000 in 2008—the year in which Fuji was placed on the tentative list for inscription as UNESCO cultural world heritage site. Thereafter the four trails have maintained a combined total of approximately 300,000 climbers per season based on beam counters located around the eighth stations (3,100 m ASL). Each of the four trails has a fifth station trailhead that ranges in altitude from Gotemba (approx. 1400 m ASL) to Fujinomiya (2400 m). The Yoshida trail on Fuji's north face has the second highest elevation (2300 m) plus proximity to the Kanto plain, resulting in the greatest footfall. It thus accounts for an increasing per cent of climbers, rising to 60% in 2015, and also attracts the most international climbers. According to sampling by visual observation on the Yoshida Route, foreign climbers accounted for just 5% to 7% of climbers in August 2009 (Jones et al., 2016). But by the 2015 season, the proportion had risen to 20% (weekends) and 30% (weekdays) based on appearance and language. Multiplying this estimated 20–30% market share by the total number of climbers suggests that some 60–90,000 foreign climbers attempt to summit Fuji each summer. Yet even this figure is likely to be an underestimate, due to difficulties in distinguishing between different nationalities for logistical reasons, such as lack of a registration system, and ethical ones, such as avoiding potential invasions of privacy by directly asking climbers their nationality.

Foreign climbers face culture shocks from the congestion and local customs, such as the mountain huts. Climbers depart huts before dawn or ascend directly overnight in time to reach the summit for sunrise. Weekends are especially congested, with Saturday nights the most crowded up to a maximum of 6,831 climbers recorded on twenty-fifth August 2012. Modern access arrangements have transformed the logistics of summiting Fuji from a week-long pilgrimage to an overnight adventure. Today, many climbers eschew a stay in a mountain hut altogether, opting to set out overnight from the fifth station trailhead (approx. 2300 m ASL) before dashing up to the summit and back again in a 'bullet climb'. Most descending climbers return promptly to Tokyo without staying another night or contributing to the region's economic impact. However, the climbers do leave behind significant environmental impacts that places severe strain on visitor facilities and flashpoints such as trails and toilets (Apollo, 2017).

The large number of climbers sparked renewed debate during the UNESCO listing process over 'cost recovery' mechanisms to mitigate such negative impacts. Although Japan's national parks charge no admission ticket per se, fees are charged for certain visitor services. For example, separate user fees have long since been charged to stay at the mountain huts, supplemented by the introduction in 1999 of a voluntary tipping system for visitors that use the toilets (Sayama & Nishida, 2001). Given the impracticality of making entrance fees compulsory, donations collected from visitors were deemed a viable alternative income stream with which to provide services. Local government units led a new system in 2013 to raise funds for conservation. The new donation system encouraged each climber to contribute ¥1000 toward conservation initiatives. These include maintenance of trails and toilets, safety measures and other visitor services. However, differences emerged between domestic and international climbers, with 71% of the latter unaware of the donation prior to climbing Fuji, compared to only 8% Japanese. International climbers with prior awareness (72%) showed significantly greater willingness to pay ¥1000 than those without (43%), suggesting that more targeted multi-lingual messages are needed to raise awareness and willingness to pay among international climbers (Fig. 3.4).

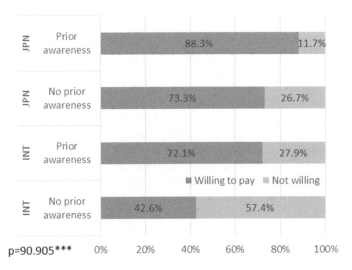

Fig. 3.4 Willingness to pay a ¥1000 donation at Mt. Fuji (Jones et al., 2016)

3.5 Discussion and Conclusion

This chapter investigated Japan's PAs from supply and demand viewpoints, exploring the 'multi-purpose' national parks from the perspectives of land ownership and zoning policy before turning to trends in nature-based tourism (NBT). The rapid global growth of NBT (Tisdell & Wilson, 2012) has relevance for an archipelago with distinct biodiversity, over two-thirds covered by mountainous forest. Prior to the peak in 1991, Japan's national parks had some of the highest per capita visitation rates in the world (Pergams & Zaradic, 2008). Today, popular destinations such as Mount Fuji continue to face management challenges such as congestion linked to 'over-tourism' (Kobayashi & Aikoh, 2008). However, overall park visitation has declined since the 1990s, and despite a recent rebound due to the rise in inbound tourism, the PA managers have struggled to adapt to changes in demand and the need for co-management (Hiwasaki, 2005).

Collaborative management goals are often undone by the fragmented realities of PAs' land ownership and governance, and Japan's national parks are no exception to global norms. After the 2008 amendments to the Nature Parks Act a 'partnership' ethos was formalized as a core management concept (Watanabe et al., 2012). The co-management concept has subsequently assumed increasing importance in line with calls for more holistic governance (Tanaka, 2014), and the need to establish a platform for regular exchange of information and multi-stakeholder dialogue (Aikoh, 2014).

However, the MOE's capacity for such coordination is complicated by the complex stakeholder networks created by national parks' 'multi-purpose' agenda and mixed ownership (Jones et al., 2018). The combination of central and local governments with private landowners results in complicated territorial mosaics that belie generic solutions through better co-management and zoning. The system has thus been criticized for its top-down approach, lack of effectiveness in biodiversity conservation and the fragmented direction of different stakeholders (Hiwasaki, 2005). The basic requirements of the collaborative approach such as openness, integration and cooperation are undermined by a park management style which offers few opportunities for exchange of information and makes no legal provision for bottom-up participation in the decision-making process (Aikoh, 2014; Hiwasaki, 2005). Consensus can be difficult to achieve even among central government agencies such as the MOE and MAFF with long-standing rivalries (Tanaka, 2014). Provision of visitor services such as trails and toilets has subsequently depended on de facto partnerships of public and private stakeholders to compensate for the lack of a 'one-stop shop' equivalent to the U.S.A.'s National Park Service (Murakushi, 2005).

Nonetheless, multi-purpose parks do own a number of advantages when compared to the 'set-aside' Yellowstone model. These include practical considerations such as minimizing the start-up capital required for PA creation by circumnavigating the need for land acquisition (Kato, 2008). Running costs, such as maintenance of roads and toilets, are largely delegated to local government. Meanwhile other visitor services such as accommodation and catering are provided by the private sector. However, this can unleash unplanned or uncoordinated development, as private companies own and autonomously operate transport systems and other tourism services including shops, hotels and mountain huts. Park management goals must compete with privately-owned, for-profit business ventures as demonstrated at Fuji.

Section 3.3 used the case study example of Mt Fuji climbers to symbolize the potentials and pitfall of Japan's multi-purpose parks in practise. In the pre- and immediate post- war periods, a series of infrastructure projects drastically altered Fuji's socio-economic access arrangements, culminating in the construction of large-scale facilities and access. One example is the Subaru line, a toll road on Fuji's north face completed in time for the Tokyo Olympics in 1964. This 30 km paved road enables access to the fifth station (2,300 m above sea level) meeting secular needs in contrast to the sacred experience once sought by the Fuji-*ko*. The pre-modern pilgrims' climbing style also differed significantly from current practise; only a few climbers were selected from amongst the members. They had ample opportunities to acclimatize during their pilgrimage, which could last a week or more. After arriving at Fuji, the pilgrims stayed overnight at the base of the mountain before beginning an ascent punctuated by numerous breaks en-route to perform ablutions and pray to Kami deities including sacred trees, stones and waterfalls. The Fuji-*ko*'s former role as cultural mediators, combined with the substantial input for regional economies invited claims of 'Edo ecotourism' (Ito, 2009), but contemporary Fuji climbs tend toward a commercialized, high-speed dash to the summit.

Recent decades have seen a growing realization of the shortcomings of this style of tourism that prioritizes speed and convenience above all else. Demand for 'slow food' and walking tours have grown along with 'green' and 'ecotourism' (Yamada, 2011). On the policy-making side, the 2008 Ecotourism Act acknowledged the need for a co-management approach to nurture the potential for NBT (Jones et al., 2018). But the parks' institutional shortcomings related to their multi-objective, multi -owner nature still undermine the capacity to respond to the wholesale changes in visitor demand. During the era of extended economic growth, steadily rising visitor numbers were able to gloss over such shortcomings, with stakeholder consensus gained via a combination of the 'stick' of central governments' command-and-control administration, coupled with frequent 'carrots', such as infrastructure and public works projects. However, more than three decades of decline in the numbers of visitors have called into question a style of administration that seems overly weighted towards restrictions—especially planning permission and access regulations—without offering fundamental solutions to promote sustainable tourism.

90 years after the passage of the original national parks act, administration remains a balancing act still seeking ways to resolve the trade-off between conservation and development goals. This chapter focused on the national parks that represent the 'gold standard' of Japan's triple-tiered nature park system and symbolize conservation even as they account for the bulk of NBT. However, it is worth noting that Japan has numerous parallel PA networks ranging from Ramsar sites to Geoparks that were excluded here due to lack of space. Other limitations include the embryonic state of English-language research, although herein lies also an opportunity to increase inbound visitation. Beyond broadening the number and range of inbound tourists' nationalities, the additional efforts to monitor and market the parks to inbounds has encouraged visitors to diversify geographically away from the urban hubs to incorporate more NBT destinations into their itineraries. For the national parks to function as effective NBT role models a co-management approach is paramount (Hiwasaki, 2005). Implementation hinges on key central and local government stakeholders, but must also include roles for local residents, landowners, tour operators and nature conservation organizations. Unlocking the potential of Japan's multi-purpose parks requires just such a kind of holistic cross-cutting partnership, with the capacity for more adaptive management in the future.

References

Aikoh, T. (2014). Issues in planning and management in the Japanese national park system: Perspectives from Daisetsuzan National Park. *Japanese Forest Economic Society, 60*(1), 14–21.

Aikoh, T., & Tomidokoro, Y. (2010). Relationship between the national park plan and the national forest plan in Shikotsu-Toya National Park. *Journal of the Japanese Institute of Landscape Architecture, 73*(5), 505–508.

Apollo, M. (2017). The good, the bad and the ugly: Three approaches to management of human waste in a high-mountain environment. *International Journal of Environmental Studies, 74*(1), 129–158. https://doi.org/10.1080/00207233.2016.1227225.

Bernstein, A. (2008). Whose Fuji?: Religion, Region, and State in the Fight for a National Symbol. *Monumenta Nipponica, 63*(1), 51–99.

Cope, A., Doxford, D., & Probert, C. (2000). Monitoring visitors to UK countryside resources: The approaches of land and recreation resource management organisations to visitor monitoring. *Land Use Policy, 17*, 59–66. https://doi.org/10.1016/S0264-8377(99)00035-6.

Cessford, G., & Muhar, A. (2003). Monitoring options for visitor numbers in national parks and natural areas. *Journal for Nature Conservation, 11*(4), 240–250.

Chakraborty, A., & Jones, T. E. (2018). Mount Fuji: The volcano, the heritage, and the mountain. In A. Chakraborty, K. Mokudai, M. Cooper, M. Watanabe, & S. Chakraborty (Eds.), *Natural Heritage of Japan*. Geoheritage, Geoparks and Geotourism. Springer International Publishing.

Critical Ecosystem Partnership Fund (CEPF). (2020). *Japan data*. https://www.cepf.net/our-work/biodiversity-hotspots/japan.

Droubie, P. (2008). *Japan's Rebirth at the 1964 Tokyo Summer Olympics*. http://aboutjapan.japansociety.org/content.cfm/japans_rebirth_at_the_1964_tokyo_sumer.

e-Stat. (2020). Ippanshoku Kokka Koumin Zaishoku Joukyou Toukeihyou (Statistics on the current state of General Civil servants). https://www.e-stat.go.jp/stat-search/files?page=1&toukei=00000002&tstat=000001134643.

Funck, C. (2013). Welcome to Japan. In C. Funck, and M. Cooper (Eds.), *Japanese tourism: Spaces, places and structures* (pp. 160–186). Berghahn Books.

Global Forest Watch (GFW). (2020). *Japan Dashboard*. https://www.globalforestwatch.org/dashboards/country/JPN.

Hadwen, W. L., Hill, W., & Pickering, C. M. (2007). Icons under threat: Why monitoring visitors and their ecological impacts in protected areas matters. *Ecological Management & Restoration, 8*(3), 177–181.

Havens, T. R. H. (2011). *Parkscapes: Green Spaces in Modern Japan*. University of Hawaii Press.

Hiwasaki, L. (2005). Toward sustainable management of national parks in Japan: Securing local community and stakeholder participation. *Environmental Management, 35*(6), 753–764.

Imura, H. (2005). Japan's environmental policy: Institutions and the interplay of actors. In H. Imura & M. A. Schreuers (Eds.), *Environmental policy in Japan* (pp. 49–85). Edward Elgar.

International Union Conservation Nature (IUCN). (2020). *Red List Japan*. https://www.iucnredlist.org/regions/japan.

Ito, T. (2009). Trail and climber management and its cost recovery from the climbers at Mt. Fuji in Edo Period. *Journal of the Japanese Forest Society, 91(2)*, 125–135.

Japan Travel Bureau Foundation (JTBF). (2014). JTBF Travel Demand Survey. In JTBF (Ed.), *Annual report on the tourism trends survey* (pp. 32). Region Ltd (in Japanese).

Jones, T. E., Beeton, S., & Cooper, M. (2018). World heritage listing as a catalyst for collaboration: Can Mount Fuji's trail signs point the way for Japan's multi-purpose national parks? *Journal of Ecotourism, 17*(3), 220–238.

Jones, T. E., & Ohsawa, T. (2016). Monitoring nature-based tourism trends in Japan's national parks: Mixed messages from domestic and inbound visitors. *Parks, 22*(1), 25–36.

Jones, T. E., & Yamamoto, K. (2016). Segment-based monitoring of domestic and international climbers at Mount Fuji: Targeted risk reduction strategies for existing and emerging visitor segments. *Journal of Outdoor Recreation and Tourism, 13*, 10–17.

Jones, T. E., Yamamoto, K., & Kobayashi, A. (2016). Investigating climbers' awareness and willingness to pay a donation: A comparative survey of domestic & international climber segments at Mount Fuji. *Journal of Environmental Information Science, 44*(5), 131–136.

Kahler, A., & Arnberger, A. (2008). A comparison of passive Infrared counter results with time lapse video monitoring at a shared urban recreational trail. *MMV4. Vienna, Austria, 485*–489.

Kanda, K. (2012). The selection process of national park landscape areas and the imaginative geographies in Taiwan during the Japanese colonial period. *Academic World of Tourism Studies, 1*, 77–87.

Karan, P. P. (2005). *Japan in the 21st century: Environment, economy, and society*. The University Press of Kentucky.

Kato, M. (2008).National Park & Protected Area Management Series. In *National Park System of Japan* (Vol. III). Kokon Shoin.

Kobayashi, A., & Aikoh, T. (Eds.). (2008).National Park & Protected Area Management Series. In *National Park System of Japan* (Vol. II). Kokon Shoin.

Kupper, P. (2009). Science and the national parks: A transatlantic perspective on the interwar years. *Environmental History, 14*(1), 58–81.

Kureha, M. (2008). Changing ski tourism in Japan: From mass tourism to ecotourism? *Global Environmental Research, 12*(2), 137–144.

Matanle, P. (2007). Organic sources for the revitalization of rural Japan: The craft potters of Sado. *Japanstudien, 18*(1), 149–180.

Matanle, P., & Rausch, A. S. (Eds.). (2011). *Japan's shrinking regions in the 21st century: Contemporary responses to depopulation and socioeconomic decline.* Cambria Press.

Meyer, J. M. (1997). Gifford Pinchot, John Muir, and the boundaries of politics in American thought. *Polity, 30*(2), 267–284.

Ministry of Environment (MOE). (2018). *Definition of National Parks.* https://www.env.go.jp/en/nature/nps/park/about/history.html.

Ministry of Environment (MOE). (2019). *Estimated inbound visits to Japan's national parks.* www.env.go.jp/nature/mankitsu-project/pdf/2019/foreign-tourists_2019.pdf.

Ministry of Environment (MOE). (2020). *Data list for each type of nature conservation.* http://www.env.go.jp/park/doc/data.html.

Mitsuda, H., & Geisler, C. (1992). Imperilled parks and imperilled people: Lessons from Japan's Shiretoko national park. *Environmental History Review, 16*(2), 23–39.

Mizuuchi, Y. (2012). Kokuritsu koen shitei ni okeru Ise Shima Kokuritsu Koen no Haikei to Ise Jingu no Kankei [Background of the Exceptional Designation of Ise-Shima as a National Park and Its Relationship with Ise Grand Shrine]. *Journal of the Japanese Institute of Landscape Architecture, 75*(5), 389–394 (in Japanese).

Murakushi, N. (2005). *Kokuritsu koen seiritsushi no kenkyū: Kaihatsu to shizen hogo no kakushitsu o chūshin ni* [The history of the establishment of the national parks: The feud between development and nature conservation]. Hosei University (in Japanese).

Muhar, A., Arnberger, A., & Brandenburg, C. (2002). Methods for visitor monitoring in recreational and protected areas: An overview. Monitoring and Management of Visitor Flows in Recreational and Protected Areas. *Institut for Landscape Architecture & Landscape Management Bodenkultur University Vienna, 2001*, 1–6.

Natori, Y. (1997). Shiretoko logging controversy: A case study in Japanese environmentalism and nature conservation system. *Society & Natural Resources, 10*, 551–565.

Nishimura, T. (2016). Nihon ni okeru Kokuritsu Koen no 'Gunjika' (1925–1944) (The 'Militarization' of Japan's National Parks). *Memoirs of Osaka Kyoiku University, 65*(1), 11–22 (in Japanese).

Oura, Y. (2008). Policy development and direction of interaction of cities and countryside after 1990. *Japanese Journal of Forest Economics, 54*(1), 40–49 (in Japanese).

Oyadomari, M. (1989). The rise and fall of the nature conservation movement in Japan in relation to some cultural values. *Environmental Management, 13*(1), 23–33.

Pergams, O. R., & Zaradic, P. A. (2008). Evidence for a fundamental and pervasive shift away from nature-based recreation. *Proceedings of the National Academy of Sciences, 105*, 2295–2300.

Polidor, A. (2007). Sacred Site Reports. www.sacredland.org/mount-fuji. Retrieved July 11, 2014.

Rainisto, S. (2003). *Success factors of place marketing: A study of place marketing practices in Northern Europe and the United States* (Doctoral dissertation). Helsinki University of Technology, Institute of Strategy and International Business.

Sayama, H., & Nishida, M. (2001). Transition of the Beautification and Cleaning Activities in Mt. Fuji Area after World War II. *Journal of the Japanese Institute of Landscape Architecture, 64*(5), 485–488.

Soshiroda, A. (2005). Inbound tourism policies in Japan from 1859 to 2003. *Annals of Tourism Research, 32*(4), 1100–1120. https://doi.org/10.1016/j.annals.2005.04.002.

Tanaka, T. (2014). Governance of Japan's national parks in 21st century: A perspective from public administration. *Landscape Research Japan, 78*(3), 226–229 (in Japanese).

Tisdell, C. T., & Wilson, C. (2012). *Nature-based tourism and conservation: New economic insights and case studies.* Edward Elgar Publishing.

Totman, C. (2014). *Japan: An environmental history.* I.B. Tauris.

Usui, R., & Funck, C. (2018). Analysing food-derived interactions between tourists and sika deer (Cervus nippon) at Miyajima Island in Hiroshima, Japan: Implications for the physical health of deer in an anthropogenic environment. *Journal of Ecotourism, 17*(1), 67–78.

Watanabe, T., Sasaki, S., Shinohe, H., & Shimomura, A. (2012). A study on the transition of resource values of the national parks in Japan and the management style. *Landscape Research Japan, 75*(5), 483–488 (in Japanese).

Weaver, D. (2001). *Ecotourism*. Wiley.

Valentine, P. (1992). *Nature-based tourism*. Belhaven Press.

Vincenot, C. (2017). *Pteropus pselaphon. The IUCN Red List of Threatened Species* 2017: e.T18752A22085351. https://doi.org/10.2305/IUCN.UK.2017-2.RLTS.T18752A22085351.en.

Woodley, S., Bhola, N., Maney, C., & Locke, H. (2019). Area-based conservation beyond 2020: A global survey of conservation scientists. *Parks, 25*(2), 19–30.

Yamada, N. (2011). Why tour guiding is important for ecotourism: Enhancing guiding quality with the ecotourism promotion policy in Japan. *Asia Pacific Journal of Tourism Research, 16*(2), 139–152.

Thomas E. Jones is Associate Professor in the Environment & Development Cluster at Ritsumeikan APU in Kyushu, Japan. His research interests include Nature-Based Tourism, Protected Area Management and Sustainability. Tom completed his PhD at the University of Tokyo and has conducted visitor surveys on Mount Fuji and in the Japan Alps.

Akihiro Kobayashi is Professor at Senshu University in Tokyo, Japan. He has conducted extensive research on issues related to planning and management of nature parks, focusing on the relationship between the natural environment and users, and the perspective of user awareness and behaviour. He is a member of Shiretoko Appropriate Use Plan Committee, Shiretoko World Natural Heritage Area Science Committee, Hokkaido Natural Environment Conservation Council etc.

Chapter 4
South Korea's System of Mountainous Protected Areas and Nature-Based Tourism

Malcolm Cooper

4.1 Introduction

The Republic of Korea (ROK) occupies the southern half of the Korean Peninsula. Its only land border is to the north, with the Democratic People's Republic of Korea (DPRK). The country has 2,413 km of coastline, and a land area of approximately 100,032km^2. Its topography is a continuation of the mountainous terrain found in the north (Fig. 4.1), and in a few places of volcanic origin. The highest peak is Hallasan (1,950 m), a volcanic cone on Jeju Island (Esperjesi, 2011; Kim, 1993), and, like its neighbor Japan, this landscape means that Korea has few extensive plains or river basins. Such basins make up only 30% of the land area of the country and are largely coastal, particularly in the west (Fig. 4.1) and in the narrow littoral plain on the east coast bordering the main mountain chain (Bong, 2007; International Park Planning Institute, 1972). The highest mountains of the peninsula are in the north and include Paektusan (Changbai Mountain, 2,749 m), Mantapsan (2,204 m), and Myohyangsan (1,909 m). These include active volcanic peaks such as Paektusan, a stratovolcano said to be the spiritual home of the Korean people (the crater is filled with water and known as Heaven Lake; Choe, 2016).

Climatically, the ROK lies in the East Asian Monsoon Region, resulting in a temperate climate with 4 distinct seasons. The winters are cold and dry, and the summers hot and humid. Spring and autumn are short intermediate seasons. Seoul temperatures in January range from −5 to −2.5 °C and summer in July sees temperatures between 22.5 and 25 °C. Because of its southern maritime location, our case study mountain park, Hallasan, has fewer extreme weather events than other parts of the country. Its temperatures range from 2.5 °C in winter to 25 °C in summer (Hong et al, 2013; MOE, 2008). These climatic influences mean that the mountainous

---

M. Cooper (✉)
College of Asia Pacific Studies, Ritsumeikan Asia Pacific University, Beppu, Japan
e-mail: cooperm@apu.ac.jp

© The Author(s), under exclusive license to Springer Nature Switzerland AG 2021
T. E. Jones et al. (eds.), *Nature-Based Tourism in Asia's Mountainous Protected Areas*, Geographies of Tourism and Global Change,
https://doi.org/10.1007/978-3-030-76833-1_4

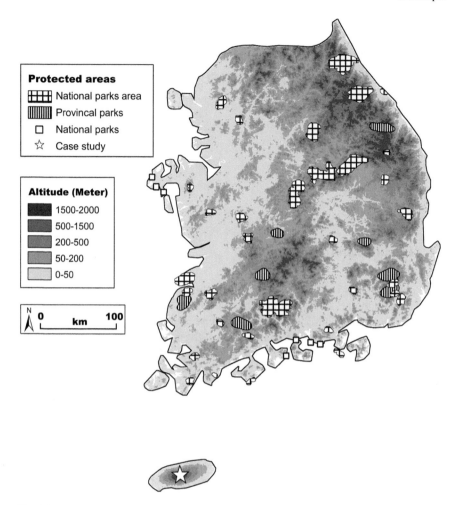

Fig. 4.1 Topography of South Korean National Parks

national parks (NPs) are accessible for much of the year, and this has allowed the rise of a significant outdoors walking culture in the country. They also mean that there is usually sufficient rainfall to support agricultural and urban life. Rarely is there less than 750 mm of rain on average in a year (MOE, 2008; UNESCO, 2007). Of course, amounts can vary, and serious droughts occur every eight years on average. There are also typhoons from the northern Pacific Ocean, and these have been increasing of late as they have in Japan although this has not (yet) been labelled as 'climate change.'

A key concept of this book is that it looks at the natural capital embodied in mountainous NPs and the social capital built on this, seeking to understand how both shape and affect natural heritage (Brondizio et al., 2019). Natural heritage is categorized as World Heritage, Global Geoparks, Biosphere Reserves, Ramsar Sites, Mountainous terrain, and other NP properties to highlight the different physical characteristics, management mechanisms, and conservation and touristic potential of each park. The natural characteristics of South Korean NP sites as tourism destinations thus range from their topography to their complex interplay of physical and biotic factors, and the interface between nature and human society (Bong, 2007; Hong et al., 2013; KNPS, 2008; Lee, 1995; You et al., 2013). This chapter seeks to explain the geologic settings, signature physical reliefs, landscaping mechanisms and ecosystems of each mountainous NP (UNEP-WCMC, 2020). The challenges facing them are analyzed, and their value for nature-based tourism is explored through themes like recent tourism trends in these areas (Oh, 1998). The chapter notes that it is worth stating an obvious message here: all these 'natural heritage' sites are different in the way they are perceived by various people—based on the processes to which society attaches particular values such as preservation or tourism development (Chape et al., 2005), and in the way the mechanisms through which they are managed are developed (Dudley et al., 2005; You et al., 2013)—but they are interrelated at the same time. Together, they are a considerable part of the natural heritage of Korea (Bong, 2007; Heo, 2007).

4.2 The South Korean System of National Parks and Protected Areas

Korea established its first national strategy for sustainable development in 2006, and premises green growth as a national priority. Acting as an important base for this policy there are 22 designated NPs and approximately 1,450 other PAs (Kim, 2013), covering a total of 23,506 km^2, or about 24% of the total land area of the country. The national parks are considered to be "areas that represent the natural ecosystem and cultural scenes of the Republic of Korea" (KNPS, 2008), while the other PAs serve a range of functions as wetlands, landscape conservation areas, nature reserves, genetic preservation areas, and marine protected areas. These are all designated by the government to protect significant natural areas but also to ensure their sustainable use, although they can be governed by a range of Ministries and this makes for a potentially fragmented system.

These parks and other protected areas are managed, with the exception of Hallasan NP (Jeju Island), by the Korea National Park Service (KNPS) (Choe et al., 2017; KNPS, 2008; MOE, 2008), and a range of Ministries (see Table 4.1). Hallasan NP is part of the autonomous *Jeju Special Self-Governing Province.* In terms of the administrative structure of governance the KNPS has been part of the Ministry of the Environment since 1998 (MOE, 2008). The designated areas are divided into Mountainous (17), Marine and Coastal (4) and Historical (1) National Parks (Fig. 4.2 and Table 4.2). Rangers are employed to control activity in the parks and Master Planning is carried out (the current plan covers the period 2012–2021). Tourism is permitted and actively promoted. The group discussed in this chapter are the 17 mountainous parks (Heo, 2007).

Table 4.1 Number of South Korean protected areas

Governing authority	Number of sites	International designation	Number of sites
Ministry of Environment	700	*UNESCO MAB*	4
Ministry of Oceans and Fisheries	10	*World Heritage sites*	1
Cultural Heritage Administration	294	*Ramsar Convention*	17
Korea Forest Service	463	*Global Geoparks Network*	2
Total	1,467	Total	24

Source IUCN (2009)

Fig. 4.2 Trail map of Hallasan National Park. *Source* Author

The KNPS was set up in 1987 to enable professional management of this resource (KNPS, 2007). Its mission was to become a "world class professional park management organization that protects nature and ensures customer satisfaction" (KNPS, 2007). The National Parks are designated areas of public land on which most forms of development are prohibited (Choe et al., 2017; Heo, 2007) and are mainly located in mountainous or coastal regions (Hee, 2007). The country's largest mountain park is Jirisan (472km^2), the first to be designated in 1967 (KNPS, 2007). The smallest is Wolchulsan, at 56.1km^2. At the beginning, policy makers in the national parks context mainly focused on providing economic activity in each area through the development of tourism (International Park Planning Institute, 1972; MOE, 2008).

Table 4.2 Area, location and designation of South Korean national parks

Name	Location	Designated	Area	Park Type
Bukhansan	Seoul	1983	80km^2	Mountainous
Byeonsan-bando	Jeollabuk-so	1988	155km^2	Marine, Coastal
Chiaksan	Gangwon-do	1984	184km^2	Mountainous
Dadohaehaesang	Jeollanam-do	1981	2325km^2	Marine, Coastal
Deogyusan	Jeollabuk-do	1975	232km^2	Mountainous
Gayasan	Gyeongsangnam-do	1972	77km^2	Mountainous
Gyeongju	Gyeongsangbuk-do	1968	137km^2	Historical
Gyeryongsan	Chungcheongnam-do	1968	65km^2	Mountainous
Hallasan	Jeju-do	1970	153km^2	Mountainous
Hallyeohaesang	Jeollanam-do	1968	545km^2	Marine, Coastal
Jirisan	Jeollanam-do	1967	472km^2	Mountainous
Juwangsan	Gyeongsangbuk-do	1976	107km^2	Mountainous
Naejangsan	Jeollanam-do	1971	81km^2	Mountainous
Odaesan	Gangwon-do	1975	304km^2	Mountainous
Seoraksan	Gangwon-do	1970	398km^2	Mountainous
Sobaeksan	Chungcheongbuk-do	1967	322km^2	Mountainous
Songnisan	Chungcheongbuk-do	1970	274km^2	Mountainous
Taeanhaean	Chungcheongnam-do	1978	326km^2	Marine, Coastal
Wolchulsan	Jeollanam-do	1988	56km^2	Mountainous
Woraksan	Chungcheongbuk-do	1984	288km^2	Mountainous
Mudeungsan	Gwangju	2012	75km^2	Mountainous
Taebaeksan	Yeongwol	2016	70km^2	Mountainous

Source KNPS (2008)

Over time has come the realization of the importance of natural capital and its need for conservation and they have become more focused on environmental impact, the promotion of public health (outdoor recreation), and efforts to ensure sustainable development (Choe et al., 2017; Dudley et al., 2005; Heo, 2007).

4.3 Public and Private Sector Involvement in the Korean National Parks

Korean PAs have a long tradition of legal protection at the national and local levels (Hong et al., 2013). This has allowed for the development of comprehensive policies, budgetary support, and effective planning, control and enforcement measures (Hockings et al., 2006). To this can be added links to the international system of environmental protection through cooperation with other government and non-government PA agencies (Chape et al., 2005; 2017; Heinonen, 2007; IUCN, 2013; Kim, 2013;

KNPS, 2007; MOE, 2008). Some problems do exist though, and these include: (i) the enforcement by police and local courts, if there are breakdowns in the responsibility system, can be weak; and (ii) there can be differences between national and local government on what level of protection is needed. The Ministry of the Environment (MOE) has jurisdiction in this situation and needs to examine ways of clarifying the system requirements for nature protection in Korea. In recognition of these problems, the country has made significant efforts recently to increase the role of international bodies (primarily the IUCN) in advising on the environmental protection and this has been successful (Bong, 2007). Assessment by the MOE itself on a regular basis is a significant part of this process and indicates to the managers of individual parks how local and international experience may be added to the national endeavor (MOE, 2008).

The main purpose of the NPs is the protection of representative ecosystems and associated cultural resources (IUCN, 2013; KNPS, 2007). There are five key requirements before designation of a park is possible (IUCN, 2013; MOE, 2008):

1. Ecosystem preservation must be possible or already in operation, or endangered species are found to be present, or there are designated National Treasures present, or plant and animal species that require specific protection have been identified;
2. The area is in a natural state, where landscapes have been preserved without significant damage or pollution;
3. If the area is also a Korean cultural landscape the associated artefacts must be integral to that area and have their own value;
4. The topography of the area is critically important, so there must be no threat from alternative development activities;
5. The proposed park must contribute to overall community aims for the preservation and governance of the local environment (IUCN, 2013; MOE, 2008).

Thus, management policies in national parks are explicitly aimed at combining both conservation and the sustainable use of natural capital under government control. In Korea, though a part of the surface area of designated national parks is private, the owners of that land have some autonomy in management decisions under the *Natural Park Act*. This situation is particularly useful when it comes to the use of these areas for tourism through public private partnerships.

Natural area values world-wide and in Korea are expected to decrease due to changes in the global environment, such as rapid climate change, without effective protection. However, natural heritage not only has environmental value in itself, but also can contribute to the revitalization of a local economy through its commodification for tourism. Therefore, any reduction in natural resources is not a simple problem, but one that is part of the organic relationship between society, economy, and environment. It is necessary to actively implement environmental policies on a macro scale to prevent any reduction in the value of natural environments (Kim, 2013; Ministry of Cultural Heritage, 2019; Mulongoy & Chape, 2004).

The management and sustainable development planning for natural capital resources in this context are normally characterized as community decisions (UNEP,

2006). When these activities are inappropriately managed, there may be significant impacts on National Park goals. Governments and the tourism industry have the duty to guarantee that they both prosper, and the future is not threatened (Dwyer & Edwards, 2010). Tourism demands on natural heritage require forms of planning and management that combines the interests and concerns of all stakeholders in a practical way (Gilligan et al., 2005; Heo, 2007). It is this situation that asks for multi-dimensional and integrative planning (Heinonen, 2007). Planning for tourism thus requires an understanding of the importance of development and control of natural capital to realize the principles of sustainable tourism. In turn, user groups must recognize the impact of tourist activity and should understand and control this, as well as the need for participative planning, consensus, and the mitigation of any conflicts between stakeholders (Dwyer & Edwards, 2010).

To achieve this outcome, the development of appropriate policies must be supported at all levels (Dudley et al., 2005; Kim, 2013). Together, stakeholders should seek the development of frameworks that include everybody when defining strategies and policies, and this is one of the tasks of the Korea Protected Areas Forum (Kim, 2013). These integrated frameworks are required for the co-creation of environmental protection and tourism, as tourism plans should have the capacity to positively influence the management of the environmental values of the national parks (IUCN, 2013; MOE, 2008).

4.4 The IUCN and Environmental Protection in Korea

The ROK is party to many of the international agreements on the protection of the environment that seek to ameliorate environmental degradation, including the UN Convention on Biological Diversity, the Paris Accords (successor to the Kyoto Protocol), agreements on Endangered Species, Hazardous Wastes, Ozone Layer Protection, and the Ramsar Convention on Wetlands (Bong, 2007; IUCN, 2013). The latter has been central to the development of Korean national parks and the conservation of natural capital in Korea (see Table 4.2).

The IUCN is a membership-based union that works through some 1,400 government and non-government organizations (IUCN, 2013). It has 6 Commissions and 11 operational regions under a Council and Secretariat. Enhancing the management of conservation areas is recognized as an important task by the communities represented by the IUCN. From 1997, this became one of the key objectives of the *World Commission on Protected Areas* (IUCN-WCPA) and forms the basis of their framework for the assessment of natural areas. This framework was originally developed in 2000 (Hockings et al., 2000) and revised in 2006 (Hockings et al., 2006). It has become the basis for many of the effective management policies in national parks worldwide, and the IUCN provides technical support for it and the resulting classification of parks (Dudley et al., 2005; IUCN, 2013). The IUCN management categories for national parks used by the KNPS to categorize and provide management guidelines are outlined in Table 4.3 (IUCN, 2009).

Table 4.3 IUCN classification of protected areas categories

Category	Objective
Ia Strict Nature Reserve (PAME, 2019)	Strictly protected areas set aside to protect biodiversity and also possibly geological/geomorphological features, where human visitation, use and impacts are strictly controlled
Ib Wilderness Area (PAME, 2019)	Usually large unmodified or slightly modified areas, retaining their natural character and influence without permanent or significant human habitation, which are protected and managed so as to preserve their natural condition
II National Park (PAME, 2019)	Usually large natural or near natural areas set aside to protect large-scale ecological processes, along with the complement of species and ecosystems characteristic of an area, which also provide a foundation for environmentally and culturally compatible, spiritual, scientific, educational, recreational, and visitor opportunities
III Natural Monument or Feature (PAME, 2019)	Set aside to protect a specific natural monument, which can be a landform, sea mount, submarine cavern, geological feature such as a cave or even a living feature such as an ancient grove. They are generally quite small protected areas and often have high visitor values
IV Habitat/Species Management Area (PAME, 2019)	Set aside to protect particular species or habitats and management
V Protected Landscape/Seascape (PAME, 2019)	A protected area where the interaction of people and nature over time has produced an area of distinct character with significant, ecological, biological, cultural and scenic value: and where safeguarding the integrity of this interaction is vital to protecting and sustaining the area and its associated nature conservation and other values
VI Protected area with sustainable use of natural resources (PAME, 2019)	Area to conserve ecosystems and habitats together with associated cultural values and traditional natural resource management systems. They are generally large, with most of the area in a natural condition, where a proportion is under sustainable natural resource management and where low-level non-industrial use of natural resources compatible with nature conservation is seen as one of the main aims of the area

Source PAME (2019), *Implementing an Ecosystem Approach to Management of Arctic Marine Environments*, 3–4; CBD and UNEP (2004) *Protected Areas and Biodiversity*, 9–10; Hockings et al. (2006) and IUCN (2013)

In addition to the protections afforded by this classification and governance system, the national parks of the ROK were also designed to assist in the development of rural infrastructure that could support the encouragement of tourism as much as they were to recognize the natural capital values of an area. While visitor limits in fragile areas within a park or restrictions on the facilities that are provided can be used to ensure that only designated trails or roads are created and used in such areas, tourism was felt to be compatible with the preservation of natural capital (Gilligan et al., 2005; Heo, 2007; IUCN, 2013). However, Forestry, farming, and hunting are usually circumscribed, and the exploitation of habitat and wildlife is generally banned. Thus, while it is important to note the importance of the IUCN classification in promoting PAs for their ecological value, the majority of ROK national parks are identified by the IUCN as a *Category II- National Park,* noted as being:

> Usually large natural or near natural areas set aside to protect large-scale ecological processes, along with the complement of species and ecosystems characteristic of an area, which also provide a foundation for environmentally and culturally compatible, spiritual, scientific, educational, recreational, and visitor opportunities. (IUCN, 2013)

This explicitly incorporates the visitation variable. Thus, only a few are classified under different categories such as Category V (Table 4.3), which is:

> A protected area where the interaction of people and nature over time has produced an area of distinct character with significant, ecological, biological, cultural and scenic value: and where safeguarding the integrity of this interaction is vital to protecting and sustaining the area and its associated nature conservation and other values. (IUCN, 2013)

The IUCN-WCPA framework used for listing under these categories is however a flexible methodology. It is recognized that standardization is not possible (Hockings et al., 2006) and that the process should therefore only provide guidance on the types and amount of evidence required to establish a reasonably comprehensive conservation system (IUCN, 2008; Lausche & Burhenne-Guilmin, 2011). The information on which this is based should include all the items in the management cycle as outlined in Table 4.4.

The ROK participated in the inaugural World Conference on National Parks in Seattle in 1962 to help build a local consensus and obtain management criteria for a conservation area system. The formal system was adopted in March, 1967, which included a National Parks Service (KNPS) and the first park (Jirisan). Today, there are 1,400 PAs of all designations and management types in South Korea, however very few of these are captured on the *World Database of Protected Areas* (Kim, 2013) or recognized by the application of the IUCN management categories.

For those that are, and any additional ones created since, the ROK KNPS proposed to the IUCN in 2006 that they jointly develop an evaluation system tailored for the Korean context (IUCN, 2013; MOE, 2008), which was implemented in 2008. The background to this was mentioned earlier in the chapter; environmental management had until the early 2000s received less recognition in Korean PAs than the peripheral (to the environment) management issues relating to visitor services and facilities. Environmental issues, such as the number of threatened species, poaching and the introduction of invasive species, are of more immediate concern to the Ministry of

Table 4.4 Management elements in the assessment of environmental values

Element of the management cycle	Assessment criteria
The natural and social capital context of the area	The area's biodiversity and social values, and their relationships
Planning and design	Protected area design and its planning and management
Resource inputs	The human and material resources required to run the protected area effectively
Management process	Management structures for the proposed park
Outputs	Whether identified work targets have been or can be met
Outcomes	Whether the overall objective of conserving biodiversity and other values can be or has been met

Source IUCN (2013)

the Environment than the experiences of tourists (KNPS, 2010). Other issues, such as the need for internal landscape renovation, the need to geographically consolidate the PAs system and the need to deal with the impact of climate change, must also be covered.

Recently a system of *specially protected zones* within certain National Parks has been set up (MOE, 2008). These are designed to protect wild animal and plant habitats, wetlands, and other areas, along with their natural resources, from natural and man-made disasters. The main policies within this new system are:

- Rotating area protection through a Rest-Year Sabbatical System (begun in 1991). Parts of a park can be 'rested' in the 6th year of a management cycle (the Rest-Year) by restricting or eliminating completely the flow of visitors. Areas accommodating endangered species that need to be protected are particularly important and can be shut off in this way;
- The areas marked for rotation were reclassified as Special Protection Zones in National Parks (effective from 2007). The decisions on rotation (exclusion of visitors) are published as a KNPS notice in the following form (KNPS, 2019) and the regulations currently specify likely exclusions out to 2026 as an indication of forward planning (Table 4.5).

4.5 Visitor Management in the ROK National Parks

As part of the Ministry of the Environment's ongoing attempts to integrate natural capital with tourism, key recommendations for improving natural capital management (UNEP, 2006) were made in conjunction with the IUCN. Table 4.5 lists the IUCN category for each Mountainous National Park and visitor numbers for some of them along with notes on their conservation value. The recommendations are:

Table 4.5 IUCN category and visitation to Korean mountainous national parks

Name	IUCN Cat.	Annual visitors	Management plan	Management notes
Bukhansan	V	5,000,000	NA	Due to the popularity of this park with hikers and Seoul residents some trails are closed for rest periods (Rest-Year Sabbatical System and local initiatives) to protect local flora and fauna. There are many trails—the most traveled routes include the Bukhansanseong fortress wall, the Baegundae peak at 837 m, and the Insu-bong peak at 810 m. The 70 km long Dulle-gil trail connects villages within the park (www.protectedplanet.net)
Chiaksan	II	NA	NA	This park protects 821 plant and 2,364 animal species. 34 animal species are endangered, including the Flying Squirrel and Hodgson's Bat. The natural forest areas are being extended (KNPS, 2019)
Deogyusan	V	1,000,000	Annual plan, 10 year 'Management directions'	This park protects 1,067 plant, 32 mammal, 130 bird, 9 amphibian, 13 reptile, 28 fish, and 1,337 insect species. Endangered species include the Flying Squirrel, Marten and Otter
Gayasan	II	NA	Annual, 10 year 'Management directions'	As a sacred site of Buddhism, Gayasan has many cultural attractions such as Haeinsa (Temple) and Palmandaejanggyeong (Tripitaka Koreana, Buddhist Scriptures). It has two peaks, Sangwangbong (1430 m) and Chilbulbong (1433 m)
Gyeryongsan	V	NA	Annual, 10 year 'Management directions'	This is a national park with important geomorphological features. There are also a number of old temples

(continued)

Table 4.5 (continued)

Name	IUCN Cat.	Annual visitors	Management plan	Management notes
Hallasan	II	2,659,776 (World Heritage site)	NA	Managed by the Jeju Special Self-Governing Province. Jeju island has three World Heritage sites and many attractions, including museums, theme parks, mountains, lava tube caves, and waterfalls. Major tourism destination
Jirisan	II	1,500,000	NA	Jirisan (The "Exquisite Wisdom Mountain") is the holiest of South Korea's mountains, as well as the first and largest land-based national park in the country. Park management is involved in biodiversity conservation (of the Asiatic black bear) and a pioneering environmental restoration program
Juwangsan	II	NA	NA	924 animal and 88 plant species are found in the park area and among them the Otter has been designated a Natural Treasure
Mudeungsan	V	NA	Annual, 10 year 'Management directions'	Mudeungsan Mountain (1,186 m) features three rock peaks called Cheonwangbong, Jiwangbong, and Inwangbong, also known as the "Jeongsang Three" (www.seoulkoreatour.net)
Naejangsan	II	NA	Annual, 10 year 'Management directions'	This park shelters 919 plant and 1,880 animal species. 12 of the animal species are classified as endangered
Odaesan	II	NA	NA	Odaesan has the famous temple Woljeongsa (Jogye Order of Korean Buddhism). Woljeongsa and Sangwon are also well-known temples located in the Park

(continued)

Table 4.5 (continued)

Name	IUCN Cat.	Annual visitors	Management plan	Management notes
Seoraksan	II	4,000,000	NA	A park noted for its floral diversity. It shelters 1,013 species of plants, and is listed with UNESCO as a tentative World Heritage site
Sobaeksan	II	NA	NA	This park has helped in the reintroduction program for the critically endangered Korean Fox through the Seoul Zoo. The target was to re-establish a local population of 50 by 2020
Songnisan	II	1,500,000	NA	This park contains one of Korea's largest temples, Beopjusa, initially constructed in 653. Songnisan is connected to adjacent mountains; from Cheonwangbong (1,058 m) in the south, there are eight peaks including Birobong, Munjangdae, and Gwaneumbong
Taebaeksan	II	NA	NA	The 22nd national park was gazetted on 22 August 2016. Taebaeksan Mountain is one of the three sacred mountains of Korea, and a notable trail destination
Wolchulsan	II	NA	NA	The park is home to 3 national treasures and several local cultural properties. The highest peak in the park is Cheong-hwang-bong, at 809 m
Woraksan	II	NA	NA	This park has 1,200 plant, 17 mammal, 67 bird, 1,092 insect, 10 amphibian, 14 reptile, 27 freshwater fish, and 118 spider species. 16 of the animal species are endangered

Source MOE (2008), UNEP-WCMC (2020). Protected Area Profile for Republic of Korea from the World Database of Protected Areas, June 2020. Retrieved from www.protectedplanet.net.

- Enhance the regional approach to conservation;
- Build up protected area system planning resources and methodologies;
- Integrate natural capital management across agencies within the ROK;
- Improve tourism and local community relations;
- Enhance visitor interaction with other stakeholders;
- Increase the level of integration between regional governments; and
- Analyze and enhance staff responsibilities in the Ministry.

Note that "Management Directions" from the MOE appear to be the preferred mechanism to guide the development and implementation of the annual management plan where one exists.

The IUCN notes that the ROK offers high class visitor services in its parks, especially those of facilities and interpretation, and that its commitment of time and resources has been impressive (IUCN, 2013). However, when dealing with the use of the PAs for tourism, high visitor numbers can be a problem. Nevertheless, the view of the Ministry of the Environment is that although impact concentrations can occur, it is preferable to find a way to manage visitors rather than limit demand and lower support for the national park system as a whole. A better range of visitor alternatives, for example developing a wide range of mountain trails for community use, organizing more guided walks (guide-based impact control), broadening the range of visitor opportunities to engage in less conflicting activities through the provision of specialist cultural heritage or wildlife experiences for the ecotourist, would assist in the management of the parks. Also, they may boost the weak and problematic local level community engagement found in the MOE studies.

Park management policy is designed to rationalize access systems through temporary closure or removal of unofficial trails, while 'hardening' a limited range of trails to accommodate higher use rates. In addition, alternative serviced rest areas are developed. These approaches give support to more controlled flow of visitor movement (UNEP, 2006), and should allow for the dissemination and acceptance of better information for visitors, local communities, and the enhancement of the skills of KNPS staff relating to tourism.

The primary forms of tourism in the ROK national parks are site seeing, mountaineering, visiting the religious/cultural sites in the parks, and hiking. Trails have appeared all over Korea recently, but the ones within the national parks are highly-rated for their length, diversity of experience and historical value (Choi, 2013). For example, the Jirisan National Park trail, known as *Jirisan Dullegil* in Korean, has a total length of 274 km. It is made up of 22 sections that range from 9 km to about 20 km. The trail was developed by linking the paths used by villagers in earlier times with relatively new roads to and from parts of the mountain. Smartphone technology plays an increasingly important part in the experience of hikers on this and other trails, some applications offer 3D maps that allow the user to get a better idea of the terrain and where small accommodation and food outlets are to be found (Choi, 2013). Our case study supporting this discussion is of Hallasan National Park, located in the province of Jeju-do, which covers the island of Jeju off the southern tip of the Korean Peninsula. This became the 9th South Korean national park in 1970. The area was designated a UNESCO Biosphere Reserve in 2002, and a World Heritage Site, *Jeju Volcanic Island and Lava Tubes,* in 2007 (UNESCO, 2007). The park is managed by the Jeju Special Self-Governing Province.

4.5.1 Case Study: *Hallasan National Park, Jeju Island*

Hallasan NP is one of Jeju Island's most important attractions; not only is Mount Hallasan the tallest in the country (1950 m), visitors can enjoy seven different walking trails and many other attractions in this park. The Park covers 153km^2 of Jeju Island, which is of volcanic origin and dominated by the Mt. Hallasan shield volcano. The island is approximately 2 million years old, made of basalt, and measures 73 km by 41 km (KNPS, 2010). The mountain supports a number of alpine plant species (1800) and animals (Esperjesi, 2011). About 12% (224km^2) of the island is covered by Gotjawal Forest. The basalt underlying the surface soil ensured that this part of the island remained largely uncultivated, a primeval forest which has developed its own unique ecology. As the lava floor is permeable the aquifer can be recharged quickly, making Gotjawal a catchment area that provides drinking water for half a million island residents and tourists. The forest is an important wetland habitat for unique plant and birdlife and it was listed as a Ramsar Site in 2011.

The Jeju Island economy has relied on primary industry such as agriculture and fishing throughout its settlement history, and is renowned for its *Haenyeo* female free divers. But in recent decades these industries have taken a backseat to tourism, as the island receives more than 15 million visitors each year. These have traditionally been mostly domestic tourists, but after 2010, many thousands of Chinese tourists have included the island and its national park in their itineraries, and their number is increasing. This trend has also been assisted by the inclusion of Jeju island in many documentaries and films made for television (for example "All-in" a TV drama from 2003; Hudson et al., 2011).

Jeju is a popular destination for Chinese tourists and commercial developers for a number of reasons, including its proximity to China (a two-hour flight from Beijing), the National Government's policy of allowing foreigners to travel to Jeju without a visa, and the willingness of Jeju officials to grant Chinese condominium owners permanent resident status (Nam, 2015; Sang-hun, 2015). Because of these attractive policies, the presence of Chinese citizens on Jeju has increased dramatically; nearly half of all Chinese tourists to the ROK visit Jeju. Additionally, Chinese investors now own 830 ha of land on the island, up from just five acres in 2009 (Sang-hun, 2015).

The most popular tourist spots are the Cheonjeyeon and Cheonjiyeon waterfalls, Mount Halla itself, Hyeobje cave, the Manjanggul and other lava tubes, some of the longest in the world. Manjanggul Cave, 30 km east of Jeju City, was designated as Natural Monument No. 98 on December 3, 1962 (MOE, 2008). The temperature inside the cave ranges from 11 °C to 21 °C, and it is biologically significant as rare species live there. The cave is considered to be a world class tourist attraction, but visitor numbers are limited to 7000 per day. Jeju Island has also recently been placed on the *Globally Important Agricultural Heritage* (GIAHS) list to protect the world's only example of black basalt stone fences, called *Jeju Batdam*, and was listed as a *Global Geopark* in 2010.

There are a variety of leisure activities for visitors in the national park, such as mountain climbing, trail walking, horse riding, hunting, fishing, and so on. Depending on the season, Jeju also hosts many festivals for tourists, including a penguin swimming contest in winter, cherry blossom festivals in spring, night beach festivals in summer, and horse riding festivals in autumn. Access to and from the island is via Jeju International Airport or by ferry. Transport on the island is by rental car. A wide range of agricultural products are sold to tourists as well as the usual souvenirs and duty-free shopping.

Mount Hallasan is in the center of Jeju Island and is the island's most distinguishing landmark, widely known for its geological and scenic values (Fig. 4.3). There are 368 parasitic volcanoes, called oreums, (scoria cone) around the main mountain, and it is famous for its vertically layered ecosystems of plants that result from the varying temperatures on the mountainside. Over 1,800 species of plants and 4,000 species of animals (3,300 species of insects) have been identified. In comparison to other national parks, Hallasan Mountain is relatively easy to hike. With hiking trails less than 10 km long, it is possible to go up to the peak and back in one day (Fig. 4.3). However, the constantly changing weather can bring a lot of wind, so visitors must be well prepared before hiking. Hiking hours are subject to change according to season and are at the discretion of park rangers. Hallasan has 7 walking trails: Eorimok (6.8 km), Yeongsil (5.8 km), Seongpanak (9.6 km),

Fig. 4.3 Crater lakes on the summit of Mt. Hallasan. *Source* Author

Seokgulam (1.5 km), Gwaneumsa (8.7 km), Donnaeko (7 km), and Eoseungsaengak (1.3 km). The longest trail is Seongpanak, which takes about 4.5 h.

4.6 Conclusions

The development of the Korean mountainous national park system in line with IUCN and other guidelines has resulted in several positive outcomes. The first is the development of an ecosystem approach to conservation and visitor management. This is a deliberate shift from species conservation to ecosystem conservation in line with global trends. In this way, communities can protect existing natural resources by developing buffer zones and 'green corridors' to facilitate species movement and counteract habitat fragmentation.

The second outcome is the building and resourcing of a coordinated approach to PA planning in the ROK. This allows policy-makers to strengthen planning through comprehensive gap analyses while including consideration of the need for protection and sympathetic options in terms of management categories and local governance types, including application of the IUCN PA categories. It also allows for management integration across agencies, the development of a legislative framework for biodiversity conservation, and the holistic integration of PA management into a single body. Hallasan provides a good example, with its Natural World Heritage site management plan "streamlined with other planning instruments" (IUCN, 2008).

Third, cooperation with a range of stakeholders in plans designed to improve local community relations and regional integration. Importantly, this includes the identification of ways to improve local support for PAs by sharing best practice examples from across Korea and elsewhere. Identification further incorporates visions of ad hoc successful outcome(s), and where necessary making changes in policies and regulations to maximize benefits for local communities. This 'co-management' model has been facilitated by promoting interactions between interest groups in both formal and informal settings. Stakeholder relations have been improved through the adoption of mutually supportive policies that have a particular emphasis on compliance, whilst acknowledging the rights of private landowners within PAs, and integrating local businesses, temples and hiking groups. In this regard, Hallasan is an outlier since there are "no residents in the park and the area is almost totally publicly owned" (IUCN, 2008).

Finally, increased management capacity has been achieved in Korea's mountainous national parks through enhancing staff effectiveness and satisfaction derived from a thorough review of recruitment policies, specialized training, more effective assignment, rotation and use of local people to reduce the rapid turnover of new staff, and bringing more local people into fulltime employment. This approach includes diversifying the parks' funding base to increase revenue and to spread risk and focus research to cover real management issues. These include, for example, greater concentration on sociological issues, cultural heritage, and climate change, while harmonizing the management of natural and cultural heritage to ensure integrated outcomes, consistent policy and good management practice.

References

Bong, S. C. (2007). Nature in Korea. In H.-Y. Heo (Ed.), *The biodiversity and protected areas of Korea*. Ministry of Environment and the Korea National Park Service.

Brondizio, E. S., Settele, J., Díaz, S., & Ngo. H. T. (Eds.). (2019). *Global assessment report on biodiversity and ecosystem services,* Intergovernmental Science-Policy Platform on Biodiversity and Ecosystem Services (IPBES).

Chape, S., Harrison, J., Spalding, M., & Lysenko, I. (2005). Measuring the extent and effectiveness of protected areas as an indicator for meeting global biodiversity targets. *Philosophical Transactions of the Royal Society B: Biological Sciences, 360,* 443–455.

Choe, S.-H. (2016). For South Koreans, a long detour to their Holy Mountain. Changbaishan Journal. *The New York Times,* September 26.

Choe, Y., Schuett, M. A., & Sim, K. W. (2017). An analysis of first-time and repeat visitors to Korean National Parks from 2007 and 2013. *Journal of Mountain Science, 14*(12), 2527–2539.

Choi, H.-S. (2013). Soaking in history and culture along the Jirisan trail. *The Korea Herald*.

Dudley, N., Mulongoy, K. J., Cohen, S., Stolton, S., Barber, C. V., & Gidda, S. B. (2005). *Towards effective protected area systems: An action guide to implement the convention on biological diversity programme of work on protected areas* (CBD Technical Series 18). Montreal: Convention on Biological Diversity.

Dwyer, L., & Edwards, D. C. (2010). Sustainable tourism planning. In J. J. Liburd & D. C. Edwards (Eds.), *Understanding the sustainable development of tourism* (pp. 20–44). Goodfellow Publishers. http://hdl.handle.net/10453/16837.

Esperjesi, J. (2011). Jeju: From peace island to war island. *Asia Times, Asia Pacific Journal Japan Focus*.

Gilligan, B., Dudley, N., Fernandez de Tejada, A., & Toivonen, H. (2005). Management effectiveness evaluation of Finland's protected areas. *Nature Protection Publications of Mets Hallitus* (A 147). Mets Hallitus.

Hee, Y. C. (2007). Birds. In H.-Y. Heo (Ed.), *The biodiversity and protected areas of Korea*. Ministry of Environment and the Korea National Park Service.

Heinonen, M. (2007). State of the Parks—Finland: Finland's protected areas and their management for 2000–2005. *Nature protection publications of Mets Hallitus*. Mets Hallitus.

Heo, H. Y. (Ed.). (2007). *The biodiversity and protected areas of Korea*. Ministry of Environment and the Korea National Park Service.

Hockings, M., Stolton, S., & Dudley, N. (2000). *Evaluating effectiveness: A framework for assessing management of protected areas*. International Union for Conservation of Nature and the University of Cardiff.

Hockings, M., Stolton, S., Leverington, F., Dudley, N., & Courrau, J. (2006). *Evaluating effectiveness, A framework for assessing management effectiveness of protected areas*. International Union for Conservation of Nature.

Hong, H. J., Choi, H. A., Byun, B. S., & Park, Y. H. (2013). Analysis of Environmental and socio-economic effect on adjustment of National Parks. *Journal of the Korean Society of Environmental Restoration Technology, 16,* 49–62.

Hudson, S., Wang, Y., & Gil, S. M. (2011). The influence of a film on destination image and the desire to travel: A cross-cultural comparison. *International Journal of Tourism Research, 13*(2), 177–190.

International Park Planning Institute. (1972). *National policy master plan for tourism, parks, recreation, and conservation—TPRC, Republic of Korea: Phase 1.* TPRC Report. International Park Planning Institute.

IUCN. (2008). *International Union for Conservation of Nature 2008 Guidelines for applying protected area management categories.* Retrieved on 27 August 2020 from www.iucn.org/pa_cat egories

IUCN. (2009). *Korea's protected areas: Evaluating the effectiveness of South Korea's protected areas system.* Report produced on behalf of IUCN, the Korean National Parks Service (KNPS), the Korean Ministry of Environment (MOE) and the Jeju Island Special Self-Governing Province. International Union for Conservation of Nature and KNPS.

IUCN. (2013). *Guidelines for applying protected area management categories.* International Union for Conservation of Nature.

Kim, S. G. (2013). *Korean protected areas in WDPA.* Korea National Park Service and Korea Protected Areas Forum.

Kim, S. Y. (1993). A study on the tourism development of Cheju Province based upon tourist area life cycle. *Tourism Geography, 3,* 85–104.

KNPS. (2007). *Annual Report 2006-2007.* Korea National Park Service.

KNPS. (2008). *National parks of Korea.* Korea National Park Service.

KNPS. (2010). *A study on national park boundary readjustment.* Korea National Park Service.

KNPS. (2019). *National parks: About the Korea national park service.* Korea National Park Service.

Lausche, B. J., & Burhenne-Guilmin, F. (2011). *Guidelines for protected areas legislation.* International Union for Conservation of Nature.

Lee, S. (1995). Institutional and legal framework of National Parks and protected areas in Korea. *International symposium and excursion on National Parks and protected areas* (pp. 111–126). Seoul.

Ministry of Cultural Heritage. (2019). *Cultural heritage 2019 by statistics.* Cultural Heritage Agency.

MOE. (2008). *A study on the feasibility investigation criteria of National Parks and improvement of the natural park system.* Ministry of the Environment.

Mulongoy, K., & Chape, S. (2004). *Protected areas and biodiversity: An Overview of key issues.* Convention on Biodiversity and UNEP-WCMC.

Nam, I. S. (2015). Chinese wealth transforms South Korea's Jeju Island. *Wall Street Journal.*

Oh, G. (1998). Studies on management systems in Korean National Parks. *Proceedings of the 21st century Korean National park policy forum* (pp. 107–126). Korean national Park Service (in Korean).

PAME. (2019). *Implementing an ecosystem approach to management of Arctic marine environments.* PAME.

Sang-hun, C. (2015). South Korean Island grows wary after welcoming the Chinese. *The New York Times.*

UNEP. (2004). *Protected areas and biodiversity.* Convention on Biological Diversity.

UNEP. (2006). *UNEP annual evaluation report.* UN Environment Evaluation Office.

UNEP-WCMC. (2020). *Protected area profile for Republic of Korea from the world database of protected areas.* www.protectedplanet.net.

UNESCO. (2007). *MAB biosphere reserves directory: Jeju Island.* UNESCO.

You, J. H., Jeon, S. K., & Seol, J. W. (2013). Flora and conservation plan of Gayasan National Park. *Journal of the Korean Society of Environmental Restoration Technology, 16,* 109–130.

Malcolm J. M. Cooper is Emeritus Professor at Ritsumeikan APU. He was the inaugural Vice President for Research at APU (2005–2012). His research interests include Protected Area Management, Leadership and Resilience, and Geotourism. He holds a Master of Laws (Environment) and has worked in the environmental planning and tourism policy areas in Australia, New Zealand, Vietnam, Sri Lanka, and Dubai.

Chapter 5
Taiwan's National Network of Protected Areas and Nature-Based Tourism

Chieh-Lu Li and Thomas E. Jones

5.1 Introduction

Taiwan consists of one main island, 12 surrounding islets (with an area over 5 km^2), and a further 153 smaller islands with a combined terrestrial area of 36,193 km^2. Two thirds of the main island consist of a mountainous backbone that contains more than 260 peaks over 3,000 m. The main island was formed approximately 4–5 million years ago at a complex convergent boundary between the Eurasian and Philippine Sea Plates. Subsequently cut off from the Asian continent, Taiwan is highly valued for its endemism and biodiversity from sea to summit, with over 1,000 km of coastline including several bays with mangrove forests (MacKinnon & Yan, 2008). Taiwan's biodiversity inventory shows 63 species of mammals (including 10 endemic); 445 species of birds (14 endemic); and over 4000 species of higher plants (1,075 endemic) (Hsu, 2014). Taiwan Forestry Bureau (2018) estimated the overall extent of forest cover at 58%, with an annual decline of 0.6%.

The first protected areas established in Taiwan were 3 national parks designated in 1937 during the Japanese colonial era (Lu et al., 2015). The designation criteria closely resembled that of national parks on the Japanese mainland, with Tatun, Tsugitaka-Taroko and Niitaka-Alishan National Parks selected as examples of magnificent mountain landscapes and sightseeing spots (Kanda, 2012). During the period after World War II, the Taiwan Authority's national policy prioritized

---

C.-L. Li (✉)
Department of Tourism, Recreation, and Leisure Studies, National Dong Hwa University, Hualien, Taiwan
e-mail: clli@gms.ndhu.edu.tw

T. E. Jones
Environment and Development Cluster, Ritsumeikan Asia Pacific University (APU), Beppu, Japan
e-mail: 110054tj@apu.ac.jp

© The Author(s), under exclusive license to Springer Nature Switzerland AG 2021
T. E. Jones et al. (eds.), *Nature-Based Tourism in Asia's Mountainous Protected Areas*, Geographies of Tourism and Global Change,
https://doi.org/10.1007/978-3-030-76833-1_5

economic development over conservation, so the National Park Act was not passed until 1972 (Lu et al., 2015).

In 1983, Kenting was designated as the first post-war protected area (PA), as well as the first national park. Four years earlier, this park on the southern tip of the island had been placed under the Tourism Bureau's jurisdiction as a National Scenic Area (Wieman, 1995). Subsequently a string of PAs was designated in the 1980s, mainly National Parks and Natural Reserves. This was followed by the Wildlife Refuges and Major Wildlife Refuges designated in the 1990s (Taiwan National Park, 2020).

In 2000, an ecological corridor was set-up in the central range and expanded. In 2015, the establishment of the Wetlands of Importance system heralded the start of the fourth wave of Taiwan's PA networks, adding 47,627 hectares of Wetlands of Importance (Lu et al., 2015). The numbers and coverage of PAs in Taiwan (1984–2015) remained flat from 1984 to 2006 in terms of total sea coverage, but there was a rapid increase in 2007 after the designation of Dongsha Atoll National Park, the biggest national park in Taiwan, added 353,489 hectares of mostly marine territory (Table 5.1).

Table 5.1 Overview of Taiwan's main systems of protected areas

PA system	No	Land	Marine	Authority	Dominant Law	IUCN	Gov't
National Parks	9	310,376	438,574	Min. of Interior	National Park Act	Cat. II	Central
National Natural Parks	1	1,123	N.A				
Nature Reserves	22	65,341	117	Council of Agriculture (Forestry Bureau)	Cultural Heritage Preservation Act	I/III	Central & local
Wildlife Refuges	20	27,144	296				
Major Wildlife Habitats	37	325,985	296		Wildlife Conservation Act	IV	Local
Forest Reserves	6	21,171	N.A			IV	Central
Total area in Ha (2014)		694,501	438,987		Forestry Act	III/IV	

Source adapted from Lee (2013). Not including Wetlands of importance ($n = 83$), 56,865 Ha (40% overlapped)

5.2 Taiwan's System of PAs and NPs

5.2.1 The National Network of PAs

21% of Taiwan's terrestrial area is covered by the 81 protected areas (MacKinnon & Yan, 2008). The two dominant central government agencies providing nature-based tourism opportunities in Taiwan's public lands are the Construction and Planning Agency (within the Ministry of Interior, responsible for national parks); and the Forestry Bureau (within the Council of Agriculture, responsible for national forest recreation areas). The PAs can be divided into six systems (Table 5.1), each comparable to the IUCN's classification of PAs. For example, according to the National Park Act, National Parks and National Natural Parks in Taiwan are similar to IUCN's Category II. However, unlike the IUCN system, there seems to be a lack of detailed laws and regulations regarding the management objective of providing a foundation for environmentally and culturally compatible scientific and educational opportunities. Likewise, according to the Cultural Heritage Preservation Act, Nature Reserves are similar to IUCN's Category Ia. However, the aforementioned difference also applies to Taiwan's Nature Reserves (Taiwan Forestry Bureau, 2005).

Running concurrently to Taiwan's networks of PAs is a system of National Scenic Areas under the Tourism Bureau (in the Ministry of Transportation and Communication). National Scenic Areas are mainly designated for tourism development purposes although conservation is also one of the stated objectives. A few areas, such as Alishan National Scenic Area, actually overlap with Alishan National Forest Recreation area. In such cases, both agencies share responsibility for administrative duties.

Except for Wildlife Refuges and a few Nature Reserves that are managed by local governments, the different PAs categories fall mostly under the jurisdiction of central government agencies (Lee, 2013). In particular, 2 central government agencies dominate PA administration: the Forestry Bureau (COA) is in charge of Natural Reserves, Wildlife Refuges, Major Wildlife Habitats and Forest Reserves (Taiwan Forestry Bureau, 2018), and the Construction and Planning Agency (MOI) has jurisdiction over National Parks, National Natural Parks and Wetlands of Importance (Lu et al., 2015).

Governance of the PAs is based on four seminal laws: the Forestry Law (1932); National Parks Law (1972); Cultural Heritage Preservation Law (1982); and Wildlife Conservation Law (1989), with other relevant acts including the Forest Law, Environment Impact Evaluation Law and Water and Land Conservancy Law (MacKinnon & Yan, 2008). The legal framework is based on four different concepts: (i) habitat preservation; (ii) forest resource conservation; (iii) protection of endangered species; and (iv) participation in international species protection (ibid.). A distinction can be drawn between Wildlife Refuges, whose overriding aim is protection of a particular species such as sea turtles, versus Nature Reserves, whose primary objective is to protect cultural assets. However, as in other chapters of this volume, considerable overlap can also be observed between the different categories due to historical and socio-political context.

The administration of PAs can be broadly divided by its function. Nature conservation work is organised by the Council of Agriculture. Management of national parks is currently under the Ministry of Interior's Construction and Planning Administration and individual National Park Headquarters. Management of national parks will be transferred to the National Park Administration, a new institute to be established within the Ministry of Interior (Li, 2017; Taiwan National Park, 2020). Environmental Impact Assessment and pollution prevention is conducted by Environmental Protection Administration, Executive Yuan Cabinet, assisted by Taiwan National Park Society, other NGOs and universities (MacKinnon & Yan, 2008).

The Tourism Bureau is also involved with the administration of Taiwan's national scenic areas. At the central level, the Tourism Bureau operates under the Ministry of Transportation and Communications (Tsao, 2002). There are 13 National Scenic Areas under the Tourism Bureau (Taiwan Tourism Bureau, 2020). Meanwhile at the local level, Tourism Sections monitor tourism business under the municipal governments' Construction Bureaus. There is also a Council for Cultural Planning and Development in charge of cultural resources (Tsao, 2002).

5.2.2 Taiwan's National Parks

Historically, selection of PAs in Taiwan was based on Japanese national park designation criteria that prioritized magnificent mountainous landscapes (Kanda, 2012). The colonial PA system also inherited the underlying tension between tourism and nature conservation goals, as epitomized by the debate over how to turn the tropical landscape of southern Taiwan into a national park (Taiwan National Park, 2020). Today, the contemporary network comprises nine national parks and one national nature park that together cover 8.6% of Taiwan's terrestrial area (see Table 5.2).

The parks include an eclectic mix of mountains, wetlands and marine areas, as well as sites of historical interest such as the Kinmen isles just off the coast of Fujian Province. This former battlefield commemorates the Battle of Guningtou that started in 1949 and continued intermittently throughout the 1950s.

The ownership of PA land belongs to Taiwan, ROC, and most parkland falls under the jurisdiction of National Forest. According to the Forestry Law, National Forest is administered by the Forest Bureau which is a long-established central government agency that pre-dates the establishment of national parks. The PA land included in national parks also comes mostly from National Forest. Hence, although the national parks agency has the legal right to administer their protected area land under the National Park Law, policy implementation in practise requires coordination with the Forest Bureau, the landowner. The National Park Law was created to preserve the nation's unique natural scenery, wild fauna and flora and historic sites and provide public recreation and areas for scientific research. According to the National Park

Table 5.2 Overview of Taiwan's nine national parks and one national nature park

Name	Area[a]	Designated	Description
Kenting National Park	332.9 km^2 (180.8 km^2 of land and 152.1 km^2 water)	1984	Taiwan's first national park; located at the southern tip; renowned for its tropical coral reef and migratory birds
Yushan National Park	1,031.2 km^2	1985	The largest territorial national park; located in the centre of the main island; contains Jade Mountain (Yushan, 3952 m) the highest peak in East Asia; Yushan is also one of the most popular mountaineering routes in Taiwan
Yangmingshan National Park	113.38 km^2	1985	A volcanic landscape of hot springs and geothermal phenomenon; located at the north of the main island
Taroko National Park	920 km^2	1986	Steep valley and marble gorge carved by the Li-Wu River; also the home of the indigenous Truku people; located in eastern Taiwan; contains the second largest national parkland and has the second highest visitation in Taiwan and the highest visitation on the east coast of Taiwan. Nanhu and Chilai Mountain are well known peaks in the park
Shei-Pa National Park	768.5 km^2	1992	Located in the centre-north of Taiwan; contains Snow Mountain, (Xueshan, 3886 m) Taiwan and East Asia's second tallest mountain; also Dabajian Mountain and the Holy Ridge are favourite mountaineering routes in Taiwan
Kinmen National Park	35.3 km^2	1995	Contains historical battlefields as well as wetland ecosystem and traditional Fujian buildings about 400 years old tracked back to Ming dynasty in China; located on an island near mainland China

(continued)

Table 5.2 (continued)

Name	Area[a]	Designated	Description
Dongsha Atoll National Park	3,536.7 km² (including 1.79 km² of land)	2007	The first oceanic national park; hosts 72 species of endemic plants and 125 species of insects; located off-shore like Kinmen National Park; due to geopolitical issues and a strict conservation policy, public access is currently prohibited
Taijiang National Park	393.1 km² (49.1 km² of land and 344.1 km² water)	2009	Located in southwest Tainan; distinctive coastal landscape comprising wetlands, tidal flats, lagoons and mangrove swamps; habitat for 205 species of shellfish, 240 species of fish and 49 crab species; also bird species, including the endangered black-faced spoonbill; giant inland sea that silted up during the eighteenth century
Shoushan National Nature Park[b]	11.2km² (2,775.0 acres)	2011	Located in south western Kaohsiung; divided into 1 special landscape area, 6 historical areas, 3 general restricted areas, and 1 recreational area; limestone caves containing stalagmites; large populations of Formosan rock macaques
South Penghu Marine National Park	358.4 km² (including 3.7 km² of land)	2014	4 remote islands located in the Penghu Archipelago, also known as the Pescadores ("fishermen" in Portuguese) that consists of nearly 90 islands altogether; clusters of Acropora coral reefs host diverse marine life including dolphins and whales; landscape known for basalt terrains and traditional low-roofed houses built from coral stone and basalt

[a] 1km² = 100 ha
[b] National Nature Park is the same as National Park but with a smaller area

Law, the zoning system in national park includes areas for conservation, special landscapes, historical preservation, general control, and recreation. Special permission must be obtained from national park administration in order to enter the conservation area (Taiwan Natural Ecological Area, 2020).

The most popular activity in Taiwan's national parks is hiking and mountain climbing, described below in a separate section. Other popular activities include biking, camping, scenic driving, forest therapy, visits to information centres, interpretative tours, nature observation, bird watching, butterfly watching and photography. However, certain activities are restricted. For example, climbing higher peaks and the use of mountaineering cabins requires a permit from the national park and forestry administration. Other examples of limits of visitor use include access restrictions on mountain bikes that are only allowed within general recreation areas in national parks (Taiwan Natural Ecological Area, 2020). Specific research has been conducted in Taiwan's National Parks. For example, the *Journal of National Parks*, published by the Construction and Planning Agency, reports the findings of investigation on natural and cultural aspects as well as human dimensions.

5.3 National Park Visitation and Nature-Based Tourism

Longitudinal trends in visitation to Taiwan's nine national parks and one national natural park from 2008–2017 show variation of visitor arrivals contingent to administrative policy (See Fig. 5.1). There was a sharp increase in aggregate national park visitation from 2012 that reflected the additional number of visitors to Shoushan (designated in 2011) and Taijiang (designated in 2009). The total visitation in Taiwan national parks reached a high record of 28.71 million visits in 2015, but this was almost cut in half in 2017, shrinking to 15.46 million visits. The main reason for this sudden decline in just two years was the loss of mainland China visitors, coupled with a cooling in domestic interest due to cheap international airfare in the Asia Pacific region (Taiwan Tourism Bureau, 2020). As a result of the national government's promotion of international tourism, the number of inbound arrivals into Taiwan increased from 10.43 million in 2015 to 10.69 million in 2016 and 10.73 million in 2017 (Taiwan Tourism Bureau, 2020).

In the peak year of 2015, the three top-ranked parks in terms of visitation were Kenting (8.05 million annual visits, accounting for 28% of the aggregate total), Taroko (6.60 million, 23%), and Yangminshan (4.53 million, 16%). Kenting showed an especially rapid increase from 4.48 (2009) to 6.34 million annual visits (2010), 7.06 (2013), and 8.16 million visits (2014). However, Kenting's visitation also showed a rapid decline, down to a record low of 3.04 million visits in 2017. Similarly, Shoushan experienced a record decline from 4.69 million visits in 2014 to 1.92 million visits in 2017, a 59% reduction. The downturn in visitation was mainly due to the decline in visitors from mainland China, as the 'New South Bound' policy re-directed them to South East Asia (Taiwan Tourism Bureau, 2020).

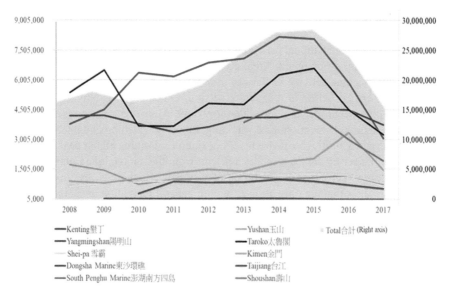

Fig. 5.1 Visitation trends to Taiwan's national parks (2008–2017). *Source* Adapted from Taiwan National Parks (2020)

Meanwhile, visitation at Taroko showed great fluctuation, from 3.68 million visits in 2011 to a maximum of 6.60 million visits in 2015, and then down to 3.26 million visits in 2017. Yangminshan and Yushan were broadly steady with around 4 million annual visits and one million annual visits respectively. Yangminshan is close to Taipei City, the capital of Taiwan that is also the most densely populated part of the country. On the other hand, Yushan is located in a relatively remote area but contains Mount Jade/Yushan, one of Asia Pacific's most iconic peaks, which attracts climbers to the park. Shei-pa showed the largest proportional increase from 826,703 visits in 2012 to 1.34 million in 2013, a 63% increase in annual visitation. Of the others, South Penghu Marine and Dongsha Marine National Parks comprise large swathes of ocean with precious little land area. Due to their restricted accessibility, these two national parks receive the lowest levels of annual visitation (Li, 2017; Taiwan Tourism Bureau, 2020).

Results from a 2018 survey of 22,165 domestic and 2,545 international respondents showed 45.3% of the respondents participated in nature-related activities (Taiwan Tourism Bureau, 2020). Among these activities, 23.7% of the respondents participated in viewing coastal and rural landscapes, wetlands, streams and water falls. Hiking on forest trails, mountaineering, camping and river tracing were selected by 14.4% of the respondents. Only 1.6% of respondents participated in watching wildlife including whales, fireflies, and birds, while 4.1% sought plants including flowers, cherry blossom, maples and big trees. Enjoying sunrise, snow scenery, and nature landscape accounted for 1.5% of the respondents. Meanwhile,

14.3% experienced cultural activities such as visiting historic sites, participating in festival activities, and watching exhibitions and shows.

5.4 Mountain Climbing in Taiwan

Taiwan contains a dense cluster of moderate-high altitude peaks that is unprecedented amongst the world's islands. There are 268 mountains above 3,000 m on the main island which is only 36,000 km^2. Many of the higher peaks lie at the heart of PAs, especially within the three high mountain national parks of Yushan, Taroko and Shei-pa. Modern mountain climbing in Taiwan emerged comparatively recently as foreign traders, missionaries and government officers made explorative forays into the mountainous wilderness at the end of the nineteenth century (Chou et al., 2018; Li et al., 2019). Systematic development of mountainous regions unfolded in the colonial era as Japanese officials and researchers penetrated the highlands further for map-making, measuring and research purposes, discovering new species and exploring local indigenous cultures in earnest from the beginning of the twentieth century (ibid.). Mountaineering was gaining popularity amongst Japanese elites who modelled their Alpine Climbers' Associations on examples from Britain and the European Alps. A Taiwanese association was established in 1926, but it was largely dominated by colonial Japanese and local elites, while the ability of less well-off sections of the population to participate in hiking activities was heavily restricted (Chou et al., 2018).

After the Second World War ended in 1945, the autonomous rule which accompanied the relinquishment of Japan's colonial claims did not immediately result in a relaxation of the strict clampdown on mountain hiking activities. Conversely, mountainous areas continued to hold strategic importance as military bases and access to protected areas was therefore tightly regulated controlled (Huang, 2015; Li et al., 2019). The 'Taiwanese Mountaineering Society', later renamed as the 'Chinese Taipei Alpine Association' (CTAA) started the 'Five Mountain Club' and '100 Mountain Club' to motivate the general public to engage in mountain climbing This led to a rapid increase in the population of mountain hikers, with four members of the CTAA successfully completing the '100 Mountain list' in 1972 to popular acclaim (Huang, 2015). Thereafter, the promotion of national parks by the government has supported the steady growth in mountain hiking culture, facilitated by relaxed access arrangements; for instance, the removal of a requirement for groups climbing above 3,000 m to be accompanied by a guide.

Data collected by Taiwan Tourism Bureau (2020) indicated that over five million people are involved in mountain hiking activities—one in four of the Taiwanese population. In 2002, a survey also conducted by the Taiwan Tourism Bureau showed that mountain hiking was the second most popular holiday activity chosen by Taiwanese (Huang, 2015). Taiwan's mountain climber market has potential for growth, since only 19% of respondents had climbed more than 50 of the 100 listed peaks, and results of a cluster analysis found only 22% to be "specialized" hikers. Nonetheless, a unique mountain hiking sub-culture has already emerged as evidenced by the set-up of 'clubs' in which more experienced hikers act as guides and hold social activities related to mountain hiking (Chou et al., 2018).

Taiwan's highest peak, Jade Mountain in Yushan National Park, is highly attractive to climbers, with over 36,000 permits issued annually. The Lodge is 8.5 km from the Tataka trail head, 2,610 m in elevation, and 2.4 km from the summit of Jade Mountain. Most of the climbers seek to reach the summit using Jade Mountain trail of 10.9 km by compliance with regulations. The number of permits for an overnight stay at the accommodation closest to the peak, Paiyun Lodge, located 3,402 m above sea level is highly restricted. The set-up of a permit system to limit the number of mountain climbers based on the physical capacity of facilities such as mountain lodges (Li, 2017). There are 116 bed berths at the Lodge. Due to Paiyun's popularity, permits are issued randomly using a lottery system, while other mountain cabins adopt a 'first-come, first-served' system. Paiyun Lodge uses solar power to generate electricity and introduced a meal reservation system since 2013 to reduce food waste (Yushan National Park, 2020).

Parks and protected areas offer a variety of other activities apart from hiking. Adventure tourism, such as rock climbing, mountain biking and canyoneering, are also proliferating, but at higher altitudes, a Taiwan Tourism Bureau survey revealed a preference for landscape appreciation and photography (Huang, 2015). Health and wellness, such as quality of life, psychological and physical well-being, are considered as an important factor behind mountain hiking's popularity. Mountain hiking to achieve health benefits is also being promoted in the context of forest/outdoor recreation and nature-based tourism (Li & Chick, 2016; Li & Haivangang, 2017; Li & Wang, 2012; 2014; Li et al., 2015, 2016; NRPA, 2018). 2020 has been designated as the Year of Mountain Tourism by Taiwan Tourism Bureau of the Ministry of Transportation and Communications to promote mountain tourism for both domestic and international hikers. The goal is to mould Taiwan into an important destination for Asian hikers focused on "smart travel and touching experiences" (Chou et al., 2018, pg. 1).

In Taiwan's second highest peak of Sheishan (3,886 m), located in Sheipa National, a study by Li and Wang (2014) on 614 hikers demonstrated a positive relationship between health/wellbeing and natural tourism such as mountain hiking. Specifically, age was shown to be a factor affecting the level of life satisfaction and perception of quality of life. Older (> 40 years), married mountain hikers had higher perceptions of life satisfaction and quality of life than single ones aged under 40 years. A positive significant relationship was found among variables, such as psychological well-being, quality of life, and life satisfaction. On the other hand, perceptions of stress showed significant and negative correlation with psychological well-being, quality of life, and life satisfaction. The result also showed that frequency of visits and mountain hiking were negatively correlated with perceived stress levels among hikers. More frequent hikers had lower perceptions of stress. Lastly, the findings showed that two factors in quality of life, i.e. the psychological and social relationship variables, along with psychological well-being, significantly explained 43% of the total variance of life satisfaction ($R^2 = 0.43$).

Taiwan is a unique high mountain island with abundant natural resources, and nature-based tourism has become increasingly popular. Where previous policies sought to limit hiker numbers for safety reason, the current calls for more mountain tourism encourage people to explore and learn from the mountains. However, a wide range of ecological and sociocultural issues have resulted from the recent promotion of mountain tourism. As mountains form the centrepiece of many PAs, there can be a corresponding lack of conservation in other ecosystem types such as the plains, coastlines, foothills and marine protected areas. Such ecosystems are already under threat from the pressure of population density combined with development and resource extraction. Specifically, MacKinnon and Yan (2008) identified the lowland forest of west-central Taiwan, Lanyu Island and other marine areas as potential PA growth areas. In addition, many of Taiwan's mountains hold a sacred place in indigenous mythology so mountain tourism must tread a tightrope between PA management and the needs of indigenous populations (Huang, 2015). To mitigate such conflicts, the Tourism Bureau that administers Taiwan's national scenic areas aims for ethical equality, for instance, promoting diverse tours that incorporate native cultures (Tsao, 2002). Several local communities are promoting the ecotourism businesses of indigenous people while preserving their unique natural and cultural resources (Tsao, 2002—see Fig. 5.2). Another way to alleviate tensions between indigenous people and national park administration is to employ more indigenous staff. A study by Li (2018) recommended that employing more Hakka language-speakers among national park staff would improve Hakka visitors' perceived service quality. Although the Hakka are not classified as 'indigenous,' these descendants of migrants from the Guangdong area comprise 14% of Taiwan's population, the

Fig. 5.2 Amis indigenous people in Hualien interpret their traditional fishery method (*Palakaw*). *Source* first author

second-largest ethnic group. In a similar vein, employing more indigenous staff in national parks could improve indigenous visitors' perceived service quality.

5.5 Case Study of Yushan

Designated in 1985, one year after the first national park at Kenting, Yushan is Taiwan's largest terrestrial national park. Covering an area of 103,121 hectares, it is located in the centre of the island within the Central Mountain Range. The park's symbol is Yushan (Jade Mountain), the tallest peak in North East Asia at an altitude of 3,952 m. However, only a few visitors climb Yushan as the trail is strictly regulated by carrying capacity controls, and hikers need to pre-register online for a permit to climb. The permit application period opens from two months to five days before the intended hike date. The permit to climb Yushan was not relaxed during the Year of Mountain Tourism in Taiwan (2020), although equivalent permit requirements were removed on other mountains (Chou et al., 2018; Li et al., 2019). A number of amenities are established to attract hikers and to enhance the climbers' safety, such as mountain cabins re-built and a WiFi station on Yushan's north peak. The policy encourages hikers to respect and gain affinity with Taiwan's mountains, and is expected to bring more visitors to the mountains.

Climbing Yushan's summit generally takes one or two days. In case a one-day arrangement is chosen, hikers are urged to self-evaluate fitness levels for safety reasons. Moreover, one-day hikers must arrive at Paiyun Lodge by 10 AM to continue up to the summit. Hikers that miss the cut-off time are told to return to the trail head. The two-day hike is normally recommended, with an overnight stay at Paiyun Lodge (3,402 m above sea level), originally built during the Japanese colonial period in 1943, and later renovated and reopened in 2013 (Li, 2017; Li & Haivangang, 2017). Located at a distance of 8.5 km from Tataka trail head and 2.4 km to Yushan main peak, Paiyun Lodge is the most popular and the last accommodation that serves meals for hikers. The permit application follows a lottery method. In 2019, the chance to win a permit was about 14% for weekends and national holidays, 58% for week days, making an overall average of 36% to win a Yushan hike permit (Table 5.3).

Paiyun Lodge has 116 beds. The building comprises a two-story steel structure, mainly uses solar power for electricity, has toilets inside and outside the lodge and a hot water supply (Fig. 5.3). Management is outsourced to a private concession. To reduce food-waste, self-catering is not permitted at the Lodge, but hikers may reserve meals and sleeping bags when applying for a permit. The fee for staying at Paiyun is NTD 480 person/night (approx. $15US). During the COVID-19 pandemic, Yushan National Park remains open. Paiyun Lodge employs a social distancing regulation and the number of beds for hikers has been reduced to 100. There are 24 bed spaces at Paiyun Lodge reserved for international climbers from Sunday to Thursday via an online booking system.

Table 5.3 Overnight stays and permits at Paiyun Lodge in 2019

Month	Hikers staying overnight at Paiyun Lodge	Average weekend (Fri, Sat & Sun) chance to win (%)	Average weekday chance to win (%)
January	1,472	57.8	100.0
February	0[a]	0.00	0.0
March	3,130	21.3	70.5
April	3,194	10.5	30.7
May	2,944	6.9	18.8
June	2,832	6.0	17.6
July	3,034	5.6	15.0
August	2,329	6.2	20.0
September	3,135	6.5	19.1
October	3,219	5.5	14.0
November	3,199	6.8	21.2
December	3,191	16.5	57.3
Total	31,676	149.5	636.6
Average	2,880	13.6	57.9
Median			35.7

½[a] "Quiet Mountain" policy was employed in February 2019 to let nature rest. All hiking activities were restricted and Paiyun permit applications temporarily suspended (Yushan National Park, 2020)

Fig. 5.3 Paiyun Lodge located at 3,402 m altitude, 2.4 km to Yushan's main peak. *Source* Yushan National Park (2020)

5.6 Discussion and Conclusion

This chapter reflects on the challenges faced by Taiwan's PA network and nature-based tourism in the mountainous areas. The historical legacy of 'magnificent mountains' remains prevalent in PA planning philosophy since the Japanese colonial era (Kanda, 2012). Thus, the positioning of mountains within the Taiwanese network of PAs reflects "emotive and historical aspects into their consideration in addition to relative or absolute height, which is the main aspect of western classification" (Huang, 2015, pg. 61). Mountainous national parks also account for much of Taiwan's nature-based tourism market, as evidenced by longitudinal trends in visitation from 2008–2017. The sharp fluctuations in aggregate national park visitation reflect evolving trends in domestic demand as well as inbound tourism policy.

However, the recent growth in mountainous nature-based tourism poses various ecological and socio-cultural challenges to conservation objectives. To realize the PAs' potential to simultaneously achieve conservation and development goals requires a more extensive legislation and use of political powers while nurturing the contribution of local communities and indigenous people in PA management. Currently, there is administrative competition between the Forestry Bureau and the Construction and Planning Agency, so it is recommended to introduce a new Ministry of Environment and Resources to take holistic charge of the PAs (Lu et al., 2015). Beyond these governance issues, a range of site-specific challenges also exist such as (Li, 2017):

- Due to congestion/crowding issues in front country settings, visitations to backcountry settings such as ecological protection zone or wilderness increased;
- Human waste and garbage problem: traditional water flush toilet causes environmental pollution. Need to promote more compost eco-toilet and 'leave no trace' practise;
- Permit lottery and queue inequality, individual users may face disadvantages drawing lots.

Although numerous challenges exist for governance of nature-based tourism in ecologically fragile areas, Taiwan benefits from several factors that favour nature conservation. Firstly, ecotourism research and educational activities are generally of a high standard and being encouraged (MacKinnon & Yan, 2008). Nature-based tourism is popular and education efforts are evidenced by the availability of, for example, over 150 higher education courses related to tourism, sport, recreation and leisure. Such programs also provide outdoor recreation, nature-based tourism and ecotourism education in PA management. Additionally, environmental education and interpretation have been a focus on Taiwan's PA management. Those practices will continue to be promoted with the boom of mountain hiking in Taiwan.

Secondly, distribution of the population is another favourable factor towards nature conservation. Li (2018) suggested that the island's population density may be already helping to moderate visitors' perceptions of congestion. He noted that Taiwan's population density is 642 people/km^2, 20 times higher than the 33 people/km^2 in the

U.S.A. The gap is even greater in densely packed megacities such as Taipei or New Taipei City whose residents would thus seems likely to have a higher tolerance for crowding and lower expectations of a solitary "wilderness" experience than American park visitors. It remains to be seen whether visitors' inherent values can predict their satisfaction with anti-congestion management strategies such as the visitation capacity control.

Thirdly, the practice of recreation carrying capacity is comparatively well-established in Taiwan's national parks and national forest recreation areas. Total visitor capacity control is deemed necessary to ensure the recreation quality. Together with the increase of outdoor recreation and nature-based tourism, more direct visitor capacity limits will be reached in some cases. For example, even in front country settings, visitors require a permit to enter popular national parks such as Yanmingshan during the annual flower event and national holiday weekends (Li et al., 2016).

Fourthly, charging user fees has long since been common practise in Taiwan's national forest recreation areas and the custom has recently been extended to the national parks. In 2016, Kenting became the first national park to charge user fees. Kenting is the most visited national park (see Fig. 5.1), and new fees include payment for car parking and admission into the visitor hubs of Kenting, including popular sightseeing spots and developed areas with facilities. Yanmingshan also started to charge interpretation fees from October 2017, at a rate of US $50 ($NTD 1,500) for domestic visitors, and US $67 ($NTD 2,000) for foreigners, including long-term foreign residents without Taiwanese ID (Li, 2017; Taiwan National Park; 2020).

Finally, recent tourism policies include the use of Chinese cultural features to promote international tourism, including major events, such as the Taipei Lantern and Chinese Food Festivals, Taroko Gorge International Marathon, Hsiukuluan River Rafting and Kayak Racing. Such events provide a unique way to add a cultural component to promote international nature-based tourism activities (Taiwan Tourism Bureau, 2020; Tsao, 2002).

References

Chou, Y. H., Ou-Yang, H. Y., & Chen, K. C. (2018). Taiwan guakwa urlinurlin yuanshifachan celue [Taiwan tourism 2020: A sustainable tourism development strategy]. *Taiwan Modern Tourism, 1*(1), 1–20 [in Chinese].

Hsu, X. H. (2014). Yuanshi taiwan: taiwan der shanudouyoungsin [Sustainability in Taiwan: Taiwan's biodiversity]. *Science Development, 501*, 44–49 [in Chinese].

Huang, M. F. (2015). Mountaineering tourism in Taiwan: Hiking the '100 mountains.' In G. Musa, J. Higham, & A. Thompson-Carr (Eds.), *Mountaineering tourism* (pp. 59–65). Routledge.

Kanda, K. (2012). The selection process of National Park landscape areas and the imaginative geographies in Taiwan during the Japanese Colonial Period. *Academic World of Tourism Studies, 1*, 77–87.

Lee, K. C. (2013). *Collaborative planning and management for IUCN category V-protected landscapes in Taiwan*. PowerPoint Presentation at the First Asia Parks Congress Working Group 4, Collaborative Management of Protected Areas.

Li, C. L. (2017, November 10). *Mountainous protected areas of East Asia: Governance, conservation & mountain tourism*. Paper presentation at the pre-conference workshop in the Asia Pacific Conference, Ritsumeikan Asia Pacific University, Beppu, Oita, Japan.

Li, C. L. (2018). Outdoor recreation in a Taiwanese national park: A Hakka ethnic group study. *Journal of Outdoor Recreation and Tourism, 22*, 37–45.

Li, C. L., & Chick, G. (2016). *Psychological wellbeing, travel distance, living circle and region*. Paper presented in the 2016 Tourism Naturally conference, Alghero, Italy.

Li, C. L., & Haivangang, B. (2017, November 11–12). *A comparative study of quality of life: The case of Yushan/Jade Mountain hikers in Taiwan*. Paper presentation in the Asia Pacific Conference, Ritsumeikan Asia Pacific University, Beppu, Oita, Japan.

Li, C. L., Liao, K. C., Haivangang, B, Jones. T., & Nakajima, Y. (2019). *Exploring mountain hiking motivations cross-culturally: The case of national park and world heritage site in Asia Pacific*. Paper presented at the International Conference on Landscape and Sustainable Development—Geopark, National Park, Satoyama Initiative, Geologically Sensitive Area, Tourism Development and Environmental Education. National Taiwan University, Taipei, Taiwan.

Li, C. L., & Wang, C. (2012). Factors affecting life satisfaction: Outdoor recreation benefit and quality of life approach. *Sports & Exercise Research, 14*(4), 407–418.

Li, C. L., & Wang, C. (2014). *The relationships of psychological and physical health and well-being for mountain hikers: An integrated ecological study*. Final report to Conservation and Research Section, Shei-pa National Park, Ministry of the Interior, Taiwan.

Li, C. L., Wang, C., Dong, E., Wu, C., & Haivangang, B. (2016). *Mountain hiking to achieve psychological well-being: A case of trails in Taiwan*. Paper presentation in the 14th LARASA World Leisure Congress (WLC 2016), Durban, South Africa.

Li, C. L., Wang, C., & Graefe, A. (2015). *Exploring recreation pattern differences among mountain hikers in Taiwan*. Paper presentation in the 2015 International Symposium on Society and Resource Management (ISSRM), Charleston, SC, USA.

Lu, D. J., Chiang, P. C., & Lin, J. (2015, October 22–25). The paradigm shift of protected area management in Taiwan. *Proceedings from the third conference of East Asian Environmental History (EAEH 2015)*, Kagawa University, Takamatsu, Japan.

MacKinnon, J., & Yan, X. (2008). *Regional action plan for the protected areas of East Asia* (p. 82). Bangkok, Thailand.

National Recreation and Park Association [NRPA] (2018). Retrieved from https://www.nrpa.org/.

Taiwan Forestry Bureau. (2005). *Analysis and evaluation of status and management of protected areas in Taiwan*. Council of Agriculture, Forestry Bureau Sponsored Conservation Research Series, no. 94-23, no. 94-00-8-08.

Taiwan Forestry Bureau. (2018). Retrieved from https://www.forest.gov.tw/EN.

Taiwan National Park. (2020). Retrieved from http://np.cpami.gov.tw/home-en.html.

Taiwan Natural Ecological Area. (2020). *Taiwan zlanshantai boufuchi* [Taiwan Natural Ecological Area]. Retrieved from https://zh.wikipedia.org/wiki/%E8%87%BA%E7%81%A3%E8%87%AA%E7%84%B6%E7%94%9F%E6%85%8B%E4%BF%9D%E8%AD%B7%E5%8D%80.

Taiwan Tourism Bureau. (2020). Retrieved from https://eng.taiwan.net.tw/. Accessed on June 22 2020.

Tsao, E. S. (2002). Ecotourism in Taiwan: Green Island. In T. Hundloe (Ed.), *Linking green productivity to ecotourism, experiences in the Asia-Pacific region* (pp. 215–220). Asian Productivity Organization.

Wieman, E. (1995). *Taiwan's national scenic areas: Balancing preservation and recreation*. Academia Sinica. 1995-06-01. Retrieved from http://www.sinica.edu.tw/tit/scenery/1295_scn1.html. Accessed on September 22 2018.

Yushan National Park. (2020). Retrieved from https://www.ysnp.gov.tw/css_en/default.aspx. Accessed on June 30 2020.

Chieh-Lu Li is with the Department of Tourism, Recreation and Leisure Studies at National Dong Hwa University and is also affiliated with National Chung Hsing University in Taiwan. Li completed his PhD at the Pennsylvania State University in USA. His research focuses on cross-cultural service marketing, ethnicity/cultural group studies, natural resource tourism, outdoor recreation, ecotourism, and human dimensions of natural resources.

Thomas E. Jones is Associate Professor in the Environment & Development Cluster at Ritsumeikan APU in Kyushu, Japan. His research interests include Nature-Based Tourism, Protected Area Management and Sustainability. Tom completed his PhD at the University of Tokyo and has conducted visitor surveys on Mount Fuji and in the Japan Alps.

Part III
Southeast Asia

Chapter 6
Indonesia's Mountainous Protected Areas: National Parks and Nature-Based Tourism

Wahyu Pamungkas and Thomas E. Jones

6.1 Introduction

6.1.1 Prologue

Indonesia is a tropical archipelago rich in biodiversity stored mostly in tropical forests. The country is one of twelve distribution hubs of plant genetic diversity known as a Vavilov centre (Endarwati, 2005 in Dunggio & Gunawan, 2009). Historically, many parts of Indonesia were home to forest-dependent peoples that practiced traditional forms of forest management (Kubo, 2008). However, the increasing demand for land and forest products has intensified the pressure for economic development, disrupting that balance and threatening both biodiversity and broader sustainability while triggering deforestation and associated conflicts (ibid.). Therefore, the authoritarian Suharto-led New Order government (1967–1998) adopted a top-down, 'science-based' Protected Area (PA) policy and expanded PAs to cover nearly 10% of the terrestrial land area in the form of national parks, wildlife sanctuaries and nature reserves (Jepson & Whittaker, 2002). As a signatory to the UN Convention on Biological Diversity (CBD), Indonesia also ratified the Aichi Biodiversity Target 11, pledging to set aside at least 17% of its territory for terrestrial PAs and 10% for coastal and marine protected areas (MPAs) (von Heland & Clifton, 2015). The bulk of this expansion was to be achieved by designating new MPAs with an interim commitment to set aside 20 million hectares by 2020 "*en route* to a wider target of setting aside at least 30 million hectares" at an unspecified date

W. Pamungkas
Forestry Agency of South Sumatra Province, Jalan Kolonel H Burlian No. 25, Palembang, South Sumatra, Indonesia

T. E. Jones (✉)
College of Asia Pacific Studies, Ritsumeikan Asia Pacific University (APU), Beppu, Japan
e-mail: 110054tj@apu.ac.jp

© The Author(s), under exclusive license to Springer Nature Switzerland AG 2021
T. E. Jones et al. (eds.), *Nature-Based Tourism in Asia's Mountainous Protected Areas*, Geographies of Tourism and Global Change,
https://doi.org/10.1007/978-3-030-76833-1_6

in the future (Soemodinoto et al., 2018). However, the first designated PAs had a distinct weighting toward mountains, and such sites still account for the bulk of human resources while providing vital ecosystem services. For example, the volcano Gunung Gede, part of an IUCN category II national park on Java, is a catchment area for Jakarta's fresh water supply (IUCN, 2008). This chapter presents an overview of Indonesia's PAs with a focus on mountainous national parks while looking at their potential for nature-based tourism. The current conditions and challenges are outlined both thematically and via the use of case study examples, especially Bromo Tengger Semeru National Park (Sect. 6.5).

6.1.2 Indonesia's National Parks

In 1980, Indonesia designated its first five national parks at Baluran, Gunung Gede Pangrango, Gunung Leuser, Komodo and Ujung Kulon. All were predominantly terrestrial, centred on iconic mountains and tropical rainforest. In 1982, after the 3rd World National Parks Congress hosted in Bali, 11 additional national parks were gazetted (Siswanto, 2017). However, the remote location of many, combined with a lack of transport or accommodation infrastructure, strongly limited visitor numbers (Cochrane, 2006). The number of PAs increased gradually and by 2004, there was a total of 20 designated national parks (Dunggio & Gunawan, 2009). By 2017, this number had increased to 54 national park units covering 16.46 million hectares, of which 654,346 hectares comprises bodies of water, including MPAs and mountainous lakes such as Rinjani, Sentarum and Zamrud (DGCNRCE, 2017). The parks are distributed across the seven main islands, with the largest number clustered in Sumatra and Java (26 national parks; 42.6% of the total). The national parks host diverse ecosystems ranging from wetlands and lakes to peat forest and savannah. Most of the popularly climbed mountains are also designated as national parks, often demarcated by the word *gunung* (mountain or mount) such as Merapi, Gede Pangrango, and Bromo Tengger Semeru (see Sect. 6.5). The parks are home to spectacular wildlife such as Komodo dragons, Sumatran elephants, tigers and rhinoceroses, orangutans and proboscis monkey, together with a vast treasure trove of less iconic but equally important species, though most of these are now under threat from development.

Understanding the specific objectives of PA designation is vital to analysing national park management effectiveness (Locke & Dearden, 2005). Designation objectives in Indonesia include: conservation of plant and animal species, maintaining life-supporting systems, and the sustainable use of natural resources and ecosystems (Putro, 2001). Meanwhile, Act No. 11/1990 on natural resources conservation claims that the main purpose of Indonesian national parks is biodiversity conservation and sustainable development. Further reasons for designating national parks include accordance with international conventions such as CITES aimed at limiting the illegal wildlife trade, in addition to political and economic considerations regarding control of nature resources (Jepson & Whittaker, 2002). In short, it can be concluded that conservation is by no means the only objective of Indonesia's national parks. Before investigating specific conservation challenges, the next section therefore explores governance and revisits the definition of national parks in Indonesia.

6.1.3 National Parks Defined

Under Government Ordinance No. 28/2011, a national park is defined as a nature conservation area which possesses native ecosystems that are managed through a zoning system to facilitate research, science, education, breeding, recreation and tourism purposes (Wiratno, 2018). According to Government Ordinance No. 68/1998, forest (and occasionally also non-forest areas) can be designated provided they fulfil the following criteria:

1. Large enough to ensure natural ecological processes;
2. Having special and unique natural resources, either plants or animals and their ecosystems;
3. Having one or more intact ecosystems;
4. An area which can be subdivided into core, utilisation, sanctuary and other zones in order to accommodate rehabilitation purposes, dependency of local people, and to support living resources and the conservation of their ecosystems (Damayanti & Masuda, 2008).

Unfortunately, the reality is that some national parks do not always live up to these designation criteria, with some converted from existing forest areas simply by altering the status and function of land classification (Damayanti & Masuda, 2008). In that sense, many parks and PAs were originally designated with little or no regard for local people (Putro, 2001). For example, during the designation of Gunung Palung National Park, a number of villages were initially excluded from the plan, although the area was later expanded to include them. It is not surprising, then, that PAs have been the site of persistent conflict regarding use rights and land ownership, as this chapter will demonstrate (Fig. 6.1).

6.2 Governance in Indonesia's National Parks

6.2.1 Land Ownership

Land ownership of Indonesian forests is classified into either national (*hutan negara*) or private forestland (*hutan hak*) where private landowners' deeds are recognized (*hutan rakyat*) (Damayanti & Masuda, 2008). National parkland is in the former category, falling under state ownership. Article 33 (3) of the Indonesian Constitution

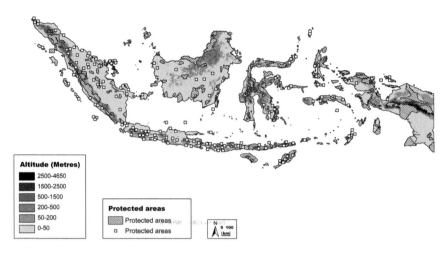

Fig. 6.1 Map of Indonesia's protected areas

declares that "the land and the waters as well as the natural richness therein are to be controlled by the state to be exploited to the greatest benefit of the people" (ibid.). Forestry law No. 41/1999 confirms that the management of forests and utilization of forest resources falls under the jurisdiction of the Ministry of Environment and Forestry (MoEF).

However, as in other countries, MoEF's PA policy has evolved in accordance with the socioeconomic and political climate of the times. In the embryonic period of Indonesia's national parks, the government adopted a 'set-aside' management approach modelled on iconic U.S. national parks such as Yellowstone, which envisage a conservation sanctuary that prioritizes protection above all else (Dunggio & Gunawan, 2009). Based on research conducted by the United Nations Environment Programme World Conservation Monitoring Centre (UNEP-WCMC) in 2018, 157 out of 733 PAs in Indonesia (21.42%) are categorized by IUCN as 'Ia' i.e. Strict Nature Reserves that are "set aside to protect biodiversity and also possibly geological/geomorphical features, where human visitation, use and impacts are strictly controlled and limited to ensure protection of the conservation values" (IUCN, 2008). But based on statistical data released by the MoEF in 2017, 293 areas are classified as IUCN category Ia in Indonesia. This discrepancy may be due to a difference between the UNEP-WCMC and the MoEF in the method of data processing of statistics. Nevertheless, both sets of data acknowledge that the largest portion of protected areas in Indonesia fall under category 'Ia' according to the Decree of Director General of Forest Conservation and Nature Protection No. 129 of 1996 (see Table 6.1).

Despite the prevalence of 'Ia' PAs, the top-down 'fences and fines' approach has been phased out in favour of more participatory policies to resource management (Kubo & Supriyanto, 2010). After the National Parks Congress in 2003, the government prioritized collaborative management via the issuance of Forestry Ministry

Table 6.1 Subdivision of Indonesian conservation areas and IUCN protected area categories

Conservation Area	Natural reserve area	Wildlife sanctuary	Ia, Ib
		Wildlife reserve	Ia, Ib
	Nature preservation area	National Parks	II
		Great forest parks	III, IV
		Natural tourism parks	V
	Hunting park		N/A
	Protected forest		VI

Source Adapted from Wiryono (2003)

Regulation No. P.19/2004 (Dunggio & Gunawan, 2009). Compromise 'win–win' solutions sought to accommodate all interests, as demonstrated by several government initiatives such as the Integrated Conservation and Development Project and Integrated Protected Area System (Dunggio & Gunawan, 2009; Sumardja & Iswaran, 1996).

Yet despite such efforts to build conservation partnerships, fragmented jurisdiction continues to plague national parks, whose core mandate often remains unclear. Apart from confusion over overlapping authorities, conflicts over ownership rights continue to be a problem as some areas of parkland overlap with timber concessions or areas where oil companies have mineral rights, along with others that are occupied or claimed by local communities. The two most frequent types of conflict are land ownership claims and encroachment on park lands by communities that live in and around the parks. The former dates back to the era of Dutch colonialism, specifically the 'Agrarian Law' (*Agrarische Wet*) passed in 1870 that paved the way for the 'domain declaration' (*Domein verklaring*) whereby the Dutch rulers claimed that 'all land not held under proven ownership, shall be deemed the domain of the state' (Kano, 2008). This principle underpinned land tenure until the 1960s, and the disputes have evolved with the changes in state forest boundaries and expansion of parkland from the 1980s to 2000s (Galudra, 2012). Meanwhile, parkland is regularly encroached upon by surrounding communities that seek to use the land for agricultural purposes, including cash crops such as vegetable, rubber and coffee, among others. For example, Yusri (2011) found that 4,830 hectares of Mount Ciremai National Park had been converted into vegetable farms. Meanwhile, the period of 2002–2011 saw a loss of 8,738 hectares of primary dryland forest in Bukit Barisan Selatan National Park mostly converted into coffee plantations (Sinaga & Darmawan, 2014).

Table 6.2 Human resources in select national parks

National Park	Area (km^2)	Human Resources		
		Rangers	Other staff	Total
Bali Barat	190	54	15	69
Bromo Tengger Semeru	503	23	20	43
Bukit Barisan Selatan	3,568	39	18	57
Halimun Salak	400	44	27	71
Kerinci Seblat	13,750	82	29	111

Source DGCNRCE (2017)

6.2.2 Administration

National parks are a form of state forest administered by the central government through the MoEF under the jurisdiction of the Ditjen KSDAE *Direktorat Jenderal Konservasi Sumber Daya Alam dan Ekosistem* (which literally translates as the Directorate General of Natural Resources Conservation and Ecosystems) via a technical implementation unit called the *Balai Taman Nasional* (National Park Bureau). This in turn consists of a head, an administrative division, a national park conservation technical section, a regional national park management section, and a functional position group consisting of rangers, or forest police officers (*polisi kehutanan*). According to the joint regulation No. NK 14/2011 between the MoEF and the state organ for human resources, rangers are the front-line employees in charge of carrying out forest protection and security as well as supervision of the distribution of forest products. Most of them are civil servants, assisted by teams of temporary employees. Aside from the rangers, there are some types of specialist positions in parks such as: forest counsellors (in charge of community development), forest ecosystem controllers (responsible for formulating forest management plans), and administration officers.

According to MoEF Decree No. P.7/2016, there are several functions held by the national park bureau, including preparation of management plans, protection and security of the area, mitigating the impact of damage to living natural resources. It can be concluded that the central government delegates considerable authority to the respective National Park Bureaus for everyday administrative purposes (Table 6.2).

6.2.3 Legislation

According to article 7 of Law No. 10/2004, the highest legal regulation hierarchy in Indonesia is the 1945 Constitution which is then followed sequentially by the provisions of the people's consultative assembly, laws, government regulations, presidential regulations, provincial regional regulations, and district regional regulations. Basically, each of these regulations should not clash with each other. In case of a conflict, the higher-ranking regulation becomes the benchmark. In addition, the principle of *lex specialis derogat legi generalis*, is also utilized wherein a special law can override a general one in the case of a jurisdictional clash.

Article 33 of the Constitution states that land, water, and everything contained therein is controlled by the state and used as much as possible for the benefit of the community. Control by the state can imply direct jurisdiction, whereby the state manages natural resources directly, for example the management of the forests in Java by state-owned enterprises. Alternatively, the state may act as a regulator that has the right to determine who will manage a natural resource, for example the appointment of concessions to manage production forests outside of Java. Either way, natural resource management should be oriented to the prosperity of the people both through the direct involvement of the community in its management and the allocation of state revenues from these natural resources for the benefit of the community.

In connection with article 33 of the Constitution, two acts are enforced, namely Act No. 5/1990 on Conservation of Living Resources and Their Ecosystems and Act No. 41/1999 on Forestry (Damayanti & Masuda, 2008). The former regulates nature reserve areas (including nature reserves and wildlife reserves) and nature conservation areas (including national parks, grand forest parks, and nature tourism parks). In article 3 of Act No. 5/1990, it is mentioned that conservation aims to realize natural resource sustainability in supporting efforts towards improved community prosperity. Furthermore, the Forestry Act categorizes forests based on land ownership into private and state forests. Yet forest can be also be sub-divided into three types, namely conservation, protection, and production forests based on their function. In terms of the categorization of a national park, one would constitute both a state and conservation forest. Article 3 of the Forestry Act explains that forest management is aimed at the maximum prosperity of the people that is just and sustainable. Technical matters regarding PAs are then regulated in legal elements derived from these two Acts such as Government Ordinance No. 36/2010 on nature-based tourism, MoEF Decree No. P.19/2004 on collaborative management of nature reserve areas, MoEF Decree No. P.76/2015 on the zoning system in national parks, etc.

Government concern for conservation areas also stems from the contribution of Indonesian forests to the global environment and attention from the international community. Therefore, conservation areas are included in the MoEF's main indicators. Despite the trade-off between conservation and economic functions of forests, the government is working towards the development of conservation areas through the allocation of areas with high biodiversity potential as Essential Ecosystem Areas (EEA). An EEA can later be designated as a conservation area after various stages of research and acts as an early preventive measure against forest exploitation. This kind of legislation has opened up opportunities for the public and private sectors to jointly utilize regions rich in natural resources. This opportunity is realized in the form of provision of utilization zones in conservation areas, collaborative or partnership management schemes, and the granting of utilization permits, among others. Policies such as implementing EEAs are intended to show the spirit of natural resources management for the prosperity of the people as mandated by the constitution.

6.2.4 The Implementation of Good Governance in Protected Areas (PAs)

Some scholars argue that governance strongly influences the effectiveness of PAs. The quality and suitability of governance type will encourage positive outcomes despite high pressure (Eklund & Cabeza-Jaimejuan, 2017). Good governance derives from decisions made with due respect for principles such as those suggested by Lockwood (2010), namely: legitimacy, transparency, accountability, inclusiveness, fairness, connectivity, resilience and adaptability. Pamungkas et al. (2018) compared state-based and community-based management and the effectiveness of good governance principles in South Sumatra Province. The results found state-based management to be better in applying good governance principles although the community-based model proved more effective for PA management from an environmental perspective as evidenced by a lower deforestation rate.

6.2.5 Zoning System

The legal and governance dimensions converge in a zoning system that seeks to map out spatial areas to accommodate the various interests outlined above. Zoning was formally embedded in Indonesian national parks following the MoEF Decree No. P.76/2015 which made provisions for four zones. In theory, conservation and community interests could be aligned through the spatial planning zoning system. In practice, zones' objectives are often difficult to realize without community involvement (Dunggio & Gunawan, 2009). The zones designated in national parks system are detailed in Table 6.3.

Table 6.3 Explanation of the zoning system used in Indonesia's national parks

Zone	Explanation
Core zones	The highest level of designated protection. Any activities that reduce or threaten park functions are prohibited along with the introduction of non-native plants and animals
Forest zones	Defined as the section of the national park that can support the conservation of the core zone and utilization zone through its location, condition and potential
Utilization zones	Parts of a national park mainly used for nature-based tourism due to convenient location, condition and natural potential
Other Zones	Divided into four categories namely, traditional; reforestation; cultural; and special zones

Source Adapted from Wiryono (2003)

6.3 Challenges Faced by National Park

National parks in Indonesia continue to face many serious challenges, especially deforestation, poaching, mining and conflicts with surrounding communities. It is common in developing countries that the local community utilizes forest resources for subsistence, and changes in control affect their livelihood if such a dependency on forest resources exist (Damayanti & Masuda, 2008). These issues often represent the accumulation of problems present since the PA designation era or earlier. Unclear boundaries, displaced settlements due to ignorance of local community rights and political and economic intervention are the roots of the problems. In addition, the magnitude of the authority of the central government over national parks and the assumption that problems within national parks are only technical in nature tend to compound the difficulties in national park administration (Dunggio & Gunawan, 2009).

6.3.1 Tenurial Conflicts

As mentioned above, disagreement and conflict between the national park authorities and surrounding communities are widespread. In Gunung Palung National Park, for example, the existence of durian plantations in the national park triggered conflicts between the community and the local national park office. The local community felt that they were not involved in determining national park boundaries. Meanwhile, unclear information about boundaries and resource utilization in Bantimurung Bulusaraung National Park has led to conflict between community and government (Wakka et al., 2015). Multiple and overlapping tenurial conflicts often exist within a single PA, such as at Mt Halimun which was reclassified as a national park in 2003 to protect lowland agriculture, mitigate erosion, and safeguard water resources for Jakarta (Lund & Rachman, 2018). Balancing interests as diverse as conservation, logging, mining and shifting cultivation require fluid, ongoing negotiations to coordinate the interests of different stakeholder groups and adjust the park plans to reflect internal zones or even modify external borders (Yusran et al., 2017).

6.3.2 Poaching

Apart from deforestation and habitat degradation, the main threat to wildlife populations comes from poaching (Dunggio & Gunawan, 2009). In 2017, for example, 87 conflicts were recorded between humans and elephants in Way Kambas National Park (Rustiati et al., 2020). Such conflicts were mostly due to poaching, a practice that also threatens deer populations in Rawa Aopa National Park (Laobu et al., 2018). Poaching also threatens the survival of the Sumatran tigers in Kerinci Seblat

National Park, both directly and via the illegal hunting of tigers' prey. From 2001 to 2016, 1,065 cases of human-tiger conflict were recorded including injuries and even fatalities on both sides (Kartika, 2016). One of the headline cases involved a female tiger named Bonita that attacked two oil palm plantation workers in Sumatra. The case stirred up local anxiety on national media as the tiger evaded capture for nearly two years, but the 'Bonita story' was symbolic of various factors that have driven the Sumatran tiger to the brink of extinction, namely: poaching of young tigers, fragmentation and reduction of tiger habitats due to human encroachment. In addition, poachers are attracted by the high economic value of tiger body parts and low levels of law enforcement, together with the lack of community participation in forest conservation, and retaliatory responses of communities to tiger attacks on livestock and humans (Yoserizal & Irawan, 2014). Policy steps have been taken to mitigate such tiger-human conflicts, namely collaborative conservation in forestry concessions, updated protocols, poaching and illegal trade monitoring, law enforcement, public awareness campaigns, education and training facilities, socialization, and prevention of livestock being preyed upon by tigers. However, the unfortunate reality is that the degradation and fragmentation of tiger habitat that are often considered to be the main driving factors in human-tiger conflict are ineffectively accommodated in mitigation policies (Fig. 6.2).

Fig. 6.2 Park rangers torch confiscated consumer goods including Sumatran tiger pelts. https://www.dailymail.co.uk/news/article-3604592/Indonesian-officials-seized-stuffed-Sumatran-tigers-elephant-tusks-endangered-animal-skins-set-fire-warning-illegal-poachers.html

6.3.3 Deforestation

Indonesia has faced severe deforestation for many decades, accounting for 61% of all deforestation in Southeast Asia from 1990 to 2010, surpassing the rate of Brazil between 2000 and 2012, and raising the problem of greenhouse gas emissions on a global level (Margono et al., 2014). To date, Indonesia's rate of deforestation remains one of the fastest in the world and requires immediate action due to illegal logging, forest fires and severe encroachment because of a lack of funds for conservation (Gaveau et al., 2013). The drivers of tropical deforestation include proximate and underlying causes (Geist & Lambin, 2002). The former includes illegal logging, agricultural expansion, infrastructure extension, and forest fires. More indirectly, underlying causes can include socio- economic factors, policy and institutional decisions and demographic development.

One example of a counter-strategy to deforestation comes from Gaveau et al. (2013) who stated that the permanent reclassification of logging concessions as forests has prevented some of these areas from being converted into oil palm plantations. Reclassification of a number of production forests into IUCN Category VI 'Protected Areas with sustainable use of natural resources' could help protect the remaining two-thirds of forested area in Kalimantan, for instance (ibid.). However, the logging concession would still be beneficial for the local community, companies and the country through the provision of financial incentives to safeguard the PAs. In other words, if logging concessions were as strictly protected from illegal encroachment as PAs are, the state would do well strategically to commit to keep natural forest timber concessions in production over the long term, alongside the PA network, to collectively conserve over two-thirds of Kalimantan's remaining forests, while at the same time providing income and employment.

6.3.4 From Timber to Tourism? Tangkahan Ecosystem Area (Mt. Leuser NP)

Gunung Leuser National Park in Northern Sumatra is home to four large and endangered Asian mammals, namely Sumatran tigers, rhinoceroses, elephants, and orangutans (Sulistyono et al., 2019). The park is listed by UNESCO as a biosphere reserve and natural World Heritage Site, reflecting its global significance. However, it has been threatened for many years by forest encroachment due to illegal logging activity. From 2001 to 2006, a series of movements headed by local youth drew attention to sustainability issues, resulting in a shift from logging towards ecotourism (Yusnikusumah & Sulystiawati, 2016). This movement began in 2001 when Tangkahan's ranger gathered logging and encroachment leaders, public figures, and village officials. This meeting resulted in an agreement to form the Tangkahan Tourism Agency to manage ecotourism. The Agency initiated a pioneering project to collaborate with the National Park Office of Gunung Leuser to maintain 1,700 hectares of core zone (ibid.). The forests still contain tropical hardwood such as *Dipterocarpaceae*, so allowing them the opportunity to regenerate enabled biodiversity to bounce back, including iconic primates such as the Orangutan and the endangered Sumatran elephant (*Elephas maximus sumatrae*), which was the main attraction for ecotourists. Other tourist activities include river rafting, jungle trekking, bathing with elephants, caving, camping, wildlife tours, and village tours (Wiranatha, 2015). This creates jobs for local guides who provide information to park visitors through several methods such as oral interpretation during tours and visual presentations. (Yusnikusumah & Sulystiawati, 2016). Tangkahan has subsequently become a benchmark for community-based ecotourism and an example for other national parks in Indonesia. However, ecotourism is still in an embryonic form and is yet to provide economic opportunities sufficient to convince all local communities to safeguard the area from illegal logging (Ginting et al., 2010).

6.4 Nature-Based Tourism

Although a handful of iconic sites such as Bali have been world-renowned destinations for decades, it is only relatively recently that Indonesia has begun to actively promote spatial diffusion of tourism by marketing PAs in under-explored, under-developed regions. In 2017, for example, a special economic tourism zone was set up in Mandalika to catch and re-distribute the Bali spill-over. In Lombok as elsewhere, the tourism industry has been growing rapidly, with "more visitors (8 million international visitors in 2012 grew to 11.5 million in 2016) [that] contribute more foreign exchange (US$12 billion in 2016), [to] an increasingly significant part of Indonesia's GDP (>4%)" (NESPARNAS, 2016 cited in Kinseng et al., 2018). The displacement effect as 'over-tourism' at Bali pushes repeat visitors to shift 'one island over' to Lombok and Mandalika in search of whiter sands and less-crowded beaches. The Mandalika project endorses many of the Asia–Pacific stereotypes of NBT development (discussed in Chap. 1) that is "often seen as large-scale and economically driven, based primarily on leisure, and characterized by resorts utilizing nature as little more than a pleasant background appealing to a wide array of holiday-makers" (Henderson, 2011 cited in Frost et al., 2014).

However, as in the aforementioned ecotourism case at Tangkahan, NBT could offer a less environmentally-damaging means of mitigating the numerous conflicts that beset national parks. Development of NBT within the parks must refer to Government Regulation No. 36/2010 that encompasses the following activities: visiting, seeing, and enjoying the beauty of nature, biodiversity, and construction of supporting facilities. Construction of supporting facilities is only allowed in the utilization zone of national parks, whose resources should be used mainly for research, education, and tourism (Putro, 2001). Tourists can be defined as all people who travel for all purposes other than residence or business. NBT to national parks comprises international and domestic segments, with the latter dominated by local or regional visits, although accurate data on national park visitation is scarce. In Table 6.4,

Table 6.4 Indonesian national parks with the highest number of tourist visits

National park name	Number of tourist visits		
	Domestic	International	Year
Bali Barat	15,346	43,902	2014
Bromo Tengger Semeru	547,445	23,713	2014
Kelimutu	18,257	26,396	2014
Komodo	13,537	67,089	2014
Rinjani	34,699	26,073	2014
Tanjung Putting	5,703	10,986	2014
Wakatobi	421	1,924	2014

Source Directorate General of Forest Protection and Nature Reserve (2015)

visit data to seven national parks reveals the NBT market to be generally dominated by domestic tourists. However, different trends are seen in the Komodo and Tanjung Puting National Parks that are dominated by international tourists due to the habitat of certain iconic, endemic species, namely Komodo dragons, orangutans, and proboscis monkey, which form major attractions for international tourists.

Visitation data is likely to be underreported, due to the ineffective method of monitoring visitor flows, although all national park visitors should enter via an official gate and buy a ticket. This procedure assists park managers in calculating the number of visits and the total receipts. Next, the data is reported to the MoEF which prepares annual statistical reports for tourist trend analysis. MoEF only releases an aggregate of tourist visits to all national parks, instead of specific data for each park. The data showed a significant increase from 2013 to 2019 (Fig. 6.3), with a growth trend that

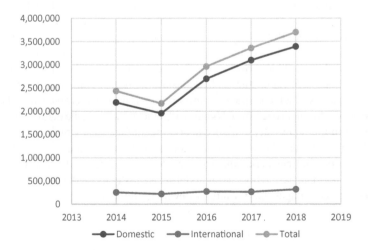

Fig. 6.3 Tourist trends to Indonesia's national parks (2014–2018)

reflects a rapid increase in domestic tourist visits, and a slight upturn in international tourists.

6.5 Case Study of Bromo Tengger Semeru National Park

6.5.1 Overview of Mount Bromo

The 2,329 m Mount Bromo is located in East Java Province. Bromo is a highly active volcano, having erupted at least 50 times since 1804 (Roscoe, 2013). It is a stratovolcano located within the Tengger caldera, comprising the three peaks of Bromo, Tengger, and Semeru. This caldera is the ancestral home of the Tengger tribe and contains unique ecosystems such as a 5,250-hectare 'sea' of fine volcanic sand created by repeated eruptions, edelweiss fields, a double-ringed caldera, 38 species of endangered wildlife and 1,025 species of orchid. Therefore, the 3 peaks were designated as a unified national park in 1982 with a total area of 50,276 hectares. Since 2015, Bromo has also been listed within the Bromo Tengger Semeru Arjuno Biosphere Reserve (Hakim & Soemarno, 2017), in recognition of its role as a biodiversity hot spot. However, in terms of visitor trends, one of the main NBT activities in Bromo Tengger Semeru National Park (BTSNP) is enjoying the morning sunrise and watching cultural celebrations of local communities. Tourists often stop off at places known as *Pedayangan*, sacred sites such as tombs, hermitages and springs that could be considered an inventory for nature-based tourism, albeit not always marketed as such. Tenggerese interpreters at the Pedayangan explain traditional concepts of nature and cultural conservation to visitors (Purnomo, 2018).

The BTSNP was also established as a spatial planning safeguard that restricts development in areas prone to volcanic eruptions, to facilitate watershed protection and as a natural space for recreational purposes along with ecological conservation objectives (Cochrane, 1997). The latter aim dominates since Mt. Bromo is one of the most accessible NBT attractions in Java due to its proximity to the main overland highway (Cochrane, 2006). Bromo can be accessed via four main gates, namely: Tumpang, Wonokirti, Ranupane, and Ngadisari, the latter being the most frequently accessed. Tourists usually use public transportation (bus and microbus), rental cars or taxis. Most visitors approach from Ngadisari/Cemoro Lawang, the gateway village to Bromo. From the main gate, tourists can continue their journey by jeep, horseback, or on foot (Cochrane, 2006). There are hotels, homestays, campsites, restaurants, souvenir shops, and car parks situated outside of the gates and accessible before entering the national park. This accommodation is mostly located outside the national park in the three villages that form the gateway to the park.

Table 6.5 Number of visits to Mount Bromo 2011–2016

Years	2011	2012	2013	2014	2015	2016
Domestic Tourists	14,022	23,472	64,294	113,611	76,873	100,120
International Tourists	4,620	4,633	7,961	14,409	12,893	26,848
Total	18,642	28,105	72,255	128,020	89,766	126,968

Source Tourism Agency of Probolinggo District in Primadianti, 2017

6.5.2 NBT at Mt. Bromo

6.5.2.1 Volcano Tourism

Due to its outstanding scenery and active volcano, tourism in Bromo has a long history dating back to tours in the 1920s and 1930s by Dutch colonial settlers and other westerners (Cochrane, 2006). BTSNP has appeared more regularly on international travellers' itineraries since the 1970s, and by the 1990s it hosted more foreign visitors than any other national park in Indonesia, becoming one of the country's most-visited NBT destinations (Petford et al., 2010) (Table 6.5).

However, as one of the most active volcanoes in Java, tourism to Bromo is also periodically devastated by volcanic events such as the eruption in 2010–2011. One of the biggest and longest-lasting in Bromo's history, it caused widespread damage to surrounding buildings, roads and farmland. The most recent eruption was in 2015, affecting visitation levels and resulting in estimated tourism revenue losses of 126 billion Rupiah. Based on historical records and seismic activity, it is predicted to erupt again sometime around 2020 (Rachmawati et al., 2018).

6.5.2.2 Pilgrimage to a Sacred Mountain

The Tenggerese people are a Javanese ethnic group located in Eastern Java and are one of the few minorities to maintain a Hindu faith (Hefner, 1985). Scattered across 30 villages, there are about 90,000 Tenggerese living in and around the park. Every summer, they make a pilgrimage to the rim of Mt Bromo's crater during the finale of a month-long celebration called *Yadnya Kasada* or the Kosodo festival (Choe & Hitchcock, 2018). Proceeding through the sand 'sea' or desert, the pilgrims stop off at the temple at the foot of the mountain, before climbing up to throw offerings into the volcano. Many travellers come to join the celebrations, with 20,000 to 25,000 visitors noted by Cochrane (2009) to account for a quarter of the park's annual visitation. The iconic ceremony is increasingly commercialized for international and domestic tourists (Cochrane, 2009), lauded by guidebooks such as Lonely Planet, which recommend the festival and list up tour agencies that offer pilgrimage packages including horseback riding and sunrise viewing.

However, the Tenggerese are the main organizers of tourism around Mt. Bromo and were early to embrace its positive economic impact as they own most of the

Table 6.6 Entrance fees at Bromo Tengger Semeru National Park in 2020

Type of fees	Domestic tourists	International tourists
Entrance fee		
a. Weekday	$1.93 (Rp. 27,500)	$15.26 (Rp. 217,500)
b. Weekend/Holiday	$2.28 (Rp. 32,500)	$22.28 (Rp. 317,500)
Additional fees		
a. Jeep driving	$45.62 (Rp. 650,000)	$45.62 (Rp. 650,000)
b. Horse riding	$10.53 (Rp. 150.000)	$10.53 (Rp. 150.000)

tourism infrastructure and organize guided tours and jeep tours to supplement their agriculture-based income (Cochrane, 1997). The local community thus benefits from additional tourism business, with services such as horse, jeep, and homestay rental providing an additional revenue stream while enabling the autonomy to resist overt outside influence and excessive development (Sutiarso & Susanto, 2018).

6.5.3 Mount Bromo PA Management

Mt. Bromo is managed by the BTSNP Office, under the jurisdiction of the MoEF. For tourism management, there is collaborative management between the national park bureau, the regional governments (province and district levels), the private sector and the local community. The government established a program to assist the national park bureau with tourism initiatives through community empowerment known by a Javanese anagram of *Pokdarwis* (Putri et al., 2015). As at other Indonesian parks, BTSNP operates dual entrance fees with international visitors paying a higher rate than their domestic counterparts. Due to the park's increasing popularity, the Mt. Bromo entrance fee was raised in 2014 from US$0.70 (Rp. 10,000) to US$1.93 (Rp. 27,500) for domestic tourists and from about US$4.00 (Rp. 57,000) to US$15.26 (Rp. 217,500) for international visitors. Entrance fees are higher on weekends and holidays and tourists pay extra for additional services such as jeep driving and horse riding (Table 6.6).

6.5.4 Tourism Management Challenges

Noviantoro (2020) described some tourism management problems at Bromo such as recognition of national park boundaries, poor governance, land encroachments, lack of data, mass tourism, lack of human resources (both in terms of quantity and quality), lack of community empowerment, lack of budget and facilities. Consequently, various regulatory issues have arisen related to tourism activities in BTSNP. The control of the number of tourists and their behaviour in and around the Bromo crater and caldera including vegetation clearing, illegal fishing and harvesting flora, collection of edelweiss flowers, graffiti and the eutrophication of Lake Regulo have all become serious problems in NBT management of BTSNP (Hakim, 2017). Unregulated tourism activities at Bromo have significant environmental impacts such as air and water pollution, effects on wildlife behaviour, habitat destruction, poor waste management leading to solid waste accumulation, erosion of the rim of the volcano, vandalism, and depleted flora due to illegal harvesting (Cochrane, 2009; Hakim, 2011). Oil spills and horse dung pollute the dirt trail around Bromo's crater, especially during religious ceremonies, when the trail is subject to extreme traffic. Furthermore, new infrastructure projects are often completed without environmental impact assessments, posing a serious threat to the national park's sustainability. Together with a lack of visible efforts from park authorities to regulate tourism activities, another ongoing issue is risk management. Potential hazards such as eruptions, earthquakes, landslides, steep slopes and unsafe modes of transportation threaten tourists' safety, and underline the need for effective early warning and evacuation systems (Meilani et al., 2018).

6.6 Discussion and Conclusions

From five national parks set up in remote mountainous regions in 1980, Indonesia has increased the number in under four decades to 54 parks covering 165 thousand square kilometres. Today's PAs cover a broader cross-section of sites, including more Marine PAs, freshwater and wetland ecosystems. Meanwhile the designation objectives have also diversified to include recognition of biodiversity and germplasm reserves, and sustainable development to support social welfare. Unfortunately, such goals are not the sole driving force behind PA designation—political and economic interests are also rife (Yusran et al., 2017). Furthermore, the gazetting of new PAs that should be based on certain criteria is, in certain cases, no more than a bureaucratic formality that simply modifies the status and function of existing state forests. This issue was further exacerbated by the neglect of local communities during the process of boundary demarcation and expansion of the national park territory over areas that had long been used by communities as a source of livelihood, as in the case of Mt. Halimun Salak National Park (Lund & Rachman, 2018). Consequently, fundamental problems faced by national park managers, especially regarding tenurial conflict cases, have simmered for decades without a clear resolution.

The National Park Bureau has been delegated considerable authority to manage the parks via existing laws and regulations such as MoEF Decree No. P.7/2016. However, in the Bromo case, Utami (2017) questioned the feasibility of the national park bureau's mandate which must call for conservation while simultaneously seeking to maximize state revenue streams. In BTSNP, as in other parks, tourism authorities propose interventions that override conservation measures (Cochrane, 2009). Moreover, PA management still faces deep-rooted problems including deforestation, poaching and tenurial conflicts, with issues of corruption, transparency, and accountability still widespread. Eklund and Cabeza-Jaimejuan (2017) stated that the quality of governance greatly effects conservation outcomes in PAs, so one way to understand the problems faced by national park management is in terms of the suboptimal quality of governance. Furthermore, the spatial planning of the PAs' zoning systems often exists only on maps, being difficult to implement in the field. This can be attributed to the rigid 'science-based' approach that is still overly emphasized by government agencies with a subsequent lack of consideration for socioeconomic aspects. The zoning system should be flexible enough to accommodate the interests of all PA stakeholders, but in reality the unclear criteria and processes currently can render the system ineffective.

Unlocking the potential of Indonesia's PAs lies in better cost-recovery from nature-based tourism, defined as a broad concept that includes activities such as ecotourism and adventure tourism, as well as conservation of nature and culture. Valentine (1992) mentioned that NBT has a lower impact on the environment, but higher social and economic benefits. From these definitions, the key aspect of NBT can be elaborated as awareness of environmental impacts. In this sense, nature is not only an object that is free to be enjoyed, but must also be respected. Using these criteria as a lens through which to evaluate the implementation of NBT in national park settings (BTSNP in particular), it can be seen that this concept is yet to be achieved. Unregulated NBT activities at Mt Bromo have resulted in negative environmental impacts that belie NBT's sustainable intentions. Furthermore, a fundamental departure from the concept of NBT can be detected in the overriding tourist profile at Mt Bromo who tend to be pleasure-seekers. This could reflect the National Park Bureau's residual focus on 'mass' tourism rather than NBT, due to goals tied to numeric quantity targets required to meet the state revenue requirements. Ironically, the respective national park offices only retain a small percent of the state revenue that they raise, an amount which is often inadequate to meet the daily needs of PA management. One proposal is thus to promote Indonesia's PAs more effectively to NBT niche markets and seek ways to improve site-specific cost recovery mechanisms that can provide sustainable revenue streams for better conservation and management.

References

Choe, J., & Hitchcock, M. (2018). 15 Pilgrimage to Mount Bromo. *Religious Pilgrimage Routes and Trails: Sustainable Development and Management, 180.*
Cochrane, J. (1997). Tourism and conservation in Bromo Tengger Semeru national park. University of Hull, Hull, UK.
Cochrane, J. (2006). Indonesian national parks: Understanding leisure users. *Annals of Tourism Research, 33*(4), 979–997.
Cochrane, J. (2009). Spirits, nature and pilgrimage: The "other" dimension in Javanese domestic tourism. *Journal of Management, Spirituality and Religion, 6*(2), 107–119.
Damayanti, E. K., & Masuda, M. (2008). National park establishment in developing countries: Between legislation and reality in India and Indonesia. *Tropics, 17*(2), 119–133.
Directorate General of Forest Protection and Nature Reserve. (2015). *Statistic of 2014*. Ministry of Environment and Forestry.

Directorate General of Natural Resources Conservation and Ecosystem (DGCNRCE). (2017). *Statistic of 2016*. Ministry of Environment and Forestry.
Dunggio, I., & Gunawan, H. (2009). Telaah sejarah kebijakan pengelolaan taman nasional di Indonesia [The historical analysis of National Park management policies in Indonesia]. *Jurnal Kebijakan Kehutanan [Forestry Policy Journal], 6*(1), 43–56.
Eklund, J. F., & Cabeza-Jaimejuan, M. D. M. (2017). Quality of governance and effectiveness of protected areas: crucial concepts for conservation planning. *Annals of the New York Academy of Sciences*.
Frost, W., Laing, J., & Beeton, S. (2014). The future of nature-based tourism in the Asia-Pacific region. *Journal of Travel Research, 53*(6), 721–732.
Galudra, G. (2012). Memahami konflik tenurial melalui pendekatan sejarah: Studi kasus di Lebak, Banten [Understanding tenure conflict through historical approaches: Case study of Lebak Banten]. *Jurnal Keadilan [Journal of Justice], 6*.
Gaveau, D. L. A., Kshatriya, M., Sheil, D., Sloan, S., Molidena, E., et al. (2013). Reconciling forest conservation and logging in Indonesia Borneo. *PLoS ONE, 8*(8), e69887. https://doi.org/10.1371/journal.pone.0069887.
Geist, H. J., & Lambin, E. F. (2002). Proximate Causes and Underlying Driving Forces of Tropical Deforestation Tropical forests are disappearing as the result of many pressures, both local and regional, acting in various combinations in different geographical locations. *BioScience, 52*(2), 143–150.
Ginting, Y., Dharmawan, A. H., & Sekartjakrarini, S. (2010). Interaksi Komunitas Lokal di Taman Nasional Gunung Leuser: Studi Kasus Kawasan Ekowisata Tangkahan, Sumatera Utara [Local communities' interaction in Gunung Leuser National Park]. *Sodality: Jurnal Sosiologi Pedesaan, 4*(1).
Hakim, L. (2011). Cultural Landscapes of the Tengger Highland, East Java. In *Landscape Ecology in Asian Cultures* (pp. 69–82). Springer, Tokyo.
Hakim, L. (2017). Biodiversity conservation, community development, and geotourism development in Bromo-Tengger-Semeru-Arjuno biosphere reserve, East Java. *Geojournal of Tourism and Geosites, 2*(20), 220–230.
Hakim, L., & Soemarno, M. (2017). Biodiversity conservation, community development and geotourism development in bromo-tengger-semeru-arjuno biosphere reserve. *Geojournal of Tourism and Geosites, 20*(2), 220–230.
Hefner, R. W. (1985). *Hindu Javanese*. Princeton University Press.
Henderson, J. (2011). "Singapore Zoo and Night Safari." In W. Frost (Ed.), *Zoos and tourism: Conservation, education, entertainment?* (pp. 100–111). Channel View.
International Union for Conservation of Nature (IUCN). (2008). *2008 Guidelines for applying protected area management categories*. www.iucn.org/pa_categories. Accessed on 27 August 2020.
Jepson, P., & Whittaker, R. J. (2002). Histories of protected areas: Internationalisation of conservationist values and their adoption in the Netherlands Indies (Indonesia). *Environment and History, 8*(2), 129–172.
Kano, H. (2008). *Indonesian exports, peasant agriculture and the world economy, 1850–2000: Economic structures in a Southeast Asian State* (p. 282). NUS Press.
Kartika, E. C. (2016). *Human-tiger conflict: An overview of incidents, causes and resolution*. Technical Report, https://doi.org/10.13140/RG.2.1.3754.2808.
Kinseng, R. A., Nasdian, F. T., Fatchiya, A., Mahmud, A., & Stanford, R. J. (2018). Marine-tourism development on a small island in Indonesia: blessing or curse? *Asia Pacific Journal of Tourism Research, 23*(11), 1062–1072.
Kubo, H. (2008). Diffusion of policy discourse into rural spheres through co-management of state forestlands: Two cases from West Java, Indonesia. *Environmental Management, 42*, 80–92.
Kubo, H., & Supriyanto, B. (2010). From fence-and-fine to participatory conservation: Mechanisms of transformation in conservation governance at the Gunung Halimun-Salak National

Park, Indonesia. *Biodiversity Conservation, 19*, 1785–1803. https://doi.org/10.1007/s10531-010-9803-3

Laobu, A., Bahtiar, B., & Sifatu, W. O. (2018). Persepsi Masyarakat Terhadap Taman Nasional Rawa Aopa Watumohai. *Jurnal Penelitian Budaya, 3*(2).

Locke, H., & Dearden, P. (2005). Rethinking protected area categories and the new paradigm. *Environmental Conservation, 32*(1), 1–10. https://doi.org/10.1017/S0376892905001852

Lockwood, M. (2010). Good governance for terrestrial protected areas: A framework, principles and performance outcomes. *Journal of Environmental Management, 91*(3), 754–766.

Lund, C., & Rachman, N. F. (2018). Indirect Recognition. Frontiers and Territorialization around Mount Halimun-Salak National Park Indonesia. *World Development, 101*, 417–428.

Margono, B. A., Potapov, P. V., Turubanova, S., Stolle, F., & Hansen, M. C. (2014). Primary forest cover loss in Indonesia over 2000–2012. *Nature climate change, 4*(8), 730–735.

Meilani, R., Muthiah, J., & Muntasib, E. K. S. H. (2018, May). Reducing the risk of potential hazard in tourist activities of Mount Bromo. In *IOP Conference Series: Earth and Environmental Science*, (Vol. 149, No. 1, p. 012021). IOP Publishing.

NESPARNAS. (2016). *Ministry of tourism and the central statistics agency (BPS): Balance of national tourism satellite (Nesparnas)*. Deputy for Institutional Development of Tourism.

Noviantoro, K. M. (2020). Evaluasi Potensi Wisata Bromo-Madakaripura Sebagai Ekowisata Dalam Meningkatkan Perekonomian Masyarakat Sekitar. *Iqtishodiyah: Jurnal Ekonomi dan Bisnis Islam, 6*(1), 49–62.

Pamungkas, W., Muluk, M. K., Rochmah, S., & Jones, T. E. (2018). Evaluating Good Governance in Preserved Forests: A Comparison Between Community-based and State-based Forest Management in South Sumatera. *Jurnal Ilmiah Administrasi Publik, 4*(3), 194–200.

Petford, N., Fletcher, J., & Morakabati, Y. (2010). On the economics and social typology of volcano tourism with special reference to Montserrat, West Indies. *Volcano and Geothermal Tourism: Sustainable Geo-resources for Leisure and Recreation*, 85–93.

Purnomo, O. A., & Nugroho, I. (2018). The sacred site: The conservation based on the local people in Tengger community and its potential as ecotourism activities. *Journal of Socioeconomics and Development, 1*(1): 7–15. https://doi.org/10.31328/jsed.v1i1.517.

Putri, T. N. T., Purnaweni, H., & Suryaningsih, M. (2015). Implementasi Program Kelompok Sadar Wisata (Pokdarwis) Di Kelurahan Kandri, Kecamatan Gunungpati, Kota Semarang. [Implementation of the Tourism Awareness Group (Pokdarwis) Program in the Kandri Village, Gunungpati District, Semarang City]. *Indonesian Journal of Public Policy and Management Review, 4*(1) 42–51.

Putro, H. R. (2001). *Participatory management of National Park and Protection Forest: A New Challenge in Indonesia*. http://www.nourin.tsukuba.ac.jp/~tasae/2001/Indonesia_2001.pdf.

Rachmawati, A., Rahmawati, D., & Rachmansyah, A. (2018). Disaster risk analysis of Mount Bromo eruption after the 2015 eruption in Sukapura District. (MATEC Web Conference 229 01015).

Roscoe, R. (2013). Volcanic cultures of risk: Photographing sites of memory. *Global Environment, 6*(11), 216–227.

Rustiati, E. L., Priyambodo, P., Srihanto, E. A., Pratiwi, D. N., Virnarenata, E., Novianasari, T., ... & Saswiyanti, E. (2020). The Essential Contribution of Captive Sumatran Elephant in Elephant Training Center, Way Kambas National Park for Wildlife Genetics Conservation. BIOVALENTIA: *Biological Research Journal, 6*(1), 38–41.

Sinaga, R. R. P., & Darmawan, A. (2014). Perubahan tutupan lahan di resort pugung tampak taman nasional bukit barisan selatan (tnbbs). *Jurnal Sylva Lestari, 2*(1), 77–86.

Siswanto, W. (2017). Pengelolaan kawasan konservasi di Indonesia: Pengelolaan saat ini, pembelajaran, dan rekomendasi. Deutsche Gesellschaft für Internationale Zusammenarbeit (GIZ) GmbH.

Soemodinoto, A., Yulianto, I., Kartawijaya, T., Herdiana, Y., Ningtias, P., Kassem, K. R., & Andayani, N. (2018). Contribution of local governments to a national commitment of the Aichi

Biodiversity Target 11: the case of West Nusa Tenggara Province Indonesia. *Biodiversity, 19*(1–2), 72–80. https://doi.org/10.1080/14888386.2018.1467790

Sulistyono, N., Ginting, B. S. P., Patana, P., & Susilowati, A. (2019). Land cover change and deforestation characteristics in the management section of national park (MNSP) VI besitang, gunung leuser national park. *Journal of Sylva Indonesiana, 2*(02), 91–100.

Sumardja, E. A., & Ishwaran, N. (1996, May 12–18). Integrated conservation and development concept and practice. Dalam Conserving Biodiversity amidst Growing Economic Prosperity. In *Proceedings of 1st Southeast Asia Regional Meeting of World Commission on Protected Area* (pp. 37–43). Cisarua dan Ujung Kulon National Park, Indonesia.

Sutiarso, M. A., & Susanto, B. (2018). Pengembangan Pariwisata Berbasis Masyarakat di Taman Nasional Bromo Tengger Semeru Jawa Timur. *Ganaya: Jurnal Ilmu Sosial Dan Humaniora, 1*(2), 144–154.

The United Nations Environment Program World Conservation Monitoring Centre UNEP-WCMC (2018).

Utami, H. S. (2017). Pengelolaan Kawasan Pariwisata (Studi di Balai Besar Taman Nasional Bromo Tengger Semeru). *Jurnal Ilmiah Administrasi Publik, 3*(1), 13–20.

Valentine, P. (1992). Nature-based tourism. Belhaven Press.

Von Heland, F., & Clifton, J. (2015). Whose threat counts? Conservation narratives in the Wakatobi National Park. *Conservation & Society, 13*(2), 154–165.

Wakka, A. K., Muin, N., & Purwanti, R. (2015). Toward collaborative management of Bantimurung Bulusaruang National Park, South Sulawesi Province. *Jurnal Penelitian Kehutanan Wallacea, 4*(1), 41–50.

Wiranatha, A. S. (2015). Sustainable development strategy for ecotourism at Tangkahan North Sumatera. *E-Journal of Tourism, 2*(1), 1–8.

Wiratno, D. (2018). *Collaborative Management Policy of Conservation Areas in Indonesia.* Collaborative National Park Management in Indonesia and Japan, Tokyo.

Wiryono, W. (2003). *Classification of Conservation Areas in Indonesia.* Policy Report: Number 11, May 2013. Centre for International Forestry Research (CIFOR).

Yoserizal, Y., & Irawan, R. E. (2014). Motif Perburuan terhadap Harimau Sumatera pada Kawasan Taman Nasional Bukit Tiga Puluh Kabupaten Indragiri Hulu (Doctoral dissertation, Riau University).

Yusnikusumah, T. R., & Sulystiawati, E. (2016). Evaluasi pengelolaan ekowisata di kawasan ekowisata Tangkahan Taman Nasional Gunung Leuser Sumatra Utara [Evaluation of ecotourism management in ecotourism area of Tangkahan of Gunung Leuser National Parks in North Sumatra]. *Journal of Regional and City Planning, 27*(3), 173–189.

Yusran, Y., Sahide, M. A. K., Supratman, S., Sabar, A., Krott, M., & Giessen, L. (2017). The empirical visibility of land use conflicts: From latent to manifest conflict through law enforcement in a national park in Indonesia. *Land Use Policy, 62*, 302–315.

Yusri, A. (2011). Perubahan penutupan lahan dan analisis faktor penyebab perambahan kawasan Taman Nasional Gunung Ciremai [Changes in land cover and analysis of factors causing encroachment on the Mount Ciremai National Park area]. Unpublished Thesis. https://repository.ipb.ac.id/handle/123456789/47612.

Wahyu Pamungkas is an analyst of planning and cooperation in the Forestry Agency of South Sumatra Province, Indonesia. His research field spans Forest Policy, Nature-Based Tourism, and Social Forestry and he completed a dual master's degree at Brawijaya University and Ritsumeikan APU.

Thomas E. Jones is Associate Professor in the Environment & Development Cluster at Ritsumeikan APU in Kyushu, Japan. His research interests include Nature-Based Tourism, Protected Area Management and Sustainability. Tom completed his PhD at the University of Tokyo and has conducted visitor surveys on Mount Fuji and in the Japan Alps.

Chapter 7
Collaborative Management of Protected Areas in Timor-Leste: Stakeholder Participation in Community-Based Tourism in Mount Ramelau

Antonio da Silva and Huong T. Bui

7.1 Introduction

Timor-Leste (East Timor in English) is the youngest independent nation in Southeast Asia. As a fragile, post-conflict state, it has some of the highest levels of unemployment, poverty, food insecurity, malnutrition and illiteracy in the world—the country faces major development challenges to improve human well-being and living standards (World Bank, 2020). Timor-Leste's population is highly dependent on natural resources (mainly oil) and subsistence livelihoods, with agriculture and fisheries being the major sources of income for the majority (94 per cent) of the population (Molyneux et al., 2012). Within the non-oil economic sector, tourism has been identified as a major government development priority for poverty reduction and to improve rural economic development (UNDP & UNWTO, 2007).

Extant studies have highlighted the potential of the natural environment for ecotourism development (Edyvane et al., 2009; Ximenes & Carter, 2000) as well as the community-based tourism (CBT) option for the country (Tolkach & King, 2015). The implementation of community-based and nature-based tourism has been highlighted as a major development priority by the Government of Timor-Leste under the Timor-Leste Strategic Development Plan 2011–2030 (RDTL, 2011) and the National Strategic Plan for Tourism 2012–2017 by Ministry of Tourism (MOT, 2012). At the intersection of these two forms of tourism, community-based ecotourism (CBET) can play a major role in community development through promoting and supporting local employment, income generation and alternative livelihood development (Edyvane

---

A. da Silva
Ministry of Tourism, Commerce and Industry, Dilli, Timor-Leste
e-mail: da19s6sn@apu.ac.jp

H. T. Bui (✉)
College of Asia Pacific Studies, Ritsumeikan Asia Pacific University (APU), Beppu, Japan
e-mail: huongbui@apu.ac.jp

© The Author(s), under exclusive license to Springer Nature Switzerland AG 2021
T. E. Jones et al. (eds.), *Nature-Based Tourism in Asia's Mountainous Protected Areas*, Geographies of Tourism and Global Change,
https://doi.org/10.1007/978-3-030-76833-1_7

et al., 2009). However, tourism planning and development in nature reserve areas are currently constrained by a lack of specific policies, plans and strategies at both national and municipal levels, and the key issues of poor institutional capacity, limited infrastructure and the unmet need of workforce training (UNDP & UNWTO, 2007). In 2017 the Timor-Leste government approved a 2030 National Tourism Policy with five thematic principles (5Ps); priority, prosperity, people, protection and partnership. These key policy goals aimed to stimulate intrepid and dynamic tourism sector growth in Timor-Leste, with an annual target by 2030 of 200,000 international inbound visitors, generating USD150 million in revenues and employing 15,000 local residents (RDTL, 2017).

The governance and management of protected areas (PAs) involves entities from both the public and private sectors (Byrd, 2007). These issues are getting more complicated in developing countries where law enforcement is unclear and weak, along with conflicts in resource and benefit sharing among different groups. Among various group of PA users, the tourism industry is an active stakeholder. This chapter brings our attention to collaborative management of national parks and nature-based tourism in the Timor-Leste. The authors centre the analysis of the nature-based tourism on Mount Ramelau, one of Timor-Leste's top-ranked national tourist attractions due to its spectacular natural scenery, religious significance, historical value and the rural community life that surrounds it. Linking the problem of resource management under stakeholder collaborative management of Mount Ramelau, the authors provide insights into the complexity of the tourism-conservation nexus, that has been relatively under-researched in the context of Timor-Leste.

7.2 Natural Conditions and the System of Protected Areas

7.2.1 A Geographical and Historical Background of the Country

Timor-Leste is the newest country in the Southeast Asian region, located to the northwest of Australia, at the eastern tip of the Indonesian archipelago. Throughout the history of the Timor-Leste, there has been an often violent and recurrent theme of struggle to achieve autonomy from foreign control, first from the Javanese kingdom of Majapahit and later from the Portuguese, Japanese and, more recently, from Indonesia. The country gained independence from Indonesia in 2002 marking the latest chapter in the country's long quest for freedom from foreign rule (Carter et al., 2001).

The country occupies approximately 15,000 square kilometers including the eastern half of the island of Timor, the enclave of Oecusse including Ataúro Island and Jaco Island. In terms of topography, the mountainous aspect and proximity to Australia produce three distinct climatic regions. The area from the coast to 600 m above sea level in the north features an annual average temperature of over 24 °C, low precipitation (500 to 1500 mm annually), and a pronounced dry period lasting five months (June to October). The mountainous zone has temperatures below 24 °C, high precipitation (2500 to 3000 mm) and a dry period of four months (July to October). The southern zone, below 600 m above sea level, has extended plains, exposed to southerly winds, average precipitation (between 1500 and 2000 mm), with average temperatures generally higher than 24 °C and a dry period of three months (August to October) (Carter et al., 2001).

The geomorphology of the Timor-Leste is different from her neighbouring islands; Timor is a continental island of Indo-Australian Plate rather than of volcanic origin. The bedrock is of limestone and other (metamorphosed) sediments and may form part of the Australian geological plate (Monk et al., 1997). In the centre of the island, it is the Ramelau mountain range with the higher elevations reaching 2964 m. In the north, the mountains stretch close to the coast (see Fig. 7.1). To the east, the rugged relief softens but is uneven, perhaps due to higher geological complexity, giving way to plains interrupted by steep cliffs. To the northeast, partially of recent volcanic origin, the land rises rapidly from the sea to 1561 m (Nipane peak). The volcanic island of Atauro emerges very steeply from the sea to form the mountain of Mano Coco (999 m) (Carter et al., 2001).

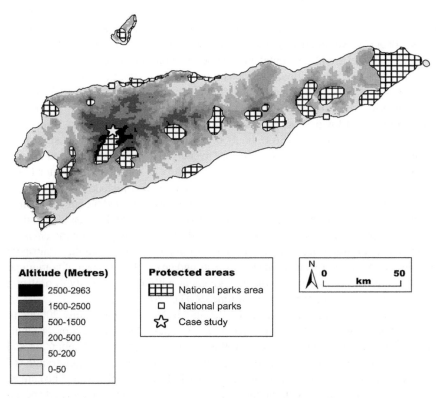

Fig. 7.1 Timor-Leste's Topography (*Source* Authors' compilation)

Although there appears to be some inherent interest in the fauna of Timor-Leste, a lack of understanding of its nature and location limits its use for tourism purposes (Carter et al., 2001). Timor-Leste's terrestrial fauna appears to have suffered along with the demise of the forest communities (Monk et al., 1997) and mammal species have suffered from 'shooting practice' (Carter et al., 2001). In contrast, the bird population is relatively diverse with more than 200, mostly non-migratory, species (Pederson & Arneberg, 1999). Timor-Leste lies in an oceanic hotspot of migratory and marine biodiversity. This region is home to 76% of the world's coral species, six of the world's seven marine turtle species, more than 3,000 species of reef fish, whale sharks, manta rays, and a diversity of marine mammals such as 22 species of dolphin, and a variety of whale species (RDTL, 2015).

7.2.2 The System of National Parks and Protected Areas

Historically, recognition of natural environmental values has its origins in Portuguese colonial rule with the establishment of the Colonial Forestry Service in 1924, driven more by an interest in commercial grade timber supply than its unique biodiversity (McWilliam, 2013). Recognition of the need for forest conservation emerged during the colonial period after WWII, but it was during Indonesian rule after 1975 that the forested Pai Cao Mountains were designated as a conservation reserve (McWilliam, 2013). Timor-Leste environmental management regulations were established during the United Nations Transitional Administration in East Timor (UNTAET) in early 2000. National policies on environment and natural resources are covered by laws and regulations formulated under UNTAET, such as laws promulgated in 2000 on the prohibition of logging operations and the export of wood from East Timor and the Law on Protected Areas (PAs). In 2006, the Timor-Leste government signed the United Nations Convention of Biological Diversity (UNCBD) and became a Party to the Convention on 8 January 2007. As a signatory to the UNCBD, Timor-Leste takes efforts to fulfil the Convention's requirements which include the Program of Work on Protected Areas. In February 2011, Timor-Leste enacted the Environmental Licensing Decree (Law No. 5) creating a system of environmental licensing for public and private projects likely to produce ecological and social impacts on the environment. There are traditional regulations and customs in Timor-Leste that contribute to conserving natural resources such as forests and crops. This system of communal protection is known as *Tara bandu*, an agreement within a community to protect a special area or resource for a period of time (RDTL, 2011/2015).

UNTAET introduced Protected Area Regulation No. 2000/19 in July 2000 with the establishment of 15 "wild protected areas" (Nunes, 2002). Protection of biodiversity requires effort and covers the following: (a) conservation of mountain and erosion-prone areas with forests; (b) conservation of coastal areas; (c) conservation of watershed areas, steep slopes and lake margins; (d) development of watershed areas; (e) conservation of areas with unique value, natural beauty and interest and special features of nature and culture; and (f) environmental impact statements. Therefore, sustainable utilization of natural resources is regulated through well-regulated activities, including (a) use of natural conservation areas for recreation, tourism, experimentation and research, (b) controlled use of wild flora and fauna, and (c) involvement of the people around conservation areas to be partners in management (Nunes, 2002).

The list of 15 sites designated as "wild protected areas" in the Timor-Leste, carried over from Indonesian designation (see Fig. 7.1), is dominated by mountain-focused areas (Carter et al., 2001). By 2016, as stated in Decree Law No. 5 (*Jórnal da República*, 2016), there are 46 protected areas including 2 marine PAs and 44 terrestrial PAs (see Table 7.1). All listed PAs and NPs officially fall under the jurisdiction of the Department of National Parks and Protected Areas (NDPA) portfolio that is responsible for management, including provisions to prohibit certain activities within the protected wild areas.

To have structured management in all protected areas in Timor-Leste, NDPA recommends major strategic needs to advance protected area planning, establishment and management. First, site designation and categorization follow alternate forms of protected area governance in Timor-Leste. Referring to the IUCN management categories, NDPA identifies four main governance types: government-managed protected areas, co-managed protected areas, privately managed protected areas, and community-conserved areas. The IUCN categories, Categories V and VI are often seen to be the most useful ones for addressing the interests of indigenous and local communities, given their emphasis on sustainable practices and the human-nature relationship that is built into their definitions (McIntyre, 2011). Second, introducing integrated land use planning incorporates indigenous or traditional systems or practices demands special consideration about social harmony and conflict minimization. Third, economic well-being of the indigenous and local communities is essential since they are the owners and are directly involved as co-managers of considerable areas of the land designated as protected areas. Thus, through participation and involvement of the traditional and local inhabitants at all stages of planning, implementation, and management, protected areas can have mutually beneficial outcomes. Fourth, protected areas can offer a wide range of employment and income opportunities for local communities.

A number of key threats to the sustainability and health of the country's natural wealth have been identified including deforestation, degradation, poaching and introduced species. Deforestation is arguably the most serious problem threatening the biodiversity of Timor-Leste, due to pressure on forests driven by the need for firewood, agriculture and out-of-control fires during land clearing or hunting. Forest cover in Timor-Leste has decreased by almost 30% in the period between 1972 and 1999 (Sandlund et al., 2001). Approximately 59% of the total land area of the country is still forested, but only 1.7% of primary forests remain, found mainly in Lautem and Covalima districts. Between 2003 and 2012, there has been a further reduction in Timor-Leste's forest cover. Approximately 184,000 ha of forest, i.e. 17.5% of the forest area of 2003, was lost in that 9-year period (RDTL, 2015). Another major threat is the gradual degradation of landscapes and up to 50% of the country is considered degraded, partly due to farmers' practicing slash-and-burn agriculture by

Table 7.1 Protected areas in Timor-Leste

Land protected area					
No	Protected area	Municipality	Post administrative	Village	Estimated size in hectares (Ha)
1	Nino Konis Santana National Park	Lautem	Tutuala	Tutuala	123.600
				Mehara	
			Lospalos	Muapitino	
				Lore 1	
			Lautem/Moro	Bauro	
				Com	
2	Mount Legumau	Lautem	Luro	Vairoke	35.967
				Afabubo	
				Baricafa	
		Baucau	Laga	Atelari	
			Baguia	Uakala	
3	Maurei Lake	Lautem	Iliomar	Tirilolo	500
		Viqueque	Uatocarbao	Irabin de Baixo	
4	Irabere Water Spring	Viqueque	Uatocarbao	Bahata	
				Irabin de Baixo	
				Irabin de Xima	
5	Mount Matebian	Baucau	Quelicai	Lai Sorulai	24.000
				Uaitame	
				Afaca	
				Namanei	
			Laga	Sagadati	
				Atelari	
				Alawa leten	
				Lavateri	
				Alawa kraik	
				Defa Uassi	
			Baguia	Osso-Huna	
				Afaloicai	
				Samalari	
				Haecono	
		Viqueque	Uatolari	Babulo	

(continued)

Table 7.1 (continued)

Land protected area

No	Protected area	Municipality	Post administrative	Village	Estimated size in hectares (Ha)
				Vessoro	
				Afaloicai	
			Uatocarbau	Afaloicai	
				Uani uma	
6	Mount Mundo Perdido	Viqueque	Ossu	Osso de Cima	25.000
				Loihuno	
				Liaruka	
				Builale	
7	Mount Laretema	Viqueque	Ossu	Uaguia	16.429
				Ua bubu	
		Baucau	Venilale	Waioli	
				Wato-Hako	
8	Mount Builo	Viqueque	Ossu	Loihuona Uaguia Ossue Rua	8.000
			Uatolari	Matahoi	
9	Mount Burabo'o	Viqueque	Uatocarbao	Afaloicai	18.500
				Uani Uma	
10	Mount Aitana	Viqueque	Lakluta	Irabin Baixo	17.000
				Ahik	
				Lalini	
11	Mount Bibileo	Manatuto	Laleia	Cairiu	19.000
		Viqueque	Lacluta	Bibileo	
				Dilo	
12	Maunt Diatuto	Manatuta	Saoibada	Fatu Makerek	15.000
				Samoro	
			Laklubar	Funar	
				Fatu Makerek	
				Mane lima	

(continued)

Table 7.1 (continued)

Land protected area					
No	Protected area	Municipality	Post administrative	Village	Estimated size in hectares (Ha)
13	Mount Kuri	Manatuto	Laclo	Uma Kaduak	18.000
14	Kay Rala Xanana Gusmao	Manufahi	Same	Holarua	
				Letefoho	
				Rotutu	
		Ainaro	Ainaro	Mauciga	
				Sorukraik	
				leolima	
15	Clere River	Manufahi	Faturberlio	Uma Berloik	30.000
				Dotik	
				Caicasa	
16	Modomahut Lake	Manufahi	Faturbeliu	Fatukahi	22
17	Welenas Lake	Manufahi	Faturberliu	Fatukahi	20
18	Maunt Manucoco	Dili	Atauro	Makili	4.000
				Vila	
				Manumeta	
				Makadade	
				Beloi	
19	Cristo Rei	Dili	Cristo Rei	Camea	1.558
				Metiaut	
20	Tasi Tolu Lake	Dili	Dom Alexio	Comoro	
21	Mount Fatumasin	Liquica	Bazartete	Metagou	4.000
				Leorema	
				Fatumasin	
22	Mount Guguleur	Liquica	Maubara	Lisadila	13.159
				Maubara Lisa	
23	Lake Maubara	Liquisa	Maubara	Vatubou	
24	Mount Tatamailau/Remaleu	Ainaro	Hatubuilico	Nunumoge	20.000
			Ainaro Lete Foho	Manutasi	
				Bobo leten	
				Katrai Kraik	

(continued)

Table 7.1 (continued)

Land protected area					
No	Protected area	Municipality	Post administrative	Village	Estimated size in hectares (Ha)
			Atsabe	Malabe	
25	Mount Tolobu/Laumeta	Ainaro			15.000
26	Mount Leolako	Bobonaro	Bobonaro	Kilatlau	4.700
			Maliana	Ritabou	
				Odomau	
			Cailaco	Riheu Atadara	
				Manapa Goulolo	
		Ermera	Atsabe	Bobo leten	
				Paramin	
27	Mount Tapoa / Saburai	Bobonaro	Lolotoe	Gildapil	5.000
				Lontas	
			Bobonaro	Oeleu Tapo	
			Malianan	Leber Sabuarai Odomau	
28	Lake Be Malae	Bobonaro	Balibo	Sanirin	
				Leolima	
				Aidaba leten	
29	Korluli	Bobonaro	Maliana	Ritabou	
				Tapo/Memo	
			Cailaco	Manapa	
30	Mount Lakus / Sabi	Bobonaro	Lolotoe	Lontas	
				Gildapil	
31	Mount Taroman	Covalima	Lolotoe	Guda	19.155
				Lupal	
			Fatul Lulik	Opa	
			Fohorem	Dato Rua	
				Dato Tolu	
				Laktos	

(continued)

Table 7.1 (continued)

Land protected area					
No	Protected area	Municipality	Post administrative	Village	Estimated size in hectares (Ha)
32	Tilomar Reserve	Covalima	Tilomar	Maudemo	7.000
				Lalawa	
				Kasabauk	
				Beseuk	
33	Cutete	Oecusse	Pante Makasar	Costa	13.300
				Nipana	
				Bobokase	
				Cunha	
				Lalisu	
34	Mount Manoleu	Oecusse	Nitibe	Usitico	20.000
				Binife	
35	Mangal Citrana Area	Oecusse	Nitibe	Binife	1.000
36	Oebatan	Oecusse	Nitibe	Suni ufe	400
37	Ek Oni	Oecusse	Nitibe	Lela-Uee, Bana Afi	700
38	Usmetan	Oecusse	Pantai Makasar	Taibako	200
39	Mak fahik	Manatuto	Barique	Manehat	
40	Mangrove Metinaro Area	Dili		Metinaro	
41	Mangrove Hera Area	Dili	Cristo Rei	Hera	
42	Hasan Foun and Onu Bot Lake	Covalima	Tilomar	Maudemu	12
				Lalawa	
				Beiseuk	
43	Bikan Tidi Lake	Ainaro	Leolina		110
44	Samik Saron	Manatuto	Barique	Barique	
			Soibada	Cribas	
			Laclubar	Manlala	

Marine protected area					
No	Protected area	Municipality	Post administrative	Village	Estimation size in hectare (Ha)
45	Natural Aquatic Reserve	Bobonaro	Balibo	Batugede	112.59

(continued)

Table 7.1 (continued)

Marine protected area					
No	Protected area	Municipality	Post administrative	Village	Estimation size in hectare (Ha)
46	Natural Aquatic Reserve	Dili	Atauro	Suco da Vila	50.85

Source Adapted from Decree Law No. 5 (*Jornal da República*)

clearing forests for new fields. Given Timor-Leste's sloping terrain and the precipitation pattern of short, intense bursts of rain, soil erosion has negative impacts on both terrestrial and aquatic biodiversity (Carter et al., 2001).

Concerning wildlife conservation, poaching of wildlife and introduced species are major problems in addition to the risk of threatened species being hunted for food, medicine, and ornaments, and also collected live for the pet trade. While there have been efforts to protect threatened species through the formulation of UNTAET regulation 2000/19, actual enforcement of this policy has been lacking (Grantham et al., 2011). It is not known how much invasive species are affecting native species but they are believed to potentially have a significant impact on native biodiversity. Recent estimates suggest that that one third of the 52 mammal species introduced on the island of Timor are thought to have accelerated the decline of some of the endemic fauna, through predation, competition, introduction of new diseases and/or consequential habitat change (Grantham et al., 2011).

Also, there is conflict of interest between the government and landowners, and between the government and communities in Malahara (subdistrict Los Palos) and Buiquira Salt Lake (sub district Laga/Baucau). The management system of established PAs is currently inadequate. Most PAs are paper parks that have been declared, but do not have management plans and lack institutional mechanisms for effective management (RDTL, 2011/2015).

7.3 Nature-Based Tourism in PAs

The Timor-Leste Strategic Development Plan 2011–2030 identifies tourism as one of the major pillars of Timor-Leste's economy (RDTL, 2011). The National Biological Action Plan (RDTL, 2011/2015) specifies nature-based tourism, community-based sustainable tourism and ecotourism for promotion. The lack of tourism development in Timor-Leste can be situated as an opportunity to build an innovative and sustainable tourism sector 'from scratch' that capitalizes on the advantages of peripherality (Weaver, 2018). UNDP and UNWTO (2007) highlight that the country has great biological and cultural diversity for ecotourism, adventure tourism, nature-based tourism and rural tourism. Terrestrial ecotourism activities (e.g., visiting historical and cultural sites, birdwatching, horseback riding, mountain climbing, hiking/trekking, trail-biking and camping) and marine ecotourism activities (e.g., snorkelling, diving, beach recreation, swimming, whale and dolphin watching, and fishing) (MED, 2011) are major assets for development.

Given the available attractions and activities, the international inbound main market for Timor-Leste tourism is adventure and dive tourism from Australia, New Zealand, the UK, Portugal and East Asian origins such as Japan, Korea, Singapore, Hong Kong and mainland China. A secondary market exists in Timor-Leste for cultural tourism based on the forts, churches, *pousadas* (guesthouses) and administration buildings. Observing Portuguese culture in Timor-Leste attracts tourism from Portugal and other former Portuguese colonies, such as Macau and Brazil, including from the Timorese diaspora (UNDP & UNWTO, 2007).

PAs are an important component of ecotourism because tourists who come to visit the areas often want to enjoy preserved nature and wildlife. The government of Timor-Leste has also supported ecotourism development by establishing the first national park and marine park, the Nino Konis Santana National Park, which is situated in Lautem District in the eastern part of the country and where the Tutuala CBET venture is now being undertaken (MED, 2011). Many ecotourism projects have been supported by local and national NGOs and international agencies in Timor-Leste. Although they have limited financial ability to support CBET development directly themselves, they have obtained financial assistance from international NGOs or international agencies such as USAID, the Portuguese Institute for Development Support, the Australian Conservation Foundation and AusAID to help local people take part in tourism development (Wollnik, 2011). CBET has been developed in several places across the country such as on Ataúro Island (Dili District), Tutuala (Lautem District), Maubara (Liquica District) and Maubisse (Ainaro District) (Edyvane et al., 2009).

Despite great potential and support for ecotourism, local communities in Timor-Leste still lack environmental knowledge and awareness to preserve their natural resources and wildlife due to poverty and lack of access to education. Biodiversity conservation management is still a relatively new concept for Timor-Leste and, due to a lack of management of the national parks, there are impacts on local community livelihoods in the surrounding areas (Cullen, 2012). The UNDP and UNWTO (2007) identify six key issues concerning tourism development including security concerns, lack of competitiveness, weak institutional systems, lack of qualified human resources, insufficient resources, and lack of tourism and environmental awareness. As tourism solutions require the collaboration of different sectors, the following section discusses the situation of nature-based ecotourism and pilgrimage in the country's most popular mountain destination, Mount Ramelau, to illustrate the importance of stakeholder collaboration for finding comprehensive solutions for sustainable visitor management.

7.4 Stakeholder Collaboration for Tourism Management in Mount Ramelau

7.4.1 Geographical, Religious and Historical Context

Timor-Leste's highest peak at approximately 2963 m above sea level, Mount Ramelau is a sacred place for both traditional and Christian believers. The highest point of Mount Ramelau named "Tatamailau" in local dialect *Mambae*; *Tatamai* means "grandfather" and *Lau* stands for "mountain"; together, the meaning of *Tatamailau* is "the grandfather of the mountains" (Visit East Timor, 2020). Ramelau is associated with the dead, with funerary rites and invisible villages of the ancestors. According to traditional belief, when someone dies, the soul is released from the body and flies to the peak of Mount Ramelau where it will wait for a minute or two and if the soul has been cleared of any capital sin, the Good Angel will descend from Heaven and lovingly hold it by the hand and guide it to heaven (Molnar, 2011). In 1997, a three-metre-high statue of the Virgin Mary was placed on the top of Mount Ramelau as a sacred symbol of Catholicism (Diocese Dili, 2018). Every year on the 7th of October, thousands of local people gather in Hatubuilico and make a pilgrimage to the peak of Mount Ramelau, commemorating the annual annunciation of the Blessed Virgin Mary.

Located 70 km south of the capital Dili in between Ainaro (to the north) and Ermera municipalities (to the northeast), the mountain is extensively covered with old-growth mountain forest that forms a habitat for endangered and rare bird species. Owing to its high altitude, the area has a relatively cool climate. In May 2017, the Department of Protected Areas and National Parks (DPA) designated Mount Ramelau (Monte Tatamailau) a terrestrial protected area with a total area of 20,000 hectares (*Jornal da República*, 2016), and upgraded it from the UNTAET Regulation 19/2000 status of a "wild protected area". The location of Mount Ramelau and the surrounding protected area is depicted in Fig. 7.1.

Being an icon for the country due to its religious significance, Mount Ramelau is one of Timor-Leste's top-ranked national tourist attractions and is popular among domestic tourists. The number of domestic visitors, mainly pilgrims, is gradually increasing. In 2017, the number of visitors to Ramelau Mountain was 21,000, a number that steadily increased to 31,000 in 2018. The number of international visitors remains quite low at less than 1,000 annually (USAID, 2018).

7.4.2 The System of Tourist Attractions and Service Stations

The description of accessibility and services in the area of Mount Ramelau is summarized from the report of the Tourism for All Project of USAID (2018). The trip to Mount Ramelau takes four hours driving from Dili while the trop from Maubisse to

Hatobuilico, the base area of the mountain takes about two hours. Maubisse's attractions include the Sacred House of Maubisse Village, a coffee and strawberry farm, and a waterfall. The village has seven guesthouses and homestay facilities supported by an Asian Foundation project for those who choose to stay overnight midway to the mountain, driving to the base village of Hatobuilico in early morning to hike to the summit. At the base of the mountain is Hatobuilico, a town that is approximately 2000 m above sea level with a population of 12,968 people. Hatobuilico is known for its naturally beautiful landscape with a view of Mount Ramelau, tranquil village life and traditional thatched-roof houses. In addition to tourism, agriculture is the primary industry of the area, with a wide range of food produced in the village. There are two private guesthouses, one government guesthouse (with price ranging from $10-$30 per night), and 32 newly added community homestays that serve basic food, costing $10 per night. There are also two restaurants and temporary vendors lining the road during the festival. No local handicrafts are available in Hatobuilico.

From the base of Hatobuilico, there are two roads that lead to the sanctuary: the old road and new road, as well as a network of trails and shortcuts. The trails are not well marked, and the old road is not suitable for vehicles. On the new road for cars and motorbikes, there are a couple of buildings used during the festival for selling food. At the sanctuary, there are seven shops selling snacks and drinks, two toilets, and five trash bins. Shop openings and an increase in trash bins are observed during the festive period (around October 7) when visitor numbers are concentrated. Cement stairs lead from the sanctuary to the chapel, located not far from the summit. The chapel is designed with a grass roof and wooden frame to reflect the natural and cultural values of Mount Ramelau and Hatobuilico. The chapel, however burned down in 2019 by natural fire, and is in the process of reconstruction (personal communication with local officers). The trails leading from the sanctuary to the chapel, and to the peak from the chapel are noticeably dangerous in the wet season, with heavy winds making it difficult to stay on the path.

7.4.3 Visitor Management

7.4.3.1 Hiking to the Summit of Mount Ramelau

Most visitors aim to hike up to the peak of Mount Ramelau, either departing from Hatobuilico in the early morning before sunrise; or making the climb in the late morning or afternoon after reaching the village and returning to Hatobuilico or Maubisse on the same day. Many foreign visitors choose to take a local trekking guide or book tour with tour operators based in Dili with an average stay of three to four days. Local hikers are mainly composed of small group ranging from 5 to 10; they usually climb without a local guide and stay, on average, for two days.

7.4.3.2 Visitor Management/Mitigation Issues

For international visitors, ratings of Mount Ramelau on Tripadvisor are quite positive, with all 15 reviews on one link being either Excellent (24%) or Very Good (76%). Most visitors recommend staying overnight in Hatobuilico and climbing the mountain in the early morning with a local guide in order to see the sunrise (USAID, 2018). However, the major problem of visitor management is with the domestic pilgrimage in peak period. The number of domestic visitors sharply increased in the first six months of 2018. The sudden increase of visitors (with many being young students) causes great concern among locals due to limited facilities, accommodations and toilets. Problems of waste left by tourists who brought their own food, their lack of awareness of local culture and religion, as well as improper behaviour at sacred sites resulted on the decision by municipal level government and Timor-Leste's Archdiocese to temporarily close the mountain. The site was finally reopened on 7 October 2018, during the annual pilgrimage of the Annunciation of the Virgin Mary.

Another major issue concerns the current capacity of the site. First, it is a concerning for local community that the base camp briefly loses its peacefulness for two days, with the influx of more than five thousand pilgrims arriving by truck, by motorbike and on foot. Second, installed trash bins are too few to manage the waste generated by the pilgrims. There's only one toilet accessible and thus not enough accessible water for the visiting masses. Trek signboards poorly inform visitors with no details on lengths of key locations along the trail and other relevant safety material.

The major issue is that the overcrowded visitors impose risk due to a lack of emergency services to assist pilgrims and trekkers in the context of a dangerous trail to the summit. There is currently no railing or barrier along the trail at the top, and the area just below the Virgin Mary structure is particularly dangerous where the trail ends at the edge of a cliff. There is no emergency protocol in place or first aid station with medical support ready and available. Visitors are exposed to the dangers of climbing and overnight stays in the mountains.

7.4.4 Stakeholders Participation in Site Management

For a mountainous area, tourism is perceived to be a positive catalyst for economic growth, but it may have a serious impact if the destination lacks long term planning which could put substantial pressure on both the fragile mountain environment and socio-cultural traditions of the locals. Identification of relevant stakeholders and the extent to which they be involved in decision-making is essential. Owing to the outstanding natural, socio-cultural and religious value of Mount Ramelau, there are different levels of governments and religious institutions that partake in tourism initiatives at the site. The roles and functions of different stakeholders are identified from an investigation by the USAID (2018) Tourism for All Project in Mount Ramelau.

At the state government level, the Ministry of Tourism, Commerce and Industry (MTCI) plays a coordinating role with Ministry of Agriculture and Fisheries for governance of protected area, with the municipal government responsible for ground operations and other stakeholders. MTCI also coordinates state investment and technical assistance from donors to the area. Revenue generation from entrance tickets and donations is earmarked to contribute to the conservation and maintenance of the site.

The Department of Protected Areas (DPA) under directory of Ministry of Agriculture and Fisheries has planned to implement zoning for community participation since 2019 along with the Planning Department. DPA identifies key threats to the protected area including burning, logging, and hunting of birds and monkeys. DPA recommends diversification of trekking routes to de-concentrate tourists around a particular area. The DPA also identifies designated areas for campfires and camping in order to mitigate the impact of tourism on forest and plants, and trains locals on recycling practices. Contribution to conservation through entrance fee and tour guiding is encouraged.

The municipal government directly manages its own tourism assets under the approval of the national government and functions of an Administrative Post. The Administrative Post has a three-fold plan for tourism development, which are (1) training people, (2) developing infrastructure for tourism and (3) creating cooperatives and supporting local people to sell local products. The Administrative Post Assembly consists of representatives of various departments including education and health, agriculture, police, three village chiefs, church heads, a women's group, and a youth coordinator. It is responsible for approving any developments regarding tourism.

The Church has been working with the municipal government and MTCI for site management and facility improvement. Entrance tickets to the site at the cost of USD 1.00 per person are planned to be implemented. In the future, the Church expects to have safety measures and improvement of trails. Facilities for pilgrimage such as water, solar panels, garden landscaping, a guesthouse, a dormitory and a kitchen are expected to be provided.

A Tourism Working Group (TWG) comprised of 15 young locals with educational backgrounds in tourism work voluntarily to improve tourism in their community. The TWG created a homestay program assigned to 32 households; they also work toward mitigating tourism impacts. Despite lacking sufficient funding to implement their working plan, TWG is working towards training for community, improving communication between stakeholders, providing support for waste collection and disposal, and conducting research for new attractions (personal communication with local officers).

The Chamber of Commerce and Industry of Timor-Leste (CCI-TL) was established in 2010 and currently has 191 members with the majority from the construction subsector, 30 from agriculture, and 37 from tourism (i.e. guesthouses and restaurants) of which four are based in Hatobuilico. The CCI-TL provided training for guesthouse owners in Maubisse, and tourism subsector training sessions in Hatobuilico. Currently, CCI-TL has had no engagement in the management of Mount

Ramelau due to limited funding. However, CCI-TL identifies that there is an urgent need to develop human resources for handicrafts, horticulture, and hospitality; to improve basic infrastructure; to provide access to microfinance and financial support to improve facilities of restaurants and guesthouses; and to support overall site management based on region-wide experience.

Local guides, homestay hosts and restaurant owners are private sector actors involved in providing accommodation, food and tourism services to visitors. There is a lack of a tour guide association and related training, resulting in a shortage of guiding skills. Except for the festive time for two or three days in October, the all-year-round regular number of guests is still quite low, with only an estimated seven to eight international guests and four to five local guests per week. Therefore, income from tourism to benefit local homestay hosts and restaurants is limited. Although, much advice on what to do and how to improve tourism is provided, there has been very little implementation and follow-through from government. Local people have not been included in tourism management discussions, are allowed to attend meetings as observers only, and there has still not been any support for the community to develop tourism.

In summary, managing tourism in a mountainous area is not only limited to the summit, the surrounding area should also be under suitable management. Currently, multiple stakeholders and agencies claim jurisdiction over the site's management. The Church claims responsibility for Mount Ramelau as one of the country's most important religious sites. MTCI is eager to protect what it rightly claims to be one of the country's most iconic tourism attractions. The municipality sees a strong role as the manager of its local tourism attractions. There is also a need to involve local businesses and guides in helping to improve and manage tourism. Hence, a participatory planning and management approach should be taken that includes local stakeholders in decision making and divides responsibilities for development and management of tourism based on each stakeholder's strengths. Community-based tourism in Mount Ramelau, however, is in its early stage. Lack of coordination and effective administration leads to the major problem of visitor management in peak season as outlined in the case above.

7.5 Conclusions

This chapter outlines the natural characteristics of Timor-Leste. Despite being a young nation, the system of laws and regulations on conservation of nature environment and biodiversity is well understood by central governmental agencies. The rich terrestrial and marine resources provide the country with potential for nature-based tourism to be developed. The development of NBT in the most attractive mountain destination in the country is still in its early stage with much need for infrastructure, resolution of safety issues, product development and provision facilities as outlined in the case of Mount Ramelau. Governance and management of NBT in Mount Ramelau is rather weak. The visitation to the mountain is mainly driven by mass domestic pilgrimage, concentrated only around the festive time in October, causing overcapacity in visitor management. There are very limited activities and programs for promotion of Mount Ramelau as a tourist attraction domestic and internationally. As it is still at the early stage of development, much work remains to make it more attractive and to maintain tourism activities in a sustainable way.

References

Byrd, E. T. (2007). Stakeholders in sustainable tourism development and their roles: Applying stakeholder theory to sustainable tourism development. *Tourism Review, 62*(2), 6–13.

Carter, B., Prideaux, B., Ximenes, V., & Chaternay, A. (2001). Development of tourism policy and strategic planning in East Timor. *Occasional Paper 2001, School of Natural and Rural Systems Management, The University of Queensland, 8*(1), 1–101.

Cullen, A. (2012). A political ecology of land tenure in Timor Leste: Environmental contestation and livelihood impacts in the Nino Konis Santana National Park. *Peskiza foun kona ba/Novas investigações sobre/New Research on/Penelitian Baru mengenai Timor-Leste*, 158–165.

Diocese Dili. (2018). Nain Feto Ramelau (Mount Ramelau). https://diocesededili.org/2018 on September 10, 2020.

Edyvane, K., de Carvalho, N., Penny, S., Fernandes, A., de Cunha, C. B., Amaral, A. L., & Pinto, P. (2009). *Conservation value, issues and planning in the Nino Konis Santa Marine Park, Timor Leste—Final report*. The Ministry of Agriculture & Fisheries, Government of Timor-Leste.

Grantham, H. S., Watson, J. E. M., Mendes, M., Santana, F., Fernandez, G., Pinto, P., Ribeiro, L. M., & da Cunha Barreto, C. (2011). *National ecological gap assessment for Timor-Leste 2010*. Byron Bay, NSW.

Jornal da República. (2016). *Decree Law No 5/2016—National system of protected areas* [Decreto-Lei No 5/2016 de 16 de Março – Sistema Nacional de Áreas Protegidas]. http://extwprlegs1.fao.org/docs/pdf/tim167551.pdf.

McIntyre, M. A. (2011). *Capacity development action plan for the programme of works on protected areas, Part 1 situation analysis, Timor Leste, 2011*. Prepared for the Department of Protected Areas and National Parks, Ministry of Agriculture and Fisheries, Government of Timor Leste with the assistance of United Nations Development Program, Timor-Leste and the Global Environment Facility. Planning for Sustainable Development Pty Ltd, Landsborough, QLD, Australia.

McWilliam, A. (2013). Cultural heritage and its performative modalities: Imagining the Nino Konis Santana National Park in East Timor. In S. Brockwell, S. O'Connor, & D. Byrne (Eds.), *Transcending the culture-nature divide in cultural heritage: Views from the Asia-Pacific Region* (p. 191). Australian National University Press.

MED. (2011). *The National Biodiversity Strategy and Action Plan of Timor-Leste*. Dilli, Timor Leste: Ministry of Economic and Development.

MOT. (2012). *National strategic plan for tourism 2012–2017*. Dili, Timor-Leste.

Molnar, A. (2011). Darlau: Origins and their significance for Atsabe Kemak identity. In A. McWilliam & E. Traube (Eds.), *Land and life in Timor-Leste* (pp. 87–116). ANU E-Press.

Molyneux, N., Da Cruz, G. R., Williams, R. L., Andersen, R., & Turner, N. C. (2012). Climate change and population growth in Timor Leste: Implications for food security. *Ambio, 41*(8), 823–840.

Monk, K., de Fretes, Y., & Reksodiharjo-Lilley, G. (1997). *The ecology of Nusa Tengara and Maluku*. Oxford University Press.

Nunes, M. (2002). *Forest conservation and fauna protection in East Timor*. Paper presented at Agriculture: New Directions for a New Nation—East Timor (Timor-Leste), Dili, East Timor.

Pederson, J., & Arneberg, M. (Eds.). (1999). *Social and economic conditions in East Timor*. Columbia University.

RDTL. (2011). *Timor-Leste strategic development plan 2011–2030.* https://www.adb.org/sites/def ault/files/linked-documents/cobp-tim-2014-2016-sd-02.pdf.
RDTL. (2011/2015). *The National Biodiversity Strategy and Action Plan of Timor-Leste (2011– 2020)—Revised edition 2015.* Dilli: Democratic Republic of Timor-Leste.
RDTL. (2015). *Timor-Leste's fifth national report to the Convention on biological diversity 2015.* Dilli: Democratic Republic of Timor-Leste.
RDTL. (2017). *Growing tourism to 2030: Enhacing a national identity.* Dilli: Timor-Leste National Tourism Policy.
Sandlund, O. T., Bryceson, I., De Carvalho, D., Rio, N., Da Silvia, J., & Silva, M. I. (2001). *Assessing environmental needs and priorities in East Timor: Issues and priorities.* Trondheim.
Tolkach, D., & King, B. (2015). Strengthening community-based tourism in a new resource-based island nation: Why and how? *Tourism Management, 48,* 386–398.
UNDP & UNWTO. (2007). *Sustainable development sector development and institutional strengthening project, Volume 1, main report.* Dilli, Timor Leste: UNDP.
USAID. (2018). *Assessement of Mount Ramelau and Hatobuilico tourism with recommendations for medium to long-term planing.* Dili, Timor-Leste: Chemonics International Inc.
Visit East Timor. (2020). *Mount Ramelau (Foho Tatamailau).* https://visiteasttimor.com/mount-ram elau/.
Weaver, D. (2018). Creative periphery syndrome? Opportunities for sustainable tourism innovation in Timor-Leste, an early stage destination. *Tourism Recreation Research, 43*(1), 118–128.
Wollnik, C. (2011). *Sustainable destination management in Timor-Leste* (Doctoral dissertation). Philipps-University of Marburg. https://www.tourism-watch.de/en/thesis/sustainable-des tination-management-timor-leste.
World Bank. (2020). *Country partnership framework for Democratic Republic of Timor-Leste.* Indonesia and Timor-Leste Country Management Unit.
Ximenes, V., & Carter, R. W. (2000). *Environmental protection and tourism: Issues for East Timor.* Paper presented at Reconstruction: Review of the Past and Perspective for the Future Conference, Tibar, East Timor.

Antonio da Silva is a Tourism marketing officer working at the Ministry of Tourism Commerce and Industry, Timor-Leste. He is a recipient of the Japan Development Scholarship (JDS) - currently enrolled in graduate study at Ritsumeikan Asia Pacific University in Beppu, Japan. He graduated from Griffith University in Brisbane, Australia majoring in tourism sustainable management.

Huong T. Bui is Professor of Tourism and Hospitality cluster, the College of Asia Pacific Studies, Ritsumeikan Asia Pacific University (APU), Japan. Her research interests are Heritage Conservation; War and Disaster-related Tourism, Sustainability and Resilience of the Tourism Sector.

Chapter 8
Protected Areas and Nature-Based Tourism in the Philippines: Paying to Climb Mount Apo Natural Park

Aurelia Luzviminda V. Gomez🄳 and Thomas E. Jones🄳

8.1 Introduction

The Philippines is an ecologically diverse archipelago located in Southeast Asia. It is the world's second largest archipelago after Indonesia, its southern neighbours in the Pacific Ring of Fire. The Philippines' land territory of about 30 million hectares is spread across more than 7100 islands in three major island groups of Luzon, Visayas and Mindanao. It is one of the world's 18 'mega diverse' countries, which together account for two- thirds of global biodiversity (Ong et al., 2002). The Philippines' tropical forests are among the most biodiverse in the world, being home to 5% of the world's flora, including at least 25 genera of plants and 49% of terrestrial animal species, making them one of the global priority targets for conservation (DENR-PAWB, 2009).

The Philippines' is also considered a biodiversity hotspot due to the high rate of destruction of natural ecosystems and the threats on endangered species, partly due to increasing population. According to the 2015 population census, the population of the Philippines stood at 101 million people with an annual growth rate of 1.7% (PSA, 2015). Based on a 2016 Labor Force Survey conducted by the same agency, 27% of the total labour force (67.2 million Filipinos aged >15 years old) was employed in the agriculture sector (including Hunting, Forestry and Fisheries (AHFF). With the subsequent dependence on natural resources as a source of livelihood, the Philippines had one of the highest rates of deforestation in the world. In the early 1900s, forest cover was about 70%, but by the late 1980s only 23% remained (DENR-PAWB,

---

A. L. V. Gomez
School of Management, University of the Philippines Mindanao, Davao, Philippines
e-mail: avgomez@up.edu.ph

T. E. Jones (✉)
College of Asia Pacific Studies, Ritsumeikan Asia Pacific University (APU), Beppu, Japan
e-mail: 110054tj@apu.ac.jp

© The Author(s), under exclusive license to Springer Nature Switzerland AG 2021
T. E. Jones et al. (eds.), *Nature-Based Tourism in Asia's Mountainous Protected Areas*, Geographies of Tourism and Global Change,
https://doi.org/10.1007/978-3-030-76833-1_8

2009). In the early 1990s, deforestation was estimated to be occurring at an annual average rate of 3.5% of forest cover.

Legislative efforts to stem deforestation have been implemented consecutively with the bigger issues of biodiversity conservation and sustainable development via the set-up of Protected Areas (PA). This chapter outlines the legal landscapes that have shaped the designation of PAs in the Philippines, with a focus on the National Integrated Protected Areas System (NIPAS) Act of 1992 and the Indigenous Peoples Right Act (IPRA) of 1997 (see Sect. 8.2). Section 8.3 uses an inventory approach to take stock of PA management in the Philippines, before Sect. 8.4 drills down into the case study example of the Mount Apo Natural Park (MANP), a Natural Park designated in 1996 that symbolizes the best intentions of the NIPAS and IPRA acts, but also the implementation challenges faced (Gomez, 2015). Implications are drawn for current and future nature-based tourism in mountainous PAs.

8.2 Legal Framework for PAs

8.2.1 *From the Brundtland Report to Protected Areas*

In 1987, the Brundtland Report defined the concept of sustainable development as that which "meets the needs of the present without compromising the ability of future generations to meet their own needs" (WCED, 1987). In the Philippines, the movement sought to marry economic growth with environmental conservation goals via decentralization efforts that were symbolized by the enactment of the Local Government Code (LGC) of 1991 (RA 7160) (Severino, 2000). Both the NIPAS and IPRA acts were closely connected to the inception of the 'sustainable development' movement that followed the Earth Summit in Rio in 1992.

The Code was the bellwether for a new set of laws intended to usher in an era of shared responsibility for local government units in natural resource conservation and management. Hitherto, natural resources in the Philippines had belonged to the state and jurisdiction over them had generally resided with central government agencies, primarily the Department of Environment and Natural Resources (DENR) (Carandang, 2012). The decentralization efforts of the LGC coincided with the passage of the NIPAS Act of 1992 (RA 7586), designed as the legal masterplan for the designation and management of the country's conservation areas. In terms of environmental conservation, NIPAS was a ground-breaking legislation (Subade, 2005), especially compared to other Southeast Asian countries (Dressler et al., 2006; UNDP, 2012). It aimed to devolve the management of forestry and other natural resources including PAs to local government units (Pulhin & Inoue, 2008; Shackleton et al., 2002).

8.2.2 The NIPAS Act (1992) and IUCN

The passage of the NIPAS Act involved the DENR, NGOs and international donors such as World Bank. It was eventually enacted in June 1992, just days before the historic Earth Summit held in Rio de Janeiro set the global stage for sustainable development targets. The primary objective of the NIPAS Act even echoes the wording of the Brundtland Report within the context of PAs, seeking *"to secure for the Filipino people of present and future generations the perpetual existence of all native plants and animals through the establishment of a comprehensive system of integrated protected areas within the classification of national park as provided for in the Constitution"* (UNDP, 2012). The Act did not create new PAs overnight, but provided a framework for the evaluation and categorization of existing and new sites. Sections 3 and 4 of the NIPAS Act defined seven categories of PAs, as shown in Table 8.1. These categories mirrored those of the International Union for Conservation of Nature (IUCN) albeit with minor differences (Corpuz & Jones, 2017).

In order to operationalize participatory management in PAs, Section 11 of the NIPAS provides for the creation of a Protected Area Management Board (PAMB), comprised of the DENR-Regional Executive Director; 1 representative from autonomous government; the Provincial Development Officer; 1 representative from the municipal government; 1 representative from each barangay inside the PA; 1 representative from each tribal community; at least 3 representatives from NGOs/POs; 1 representative from other relevant departments/national agencies. As of 2014, DENR-BMB reported that 177 PAs had organized management boards,

Table 8.1 Categories of Philippine PAs and aggregate area as of 2012

NIPAS category (aggregate area in ha)	Number	%[a]	Comparison with IUCN global benchmarks
(a) *Wilderness areas* (total 430 ha in Philippines)	12	5	Most restrictive category under NIPAS akin to IUCN cat Ib. that restricts access
(b) *Natural parks/ national parks* (total 1,332,268.08 ha)	61	25	Similar to IUCN cat. II, but "national park" also implies a particular category of public lands that includes all PAs
(c) *Natural Monument* (total 24,206.16 ha)	4	2	Essentially the same as IUCN cat. III (Natural Monument)
(d) *Game refuge Bird/wildlife Sanctuary* (total 1,233,946.75 ha)	14	6	Essentially the same as IUCN cat. IV (Wildlife Sanctuary)
(e) *Protected landscape/seascape* (total 1,805,375 ha)	64	27	Similar to IUCN cat. V, but NIPAS emphasizes opportunities for NBT
(f) *Natural Biotic Area* (total 12,156.35 ha)	4	2	Similar to IUCN cat. VI, but NIPAS emphasizes role of indigenous peoples

[a] per cent of the total aggregate area of 5,452,159 ha
Source Adapted from UNDP (2012, p. 9) and DENR-PAWB (2013)

of which 98 areas with a PAMB were proclaimed under NIPAS. To implement the NIPAS Act, the Implementing Rules and Regulations were enforced through DENR Administrative Order No. 25 on June 29, 1992. However, the IRR was revised pursuant to DENR-A.O. No. 26 on December 24, 2008 to incorporate and integrate all existing regulations including the Wildlife Resources Conservation and Protection Act (RA 9147); Caves and Cave Resources Management and Protection Act (RA 9072); Philippine Mining Act (RA 7942).

8.2.3 The IPRA Act (1997)

Five years after the NIPAS Act of 1992, the Indigenous People's Rights Act of 1997 (RA 8371) was passed, laying a foundation for the involvement of indigenous people in the management of PAs and on sensitive issues of ancestral domain. Due to the scattered topography of the Philippine archipelago that contains over 7100 islands, the marginalized and often neglected Indigenous Peoples are diverse in terms of ethnicity and tribal groups. The IPRA Act of 1997 was intended to improve the living standards of the estimated population of 14–17 million that have faced perennial struggles to protect their rights (UNDP, 2012). IPRA further established the National Commission on Indigenous People (NCIP) with a mandate to "protect and promote the interest and well-being of the ICCs/IPs with due regard to their beliefs, customs, traditions and institutions" and ancestral domains claims were ceded from DENR to NCIP.

IPRA was intended to safeguard IPs' claims to ancestral domains over land, water and natural resources. The legal concept of ancestral domains transcends physical and residential territories to include areas of spiritual, cultural and traditional practices. IPRA contains a whole chapter detailing the IPs' rights to ancestral domains, including the right of ownership; right to develop lands and natural resources; right to reside in the territories; right in case of displacement; right to regulate entry of migrants; right to safe and clean air and water; right to claim parts of reservations; and the right to resolve conflict. In the case of ancestral lands, IPs have the right to transfer land/property rights to/among members of the same ICCs/IPs and the right to redeem the property in case of transfers that raise questions on consent given by IPs and transfers made with unjust considerations and/or prices.

IPs whose ancestral domains have been officially delineated and determined by NCIP are issued a Certificate of Ancestral Domain Title (CADT) in the name of the community concerned. Notwithstanding cases of conflicts, the recognition of claims on ancestral lands or waters and rights to livelihood opportunities as legitimized by Local Government Code and IPRA proved to be a favorable result of devolution (Dressler et al., 2006). Another key aspect of the law in relation to the management of PAs is outlined in Chapter 4: "Right to Self-governance and Empowerment". Self-governance guarantees the rights of IPs in the pursuit of their cultural, economic and social development. Sections 16 and 17 provide for the right to participate in decision making and the right to determine and decide priorities for development. In short, these provisions legitimized the involvement of IPs in PAMBs, 10 years prior to the UN's Declaration on the Rights of Indigenous Peoples (Simon, 2013).

8.3 Protected Area (PA) Management in the Philippines

Despite the passage of NIPAS in 1992 and IPRA in 1997, PA administration in the Philippines remains problematic with threats to biodiversity unresolved. Under the NIPAS Act, a protected area (PA) is defined as "identified portions of land and water set aside by reason of their unique physical and biological significance, management to enhance biological diversity and protected against destructive human exploitation" (Section 4b). The designation of PAs is the centrepiece of the Philippines' strategy to conserve the country's biodiversity, "with the aim of achieving economic growth without depleting the stock of natural resources and degrading the environment" (DENR-PAWB & GIZ, 2011, p. 9).

As of 2012, the DENR Protected Areas and Wildlife Bureau (currently known as the Biodiversity Management Bureau), reported a total of 240 PAs in the country under the NIPAS Act, covering an area of 5.44 million hectares. 75% (or 170 of the PAs in Table 8.1.) were accounted for by terrestrial areas and 25% (70 PAs) are marine ecosystems, representing 13.6% of the Philippines' total land area and 0.6% of marine territory respectively (DENR-PAWB, 2012). By 2014, the DENR reported that of the 240 protected areas, 113 had been proclaimed by the President under NIPAS covering 3.57 million hectares (see Fig. 8.1). 71 of the 113 areas were initially identified as components with a combined area of 2.0 million hectares and 42 additional areas with combined area of 1.57 million hectares. Natural parks/national parks comprised 25% of all PAs. Of the 113 designated PAs, 29 were marine with an area of 1.37 million hectares while 84 were terrestrial with a total area of 2.20 million hectares. In addition to Table 8.1, the following PAs were left unclassified total of the 240 PAs: 2 'resource reserves' covering 176,000 ha; 56 'watershed forest reserves/areas' covering 834,632 ha, and 23 'mangrove swamp forest reserves' covering 33,143 ha.

PAs designated under the NIPAS act are managed by a multi-sectoral Protected Area Management Board (PAMB). The aforementioned PAMB member structure is headed by the regional executive director of the DENR. The principal role of each PAMB is to guide PA management, e.g. by granting permission for specific activities to be conducted within the PA. Each PAMB has a body of staff, headed by the PA superintendent serves as secretariat of the PAMB and reports to the regional executive director of the DENR (DENR-PAWB & GIZ, 2011).

However, despite the best intentions of PAMB legislation, insufficient funding is a major challenge in PA management in the Philippines. The majority of the funds for PA management come from government coffers and, in some instances, external sources. From 2005 to 2009, for example, approximately 74 and 11% of funding for PA management came from government grants and international sources, respectively. The remaining 15% came from other sources, such as private donations and fees/concessions (DENR-PAWB, 2012). External funding sources included the

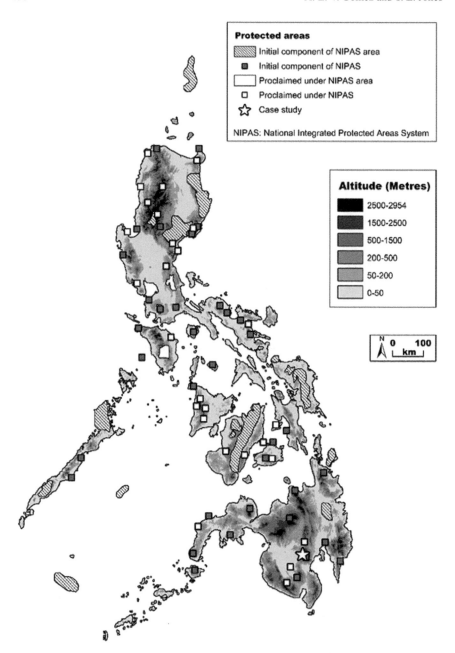

Fig. 8.1 Location of the Philippines' Protected Areas (PAs)

Global Environment Facility of the World Bank, the European Union, and the United Nations Development Fund. These organisations provided funds for activities to support PA establishment and management, mainly in the form of technical assistance and capacity-building projects.

Despite such overseas aid, the DENR estimated that as of 2008, there was an estimated shortfall of 1478 personnel and an operating expense equivalent to about PhP350 million (approximately US $7.8 million) (DENR-PAWB, 2012). Between 2010 and 2013, the proportion of the DENR budget allocated for PA management increased from PhP 173.41 million to PhP 223.38 million (approximately US $3.87 to 4.98 million), or an average increase of 10% per year. Of the amount allocated for PA management, more than 60% was for personnel services. Between 2010 and 2013, inflation increased at annual average rate of 3.6%.

The NIPAS Act provides for a fee system whereby charges can be imposed on users of PAs with revenues channelled back to management activities. However, in 2005–2009, fees collected contributed only 11% of PA management expenditure. With a shortage of funds from various sources, the DENR recognises the necessity of exploring a system of sustainable financing based on payment for environmental services (DENR-PAWB, 2012), an approach that will next be investigated using the case study of the MANP.

8.4 The Mount Apo Natural Park (MANP)

8.4.1 Location, Designation and Population

Mount Apo is a dormant stratovolcano with three peaks on the island of Mindanao, 45 km southwest of Davao City. The MANP was declared a natural park by Presidential Proclamation No. 882, passed on 24 September 1996. In accordance with the provisions of the NIPAS Act, the MANP was one of the initial components laid out by NIPAS in 1992. The MANP was legally declared as a PA by virtue of Republic Act 9237, otherwise known as the Mount Apo Protected Area Act of 2003.

Natural Parks are one of the PA categories established under the NIPAS Act, defined as a "relatively large area not materially altered by human activity where extractive resource uses are not allowed and maintained to protect outstanding natural and scenic areas of national or international significance for scientific, educational and recreational uses" (Section 4, h, NIPAS Act). Like MANP, most of the Philippines' 61 Natural parks were designated shortly after the NIPAS act, many of them in mountainous areas. This category thus symbolizes the intention to promote PAs as the flagship of sustainable development. In addition, the MANP has an earlier history of designation as a national park in 1936 and is thus considered to be one of the country's priority conservation area (Ong et al., 2002). MANP is an important habitat for birds (Mallari et al., 2001) and a listed ASEAN heritage site (DENR-PAWCZMS, 2013). Along with biodiversity outlined below, the defining feature is

Mt Apo, the Philippines' highest mountain with an estimated maximum altitude that ranges from 2954 to 3144 m above sea level (UNESCO, 2011; Apollo et al., 2020).

Despite being designated as a natural park under the NIPAS Act, at least 44 *barangays* are located within the MANP, including 32 in Davao del Sur and 12 in North Cotabato (DENR-PAWCZMS, 2013). *Barangay* (literally meaning a village, this is the basic local government unit in the Philippines) are permanent settlement areas that are inhabited by indigenous peoples and non-indigenous peoples alike. Although the exact population is unknown, at least 120,000 people distributed in at least 27,000 households were living within the boundaries of the MANP based on the 2010 census conducted by the National Statistics Office. The population is composed of indigenous peoples alongside migrants from other parts of Mindanao or the Visayas. Their major source of livelihood is subsistence agriculture, with some commercial agriculture. The DENR nominated Mt Apo for inclusion on the UNESCO world heritage list in 2009, but it was subsequently removed from the tentative list in 2015 following recommendations to ease the pressures from logging and poaching etc.

8.4.2 Biodiversity of the MANP

The flanks of Mt Apo are dominated by forests, such as the lowland evergreen, lower montane, upper montane mossy or cloud and subalpine forest zones. There are also grasslands consisting of scrub and open areas that cover almost 36% of the MANP. The lowland evergreen rainforest that ranges from the foot slopes to around 1200 m was extensively commercially logged from the 1950s to 1970s. The lower montane rainforest is the most extensive ecosystem and ranges from 1200 m up to 1800 m. This zone hosts several rare, endemic bird species found only on a very few higher mountains in Mindanao, such as the Slaty-backed Jungle-flycatcher (*Rhinomyias goodfellowi*), the Red-eared Parrotfinch (*Erythrura coloria*) and the Apo Myna (*Basilornis Miranda*) (Mallari et al., 2001). The upper montane mossy or cloud forest extends between 1800 and 2600 m in elevation (DENR-PAWCZMS, 2013). This forest ecosystem has steep slopes, rugged terrain, a moist climate, constant precipitation and high humidity due to dense cloud formations. The harsh climatic conditions, thin soils and the scarce and acidic plant nutrients at these elevations inhibit the succession of shrubs into forests. The MANP is a watershed of several river systems such as the Marbel and Matingaw Rivers, and a major tributary of the Mindanao, Sibulan and Digos Rivers, with the latter discharging in the Davao Gulf.

The MANP is a priority site for conservation and research of arthropods, amphibians and reptiles, birds and terrestrial mammals (Ong et al., 2002). It is also listed as an Important Bird Area (Mallari et al., 2001). However, there is no comprehensive monitoring on the flora and fauna other than a baseline study commissioned by the Energy Development Corporation (Dames & Moore Inc, 1994).

8.4.3 Governance and Management of the MANP

The MANP covers an area of 64,053 hectares, composed of 54,975 hectares (85.8%) of PA and 9078 hectares (15.2%) of buffer zones. The PA is further classified into core and multiple-use zones, as shown in Fig. 8.2. The NIPAS Act itself does not prescribe a 'core zone,' but only specifies the 'protected area' (designated for biodiversity, as mentioned in 2.1 above) and 'buffer zones'. The latter are "identified areas outside the boundaries of and immediately adjacent to designated protected areas...that need special development control in order to avoid or minimize harm to the protected area" (Section 4.c).

The preparation of the MANP general management plan for 2010–2030 highlighted the enormous challenge of enforcing the standards of the NIPAS Act due to several factors. First, over half of the total land area of the MANP is covered by ancestral domain titles that grant the indigenous peoples a certain degree of autonomy in managing the land. Second, before designation as a PA, some parts of the MANP already had permanent settlements where people engaged in various forms of agricultural activities and could qualify for land tenure. Thus, the PA is further divided into two zones: the core zone and multiple-use zone. The de facto core (40% of the PA) is managed primarily for biodiversity conservation, but may also be used for research, as well as religious and ceremonial activities. The multiple-use zone covers the remaining 60%, where human settlement, agricultural development and

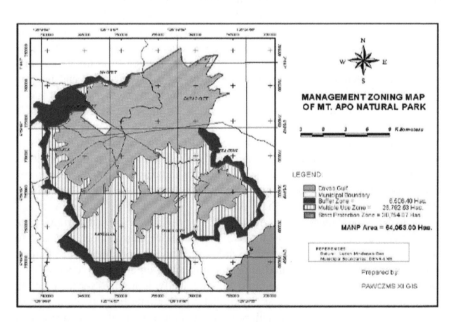

Fig. 8.2 Management zones of the MANP (*Source* DENR XI)

other land uses pre-date PA designation. The major goal of the zoning is the "effective management of the protected area and buffer zones and [to] promote sustainable development of all legitimate stakeholders" (DENR-PAWCZMS, 2013, p. 20).

MANP stakeholders include a mix of government agencies, local government units, non-government organisations, indigenous peoples, water utilities, private energy corporations, water users, mountain climbers, other users and beneficiaries of the ecosystem services. As the primary agency for environment and natural resources management, the DENR is a major stakeholder, but other government institutions are also involved including—but not limited to—the Departments of Agriculture, Energy, and Tourism. In addition, NIPAS theoretically empowers local government authorities with shared responsibility, so another set of stakeholders are overlapping LGAs such as the provinces of Davao del Sur and Cotabato; the cities of Davao, Digos and Kidapawan; the municipalities of Santa Cruz, Bansalan, Makilala, and Magpet; and the 44 barangays within the boundaries of the MANP.

Two major communities also represent significant stakeholders. One group is the indigenous communities that consider the MANP as their ancestral domain. Their main agenda is to claim certificate of ancestral domain titles to ensure their human and cultural wellbeing. Another group are the tenured migrants that have continuously occupied public lands within the PA from before 1st June 1987 and are substantially dependent on the MANP for their livelihood (Section 3, w, RA 9237).

Another set of stakeholders are NGOs and other civil society groups including the Mount Apo Foundation Inc. (MAFI), the Philippine Eagle Foundation, the Kapwa Upliftment Foundation, and local mountaineering groups. The MAFI was established as a legal requirement for obtaining an environmental compliance certificate of Energy Development Corporation for its geothermal energy operation in Mt Apo. MAFI receives one-centavo per kilowatt of power generated from the geothermal plant to fund community development initiatives on the reservation and up to a 10 km-radius of the geothermal plant. Utility companies are another group of stakeholders. This group includes the various water utilities that serve areas in and outside the MANP, the Energy Development Corporation that operates the aforementioned geothermal plant, and Hedcor Sibulan Inc., a private enterprise that is currently developing hydropower projects in and around the MANP. The final group of stakeholders are water users such as farmers, households, commercial and industrial water users, and tourists. In sum, the general public and wider community also benefit from the ecosystem services provided by the MANP.

The Mount Apo Protected Area Act (RA 9237) provides the legislative framework for the management of the MANP. RA 9237 mandates the management of MANP to ensure the conservation of biodiversity, while respecting and promoting the interests of indigenous peoples and tenured migrants along with other stakeholders through sustainable and participatory development. Specifically, it provides for the creation of the multi-sectoral PAMB headed by the DENR Regional Executive Director of Region XI and including representatives from all of the aforementioned government agencies and one representative from each of the three sub-tribes (Jangan, Ubo and Tagabawa) in the MANP by an independent institution or academe and validated by the NCIP; along with up to eight representatives from NGOs. Under RA 9237, the

PAMB has the following major functions: reviews, approves and adopts proposals, management plans and development programs for the MANP; approves proposals for the budget and exercises accountability over donations, budget allocations and all other funding that may accrue; adopts rules and procedures in the conduct of business. The PA superintendent, as the chief operating officer of the MANP, is responsible for implementing the management plan as detailed in the annual work program. He also maintains peace and order, and has the right to arrest anyone found to have violated provisions in RA 9237 or confiscate any forest resource or instruments that are obtained or used in violation of PA laws.

The general management plan contains five programs aimed at effective management via: (1) biodiversity research and rehabilitation; (2) indigenous peoples' affairs and cultural programs; (3) community-based resource management; (4) ecotourism; and (5) institutional strengthening, partnership and co-management (DENR-PAWCZMS, 2013). However, implementation faces challenges linked to the lack of comprehensive baseline information on biodiversity, or the socio-demographic and cultural profiles of residents of the MANP. Another issue is the lack of a sufficient funding despite climber entry fees described below.

8.5 Nature-Based Tourism in the MANP

The principal NBT activity in the MANP is climbing and several trails lead to the summit of Mt Apo. The established trails include the Agco Trail in Kidapawan City, New Israel trail in Makilala, Bongolanon trail in Magpet, Kapatagan trail in Digos City, Sibulan trail in Sta. Cruz, and Bansalan trail (Fig. 8.3). Reliable records of climber numbers were not available, but trends obtained from municipal tourism offices indicate that, between 2006 and 2013, around 3500 climbers per year aimed to summit Mt Apo (Gomez, 2015). This does not include other NBT users attracted by opportunities for hot and cold thermal springs, nor the unknown number of river rafting participants involved with water-based recreation.

8.5.1 Mount Apo Climber Survey

A climber survey was conducted during the climbing seasons of April–May and October 2010. Survey implementation was assisted by enumerators who had prior experience conducting surveys, either with the University of the Philippines Mindanao or with the National Statistics Office. Enumerators were briefed with the background of the research and were trained in intervention procedures. Each interview was preceded by a plain English language statement about the research, supplemented when necessary by an explanation in the local dialect. It was emphasised that the research was not commissioned by any government agency, nor was it part of any existing plan to increase fees for climbing. Periodic feedback sessions were

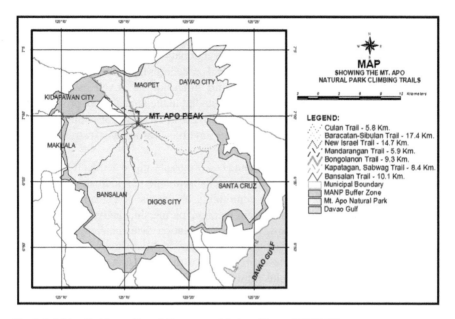

Fig. 8.3 Main climbing trails and distances to Mt Apo (*Source* DENR XI)

conducted after the completion of 15–20 interviews. Climber respondents were interviewed at local tourism offices after completing all the requirements for climbing. The interviews were conducted after respondents had given their consent by signing consent forms. 431 valid responses were collected from a cross-section of climbers that used the official trekking trails located in the four municipalities and two cities included in the MANP.

8.5.2 Mount Apo Climber Profile and Motivations

Climbing Mount Apo is the principal recreational activity in the core of the MANP, and therefore climbers suitably encapsulate the recreational use value of the protected zone. The survey obtained responses from 431 climbers (Gomez, 2015); 43.2% resided outside the MANP, but 32.5% were living in Davao City. Climbers tended to be male (68.4%) and young, with over 85% aged under 36 years old. Only three respondents were older than 55 years. The majority (93.7%) had at least some college education and more than half of all climber respondents had finished college. Almost 10% had postgraduate education. More than half of the respondents had a monthly personal income of less than PhP 8333 (US $186), while approximately 10% earned more than PhP 25,832 per month (US $576).

Climbers tended to climb Mt Apo in the company of friends (42.5%) or as a member of an organised group (36.9%), whose number ranged from two to 100 persons (for safety, solo climbs are not permitted). Mt Apo trailheads are accessible by different modes of transport. Almost half of the climber respondents had utilised only public transportation. Other respondents travelled either by private vehicle only (25%) or through a combination of public and private transport (26%). Public transportation in the area usually entailed a combination of bus, jeepney and motorbike. Climbers from outside Mindanao included a flight as one leg of their expedition. Respondents were asked about their previous experience as well as future plans to climb Mt Apo again. The majority of respondents (67%) had climbed Mt Apo at least once before. On average, respondents had climbed Apo 2.3 times prior to the survey. A small number (3.2%) had climbed Mt Apo over 10 times previously. The majority (81%) indicated that they planned to return. The average duration of respondents' time away from home for the climb was 4.8 days, including camping in designated camp sites en route to the peak. Reaching Apo's summit from the various trailheads typically takes from eight hours to two days, depending on the trail used and the climber's level of fitness. Mt Apo attracts climbers predominantly because it was the highest mountain in the Philippines (Fig. 8.4.). More than two-thirds of respondents indicated that 'sharing nature experience with family and friends' strongly motivated them. Almost half of the respondents 'just wanted to climb Mount Apo'. A chance to see a Philippine eagle, *Pithecophaga jefferyi*, or other rare plants and animals, and 'Mt Apo being a sacred mountain', were less frequent motivators. When asked about their likely activities if they were not climbing Mt Apo, almost half replied that they would be working or visiting other tourist destinations.

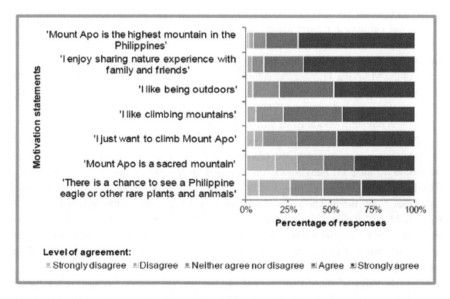

Fig. 8.4 Agreement of climber respondents with motivations for climbing Mt Apo ($n = 431$) (*Source* Gomez, 2015)

The average entrance fee paid by climber respondents was PhP 596.50 (US $13), which represented about 12% of all travel expenses. The biggest median cost was Transport (PhP 975.2, US $22). Climbers from outside Mindanao, such as those from Luzon and overseas, incurred accommodation expenses. Less than half of the respondents paid for a guide, but another possible expense was the rental or purchase of camping gear (Gomez, 2015).

8.5.3 Use and Non-use Values of the Mt Apo Natural Park

Calculating the economic value of ecosystems has long been recognized as a powerful incentive for conservation (Balmford et al., 2002; Clawson & Knetsch, 1966). Total economic value (TEV) estimates are useful in describing the current status of natural resources and possible losses that may be incurred if the resources are not properly managed and conserved (Stoeckl et al., 2011). Gomez (2015) used a TEV framework to estimate the use and non-use values of the MANP, based on consumer surplus derived from willingness to pay estimates from anchored open-ended bids (Greiner & Rolfe, 2004). The methodology combined (i) use values associated with water provision and climbing; (ii) option and non-use values of watershed protection; and (iii) option value of climbing and non-use value (bequest and/or existence) of biodiversity. Results from Gomez (2015) showed that the lower bound estimate of the TEV was PhP 6482.3 million (net present value for the year 2010 being approximately US $144.6 million).

The use value of domestic water provision and recreation (climbing) accounted for 41% of the estimated TEV; 59% was accounted for by option and non-use values of watershed protection, climbing, and biodiversity conservation. The values of components of the MANP's TEV supports the idea that societies generally favour ecosystem services that provide direct benefits (Rodríguez et al., 2006). The relatively high option value of biodiversity (47% of TEV) indicates that individuals place high value on ecosystem goods from which they may not directly benefit.

The TEV estimate for the MANP is a lower bound estimate for a number of reasons. For instance, not all values were measured, and also, the MANP is irreplaceable because Mt Apo has an iconic status as the Philippines' highest mountain. Third, the estimates were based only on the population of the municipalities and cities that cover the MANP, which was roughly 23% of the combined population of Region 11 and Region 12, and only 2% of total Philippine population. Fourth, the use value for water provision is based on the consumer surplus of household water users only, but in fact human consumption of water is comparatively smaller than other uses such as agriculture.

The use value of recreation is likely to be underestimated because other recreational activities such as hot springs, waterfalls and river rafting were excluded. Finally, the true option and non-use values are likely to exceed the current estimate for several reasons. The option value for climbing may be higher for climbers from areas outside the MANP, including from Manila and overseas. This is because analysis of the climber sub-sample indicates that climbers who reside outside the periphery of the MANP were willing to pay 74% more to climb Mount Apo than climbers living on the periphery of the MANP. This is consistent with prior findings that climbers who travel longer distances are willing to pay a higher amount to enter a tourist site (Schroeder & Louviere, 1999). Higher household income also had a significant positive influence on willingness to pay for the conservation of marine turtles (Jin et al., 2010).

8.6 Discussion and Conclusions

This chapter focused on two legislative landmarks, the NIPAS (1992) and IPRA (1997) that symbolize the pre-millennial policy shift in designation of PAs in the Philippines. In particular, NIPAS has been a landmark in the evolution of the Philippines's conservation policies, paving the way for the 'sustainable development' principles encapsulated in the 1992 Earth Summit to be implemented in the designation and management of PAs.

Transformative legislation was especially needed in the Philippines due to the rapid rates of deforestation. However, neither NIPAS nor IPRA, nor the new Local Government Code that preceded them in 1991, could immediately resolve the problem of illegal logging and land conversion. Instead, forest statistics indicate a gradual improvement, with an increase in forest cover across the Philippines from 6,570,000 ha in 1990 (21.9% of total land mass) to 7,665,000 ha in 2010 (25.6%) (FAO, 2010). Even this reported 'increase' is attributed mainly to changes in the international definition of forest adopted from the FAO with Forest is defined as "land spanning more than 0.5 hectare with trees higher than 5 meters and a canopy cover of more than 10 percent" (FAO, 2015), although there has been some natural regeneration and afforestation on both private and public lands. An alternative means of boosting biodiversity was to protect existing stocks, but La Viña et al (2010) found that PAs in the NIPAS Act account for less than half of the priority sites identified by a 2002 report commissioned by DENR and delivered by the Foundation for the Philippine Environment in collaboration with Conservation International Philippines. The same authors also showed that initial components incorporated into NIPAS cover additional areas that are not considered strategically important for biodiversity conservation, exacerbating the gap between preferential and actual conservation sites. In fact, only 82 of the 240 PAs listed by NIPAS (63 terrestrial and 19 marine) correspond to the identified key biodiversity areas, accounting to 34.8% of the total key biodiversity areas in the country (UNDP, 2012). Moreover, as the NIPAS Act does not specifically differentiate between core or buffer zones, designated PAs run the risk of being 'paper parks' that exist on a map but are in reality face sub-optimal administration that belies the Aichi Target 11 goals (Mallari et al., 2016; see also

Chapter 14 of this book). Nonetheless, one of the key methods to counter deforestation was through expanding the national network of PAs and modifying the management style from 'top-down' to the holistic co-management approach epitomized by NIPAS and the cross-cutting councils called PAMBs.

Senga (2001) pointed out that the structure of PAMB could be "cumbersome and unwieldy," as our case study confirmed with the long list of MANB stakeholders, but still believes that the Board could be the embodiment of an effective administrator. However, the PAMB's effectiveness relies on consensus-building among members, which is an admirable aim but can be problematic when discontent or vested interests emerge (Cortina-Villar et al., 2012). Dressler et al. (2006) also cited the "ambiguous management relationship" in a case from Palawan where a local politician appointed members of the Board from the same political party, casting doubts over the impartiality of the PAMB's decision-making.

As in other chapters of this book, the funding of PAs is of paramount importance, but results from field work are not encouraging. At Mt Apo, a portion of the climbing fee is supposed to be remitted to the through the DENR to finance conservation and management of the MANP (DENR, 2008). DENR data on the 2011 revenue generated by PAs indicated that the MANP was not among the Philippines' top ten revenue generating PAs (DENR-PAWB, 2012). However, available data on climbers to Mount Apo indicate that total nominal income from 2008 to 2011, based on the climbing fee of PhP 500 (US $11.15), should have been at least PhP 5.154 million (approximately US $110,000). This amount was 12% higher than that of the 10th highest earning PA in the Philippines indicated in the DENR-PAWB list. In spite of the ID cards issued to Apo climbers upon registration, this lack of transparency hints at deeper issues surrounding the collection and utilisation of climber fees. The upshot is a lack of basic visitor services, from trails to toilets, that entails negative environmental impacts (Apollo, 2017) and undermines future willingness to pay. Based on personal observation, there is an urgent need to rehabilitate and improve certain trail sections on Mt Apo to ensure the safety of climbers and to prevent further damage from erosion (Gomez, 2015).

As in other chapters of this volume, PAs in the Philippines suffer from underfunding due to an over-reliance on government coffers. Insufficient funding undermines the effectiveness of PAMB management. This research provides empirical evidence for the MANB, suggesting that there is ample scope to widen the funding base for PA management. Gomez (2015) estimated the consumer surplus for water provision, biodiversity conservation and recreational use, and the lower-bound of the total economic value (TEV) of the MANP. Consumer surplus estimates for Apo climbers suggests that climbing fees could be doubled without a long-term reduction in climber numbers. As with other iconic mountains investigated in this book, Apo's irreplaceable status as the Philippines' highest peak ensures that it will continue to attract climbers. In addition, the demand for unique natural sites that involve long travel distances also tends to be price inelastic (Clawson & Knetsch, 1966). This implies that the cost of visiting an area does not have much influence on the decision to visit, especially as the current entrance fee is such a small proportion of the total cost of climbing Apo (Gomez, 2015). It also reflects prior research suggesting that visitors' willingness to pay for access to national parks adjusts as the entrance fee increases (Chase et al., 1998).

Beyond the tangible revenue of tickets and entry fees, PAs are a cornerstone of biodiversity conservation, and also provide a range of ecosystem services, safeguarding benefits from water provisions to nature-based tourism. The TEV of the MANP was estimated at PhP 6482 million (approximately US $144.6) at 2010 values (Gomez, 2015). The use value of water provision for domestic use accounts for 38% of TEV, while the non-use value of biodiversity accounts for 47% of TEV. The remaining 15% of TEV is accounted for by option and non-use values of watershed protection, use value for climbing and the option value of climbing. The estimated TEV of the MANP contributes to around 2% of the combined gross regional domestic product of Regions 11 and 12 in Mindanao. However, the actual TEV is likely to be higher if the estimate includes all users and beneficiaries of water from the MANP watershed areas, as well as users of recreational opportunities other than climbing. The research also finds that climbers who reside further from the MANP were willing to pay significantly higher amounts for climbing. The empirical research findings clearly establish the economic potential for PES schemes in watershed protection, biodiversity conservation, and mountain climbing in the MANP. PES schemes could address the revenue shortfall of PA management and generate significant supplemental funds for management and conservation of the MANP.

These findings show that properly designed, targeted PES schemes have the potential to help find the following key management programs for the MANP as identified in the general management plan: (1) biodiversity research, conservation and environmental restoration programs; (2) indigenous peoples' affairs and cultural programs. Although the MANP is a government-owned PA, a significant portion of the land is dependent on ancestral domain claims of indigenous peoples who have a communal right to manage the land. Meanwhile non-indigenous long-term residents can also apply for tenured migrant status, ensuring use rights although they cannot sell the land on; (3) community-based resource management programs; (4) participatory and community-based ecotourism programs that use NBT tools to help incentivize sustainable participation of locals, for example by generating employment opportunities during the climbing season for guides, porters, cooks and mountain rescue. Based on this research, less than half of Apo climbers currently employ a guide. PA managers could help raise the rate by offering capacity-building programs to improve Guides' conversational English-speaking skills, knowledge of the flora and fauna and communication abilities; and (5) institutional strengthening, partnership and co-management programs to improve the effectiveness of the PAMB.

In conclusion, this research investigated the seminal legislative changes in the Philippines' post-Rio PA policy-making. The chapter summarized Gomez's (2015) consumer surplus estimates, thus making a contribution to the knowledge and understanding of the values to users and beneficiaries of ecosystem goods and services, via the case study of Mount Apo. The 1992 NIPAS Act pioneered a more participatory style of governance (La Viña et al., 2010) by formally recognizing multi-stakeholder involvement including acknowledgment of the role of local communities. The rights of indigenous peoples living in the area that depended on the resources for their livelihood were also recognized and subsequently legitimized through the enactment of the Indigenous Peoples Right Act (IPRA) in 1997. However, more efforts are needed to ensure that fees from climbers and other types of NBT can achieve their potential to become a viable revenue stream for effective conservation in the Philippines' PAs.

References

Apollo, M. (2017). The good, the bad and the ugly–three approaches to management of human waste in a high-mountain environment. *International Journal of Environmental Studies, 74*(1), 129–158.

Apollo, M., Mostowska, J., Maciuk, K., Wengel, Y., Jones, T. E., & Cheer, J. M. (2020). Peak-bagging and cartographic misrepresentations: A call to correction. *Current Issues in Tourism.* https://doi.org/10.1080/13683500.2020.1812541

Balmford, A., Aaron, B., Cooper, P., Costanza, R., Farber, S., Green, R. E., Jenkins, M., Jefferiss, P., Jessamy, V., Madden, J., Munro, K., Myers, N., Naeem, S., Paavola, J., Rayment, M., Rosendo, S., Roughgarden, J., Trumper, K., & Turner, R. K. (2002). Economic reasons for conserving wild nature. *Science, 297*(5583), 950–953.

Carandang, A. P. (2012). Assessment of the contribution of forestry to poverty alleviation in the Philippines. In FAO, *Making forestry work for the poor: Assessment of the role of forestry on poverty alleviation in Asia and the Pacific* (pp. 267–292). Regional Office for Asia and the Pacific, FAO.

Chase, L. C., Lee, D. R., Schulze, W. D., & Anderson, D. J. (1998). Ecotourism demand and differential pricing of national park access in Costa Rica. *Land Economics, 74*(4), 466–482.

Clawson, M., & Knetsch, J. L. (1966). *Economics of outdoor recreation.* Johns Hopkins University Press.

Congress of the Philippines. (1997). Indigenous Peoples Right Act (Republic Act 8731).

Corpuz, J. D., & Jones, T. E. (2017). Common goals, different roles in devolved management: A review of conservation policies in Palawan Protected Areas. *Governance Policy Research Network Bulletin, 12*, 69–97.

Cortina-Villar, S., Plascencia-Vargas, H., Vaca, R., Schroth, G., Zepeda, Y., Soto-Pinto, L., & Nahed-Toral, J. (2012). Resolving the conflict between ecosystem protection and land use in protected areas of the Sierra Madre de Chiapas Mexico. *Environmental Management, 49*(3), 649–662.

Dames & Moore Inc. (1994). *Study on the biodiversity and cost benefit analysis of the Mt. Apo geothermal project.* PNOC-Energy Development Corporation.

DENR. (2008). *DENR Administrative Order No. 2008-26: Revised implementing rules and regulations of Republic Act No. 7586 or the National Integrated Protected Areas System (NIPAS) Act of 1992.* Department of Environment and Natural Resources, Quezon City, Philippines.

DENR-PAWB. (2009). *Assessing progress towards the 2010 biodiversity target: The 4th national report to the Convention on Biological Diversity.* Protected Areas and Wildlife Bureau, Department of Environment and Natural Resources, Quezon City, Philippines.

DENR-PAWB. (2012). *Communities in nature: State of protected areas management in the Philippines.* Protected Areas and Wildlife Bureau, Department of Environment and Natural Resources, Quezon City, Philippines.

DENR-PAWB. (2013). *Technical bulletin number 2013-01: List of protected areas under the National Integrated Protected Areas System (NIPAS).* Protected Areas and Wildlife Bureau, Department of Environment and Natural Resources, Quezon City, Philippines.

DENR-PAWB & GIZ. (2011). *An in-depth review of the NIPAS Law and related statutes on the establishment and management of protected areas in the Philippines: A final report.* Protected Areas and WIldlife Bureau, Department of Environment and Natural Resources and the Deutsche

Gesellschaft für Internationale Zusammenarbeit (GIZ) GmbH, Manila, Philippines. http://www.enrdph.org/docfiles/20130228_NIPAS%20Review_FINAL_[web].pdf.

DENR-PAWCZMS. (2013). *The Mount Apo Natural Park updated general management plan 2013–2033*. DENR-PAWCZMS 11, Davao City, Philippines.

Dressler, W. H., Kull, C. A., & Meredith, T. C. (2006). The politics of decentralizing national parks management in the Philippines. *Political Geography, 25*, 789–816.

FAO. (2010). *Global forest resources assessment 2010 country report: Philippines*. FRA2010/164. Food and Agriculture Organization of the United Nations, Rome, Italy.

Forest Resources Assessment Programme (FAO). (2015). Global forest resources assessment 2015. http://fao.org/3/a-i4808e.pdf.

Gomez, A. L. V. (2015). *Identification and valuation of ecosystem services in the Mount Apo Natural Park, the Philippines, as a basis for exploring the potential of 'payments for environmental services' for protected area management* (Unpublished PhD thesis). Charles Darwin University, Australia.

Greiner, R., & Rolfe, J. (2004). Estimating consumer surplus and elasticity of demand of tourist visitation to a region in North Queensland using contingent valuation. *Tourism Economics, 10*(3), 317–328.

Jin, J., Indab, A., Nabangchang, O., Thuy, T. D., Harder, D., & Subade, R. F. (2010). Valuing marine turtle conservation: A cross-country study in Asian cities. *Ecological Economics, 69*(10), 2020–2026.

Mallari, N. A. D., Tabaranza, B. R., & Crosby, M. J. (2001). *Key conservation sites in the Philippines: A Haribon Foundation and BirdLife International directory of important bird areas*. Bookmark.

Mallari, N. A. D., Collar, N. J., McGowan, P. J. K., & Marsden, S. J. (2016). Philippine protected areas are not meeting the biodiversity coverage and management effectiveness requirements of Aichi Target 11. *Ambio, 45*, 313–322.

La Viña, A. G. M., Kho, J. L., & Caleda, M. (2010). Legal framework for protected areas: Philippines. *IUCN-Environmental Policy and Law Papers, 81*. https://www.iucn.org/downloads/philippines.pdf.

Ong, P. S., Afuang, L. E., & Rosell-Ambal, R. G. (Eds.). (2002). *Philippine biodiversity conservation priorities: A second iteration of the national biodiversity strategy and action plan*. DENR-PAWB, CI Philippines, UP-CIDS & FPE, Quezon City, Philippines.

Philippine Statistics Authority (PSA). (2015). *Census of Population, computed from 2010–2015 growth rate*. www.psa.gov.ph.

Pulhin, J. M., & Inoue, M. (2008). Dynamics of devolution process in the management of the Philippine forests. *International Journal of Social Forestry, 1*(1), 1–26.

Rodríguez, J. P., Beard, J. T. D., Bennett, E. M., Cumming, G. S., Cork, S. J., Agard, J., Dobson, A. P., & Peterson, G. D. (2006). Trade-offs across space, time, and ecosystem services. *Ecology and Society, 11*(1). http://www.ecologyandsociety.org/vol11/iss1/art28/.

Schroeder, H. W., & Louviere, J. (1999). Stated choice models for predicting the impact of user fees at public recreation sites. *Journal of Leisure Research, 31*(3), 300–324.

Senga, R. G. (2001). Establishing protected areas in the Philippines: Emerging trends, challenges and prospects. *The George Wright FORUM, 18*(1), 56–65.

Severino, H. (2000). The role of local stakeholders in forest protection. In P. Utting (Ed.), *Forest policy and politics in the Philippines: The dynamics of participatory conservation* (pp. 84–116). Ateneo de Manila University Press.

Shackleton, S., Cambell, B., Wollenberg, E., & Edmunds, B. (2002). Devolution and community-based natural resource management: Creating space for local people to participate and benefit? *Natural Resource Perspectives, 76*, 1–6.

Simon, S. (2013). Of boars and men: Indigenous knowledge and co-management in Taiwan. *Human Organization, 72*(3), 220–229.

Stoeckl, N., Hicks, C. C., Mills, M., Fabricius, K., Esparon, M., Kroon, F., Kaur, K., & Costanza, R. (2011). The economic value of ecosystem services in the Great Barrier Reef: Our state of knowledge. *Annals of the New York Academy of Sciences, 1219*(1), 113–133.

Subade, R. F. (2005). *Valuing biodiversity conservation in a world heritage site: Citizens' non-use values for Tubbataha Reefs National Marine Park, Philippines*. Research Report No. 2005-RR4, Economy and Environment Program for Southeast Asia, Singapore. http://www.eepsea.net/pub/rr/11201034921RodelRR4.pdf.

United Nations Development Program (UNDP). (2012). *Communities in nature: State of protected areas management in the Philippines* (48 pp.). PAs & Wildlife Bureau.

UNESCO World Heritage Centre (WHC). (2011). *Tentative lists: Mount Apo Natural Park*. https://web.archive.org/web/20110311220211/https://whc.unesco.org/en/tentativelists/5485/.

WCED (1987). *Our common future* (G. H. Brundtland, Ed.). The World Commission on Environment and Development. Oxford University Press.

Aurelia Luzviminda V. Gomez is Assistant Professor in the School of Management at the University of the Philippines Mindanao in Davao City, Philippines. Her research interests include nature-based tourism, protected area management, and socio-economic aspects of natural resource management. She obtained her PhD at Charles Darwin University in Australia; her PhD research focused on ecosystem services valuation at Mount Apo Natural Park, Philippines.

Thomas E. Jones is Associate Professor in the Environment & Development Cluster at Ritsumeikan APU in Kyushu, Japan. His research interests include Nature-Based Tourism, Protected Area Management and Sustainability. Tom completed his PhD at the University of Tokyo and has conducted visitor surveys on Mount Fuji and in the Japan Alps.

Chapter 9
Governance and Management of Protected Areas in Vietnam: Nature-Based Tourism in Mountain Areas

Huong T. Bui, Long H. Pham, and Thomas E. Jones

9.1 Introduction

Vietnam extends more than 1,650 km from north to south, from 23°30'N to 8°30'N. Three quarters of the country is hilly or mountainous, with Hoàng Liên in the north and Trư`ờng Sơn in the west being the country's two major mountain ranges. The lowland areas include two major river deltas: the Red River Delta in the north and the Mekong River Delta in the south. A narrow, coastal plain runs along much of the country's coastline. Vietnam's climate is tropical monsoonal, dominated by the southwesterly monsoons from May to October and northeasterly monsoons during the winter months. Annual rainfall averages between 1,300 mm and 3,200 mm, but can vary from as much as 4,800 mm to as little as 400 mm in some areas. The country's flora and fauna combine influences from the Palaearctic realm's Himalayan and Chinese sub-regions with the Indo-Malayan realm's Sundaic subregion (ICEM, 2003).

Vietnam is ranked as the 16th most biodiverse country in the world, hosting 110 Key Biodiversity Areas (Mittermeier et al., 2004). UNESCO Vietnam recognizes that "Vietnam is one of the most important biodiversity ecosystems in the world, both in the marine and terrestrial ecosystems, especially in forest ecology and mangrove forests" (Van, 2019). It is home to several of the world's most iconic species, including the Indochinese tiger (*Panthera tigris*) and the Asian elephant

(*Elephas maximus*). Since 1992, four mammal species have been discovered that were previously unknown to science (Timmins & Duckworth, 2001). Overall, 109 large mammals and about 850 bird species have been recorded in Vietnam, which is also home to between 9,600 and 12,000 plant species (Vo, 1995). This level of diversity is remarkably high for a relatively small country of only 33.12 million hectares. Besides the global importance of its natural biodiversity, Vietnam is also known as one of the richest countries in terms of agricultural biodiversity with over 800 plant species cultivated in diversified agroecological production (USAID, 2013).

Forty-five years after the end of the Vietnam War unifying the country and after more than three decades under dramatic economic reform, governance and natural resources management in Vietnam have experienced dramatic changes alongside the changes in economic policies. In the meantime, many nature-based destinations in mountain areas have opened to tourism activities. However, research on nature-based tourism in the mountain areas of Vietnam remains in its infancy. This chapter aims to provide an overview of mountainous protected areas in Vietnam with reference to classification, governance and management issues. Having analyzed tourism-led development strategy in the northern mountain region of the country, our work opens up a discussion on the role of tourism within the broader context of natural resource governance and policy in Vietnam.

9.2 Protected Areas and National Parks

9.2.1 *Classification*

Using the 1994 International Union for Conservation of Nature (IUCN) Guidelines document as a template, Vietnam developed its own system for the categorization of protected areas (PAs), aiming to deal with the problems raised by the current categorization system, whilst maintaining the values inherent to the IUCN system (Stolton et al., 2004). The Vietnamese system consists of four categories of protected areas:

- *Category I*: national park, a PA managed mainly for ecosystem protection, research, environmental education and recreation, equivalent to IUCN Category II (national park).
- *Category II*: nature reserve, a PA managed mainly for ecosystem or species protection, research, monitoring, recreation and environmental education. There is no direct equivalent to an IUCN category.
- *Category III*: habitat and species management area, a PA managed mainly for environmental and biodiversity conservation through management interventions (with increased provisions for the co-management of resources). It is equivalent to IUCN Category IV (habitat/species management area).

- *Category IV*: protected landscape/seascape, a PA managed mainly for landscape or seascape conservation and recreation. This category is equivalent to IUCN Category V (protected landscape or seascape).

In the Vietnamese vernacular, the forerunners of PAs were initially understood as 'prohibited forests' and from 1960 they were termed 'special-use forests' (SUF) (Stolton et al., 2004). Under the SUF system, Cuc Phuong Protected Forest (now Cuc Phuong National Park) was the first designated PA in Vietnam. Following reunification in 1975, attention focused on identifying and surveying potential protected forests throughout the country. In 1994, Vietnam ratified the Convention on Biological Diversity (CBD), followed by the creation of a Biodiversity Action Plan with a recommendation to strengthen the SUF system. The current SUFs include national parks, nature reserves, species and habitat conservation areas, landscaped protected areas, and experimental and scientific research areas (USAID, 2013) as outlined in Table 9.1.

Table 9.1 Protected areas in Vietnam

Type of protected area	Total number	Total protected area (ha)
National protected areas		
National parks	33	1,093,982
Nature reserves	57	1,044,213
Species/habitat conservation areas (SHCA)	12	38,777
Landscape protection areas (LPA)	53	78,129
Experimental and scientific research areas[a] (ESRA)	9	10,653
Total special-use forest[b]	**164**	**2,265,754**
Marine protected areas[c]	9	172,577 (including 104,098 ha. of Marine area)
Internationally-recognized conservation areas		
Ramsar Wetlands of International Importance	5	84,982
UNESCO Biosphere Reserves	8	n.d.
UNESCO Natural World Heritage Sites	3	n.d.
ASEAN Heritage Parks	4	n.d.
Important Bird Areas	65	n.d.

[a]This is considered to be a type of landscape protection area under the Law on Forest Protection and Development, but is recognized as a separate category of PA under the more recent Law on Biodiversity
[b]Special-use forest includes national parks, nature reserves, SHCAs, LPAs and ESRAs
[c]Three categories of MPA are recognized under the law on Fisheries: national park, aquatic natural reserve, and species and habitat conservation areas. However, six of the 9 existing MPAs are currently established as SUF
Sources USAID (2013) and GIZ and VN Forest (2019)

Forest coverage of Vietnam has changed dramatically under the pressures of prolonged wars and post-war economic development. In 1943, statistics by Paul Maurand (1943, cited in Nguyen & Vu, 2007) estimated the forested area of Vietnam to be 14.3 million ha, covering 43% of the country's territory. From 1943-1975, forest areas declined to 11.2 million ha with a territorial coverage of 34% (FIPI, 1976 cited in Nguyen & Vu, 2007), as much of the center and north of the country was heavily defoliated by aerial application of herbicides during the Vietnam War, with the side effects of these pollutants still being experienced today. From 1976 to 1990, forests were excessively exploited for post-war reconstruction, leading to further reduction of the forest area by two million ha, with the reduced coverage ratio remaining at 27.8%. Since 1990, with the policy of reforestation and incentives by the central government, the forest coverage has increased. Statistics from 2005 show the entire country's forest area has reached 12.6 million ha, covering 37% of the territory (Nguyen & Vu, 2007), with the growth attributed to the efforts of reforestation programs. However, the reforestation program focused almost entirely on plantations of exotic species, which have very low biodiversity value, and remaining forests are often fragmented and degraded. The Forest Investigation and Planning Institution (FIPI) of Vietnam estimates that only 6–8% of remaining cover is primary forest, i.e. old growth forest untouched by human activities (USAID, 2013). Due to the continued reforestation program, forest cover reached 14.45 million ha in 2018, representing some 41.65% of Vietnam's landmass and comprising 10.3 million ha of natural forests and 4.15 million ha plantation forests, respectively (GIZ & VN Forest, 2019).

There are 164 PAs nationwide, including 33 national parks (NPs) (6 under the Vietnam Administration of Forestry and 27 under a Provincial People's Committee or Department of Agricultural and Rural Development), 57 nature reserves, 12 species and habitat conservation areas, 53 landscaped protected areas and 9 experimental and scientific research areas (GIZ & VN Forest, 2019) (see Fig. 9.1). Three quarters of PAs in Vietnam are located in mountainous regions or include mountain peaks. Among the 33 NPs, 22 are designated to be in mountainous regions (see Table 9.4).

The Trư`ờng Sơn (Annamite Range) is the longest mountain range in eastern Indochina. It extends approximately 1,100 km, parallel to the Vietnamese coast, through Laos, Vietnam, and a small area in northeastern Cambodia. The Trư`ờng Sơn range naturally borders Laos to the west and Vietnam to the east, although the Tây Nguyên (Central Highlands) region lies west of the divide; populated with ethnic minorities, the region joined the territory of Vietnam in the mid- twentieth century. As shown in Table 9.4, five NPs in the Central Highland area as well as five other NPs located in the North Central Coast are mountainous. Two NPs (Cát Tiên and Bù Gia Mập) are located on the southeast end of the Trư`ờng Sơn Range. Among the 13 NPs along the Trư`ờng Sơn Range, nature-based tourism has developed in four NPs, namely Phong Nha-Kẻ Bàng, Bạch Mã, Yordon and Cát Tiên.

The Northeast and Northwest regions of Vietnam are also home to the Hoang Lien Mountain Range that borders Vietnam and China to the north, and Laos to the West with the highest peak in Indochina, Fansipan. The Tam Đảo Range marks the southern border of the Northeast region. Additionally, the Sông Mã Range marks the western end of the northwest region. Table 9.4 shows that six out of seven NPs in the

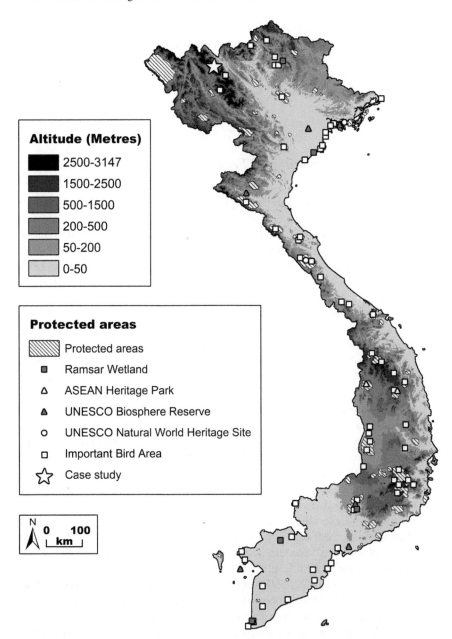

Fig. 9.1 Distribution of protected areas in Vietnam. *Source* Author compilation based on data from USAID (2013)

Northwest region are mountainous NPs. Three NPs in the Red River Delta are also bordered by mountains with well-established tourism activities; these NPs are Ba Vì and Cúc Phương and Tam Đảo owing to proximity to the capital city, Hanoi. In the Northeast region, Hoàng Liên and Ba Bể National Parks have unique nature-based tourism activities, but are not easily accessible from the major tourist hub, the capital city of Hanoi.

9.2.2 Legal and Institutional Framework

Vietnam has a comprehensive legal and regulatory framework for biodiversity conservation. At the national level, it consists of laws, decrees, decisions and resolutions. In addition, each ministry may develop its own sector-level circulars and decisions to guide the implementation of national legislation (USAID, 2013). The passage of the Biodiversity Law in 2008 represented a milestone for conservation, elevating biodiversity to the same legal status as other sectors. Prior to this law, the only references to biodiversity within Vietnam's legal framework were in sectoral laws, such as the Law on Water Resources (1998) and the Law on Forest Protection and Development (2004), as outlined in Table 9.2. Vietnam is a signatory to a number of international agreements relevant to biodiversity and conservation including the Ramsar Convention in 1989, in addition to the Convention on Biological Diversity, the Convention on International Trade in Endangered Species of Wild Fauna and Flora, and the UN Framework Convention on Climate Change, all agreed to in 1994.

The implementation of laws and compliance, however, reveals major problems as identified in a report by USAID (2013). Overlap and conflict between institutional mandates occur due to the fact that laws and decrees are prepared by different ministries. For example, Vietnam's PAs are classified into four types by the Law on Biodiversity; five types by the Environmental Protection Law; five types by the Law on Forest Protection and Development; and three types by the Law on Fisheries. Only two of these categorical types are common across all of the guiding regulations. This inconsistency makes the application of standard methodologies, strategic development and management of PAs problematic. While the complex legal framework is indicative of the state's concern about environment, this interest still needs to be translated into action at local levels of government.

The legal framework delegates PA conservation responsibilities to two ministries. At the national level, the Ministry of Natural Resources and Environment (MoNRE) takes the lead for biodiversity. However, the Ministry of Agriculture and Rural Development (MARD) and its provincial departments are responsible for all SUFs. Meanwhile, the Ministry of Fisheries is responsible for developing a system of marine PAs in Vietnam. While MoNRE is legally responsible for biodiversity conservation, in fact MARD is responsible for the management of SUFs and the provision of technical support services to Provincial People's Committees (PPCs) and lower level administrative units, such as Community People's Committees (CPC) (USAID, 2013). The agencies responsible for managing these areas in the provinces are the Departments

Table 9.2 Environmental laws and regulations

Number	Description	Date
LAWS		
16/2017/QH14	Law on Forestry	15 November 2017
17/2003/QH11	Law on Fisheries Resources Protection	26 November 2003
29/2004/QH11	Law on Forest Protection and Development	14 December 2004
52/2005/QH11	Law on Environmental Protection	12 December 2005
20/2008/QH12	Law on Biodiversity	28 November 2008
DECREES, RESOLUTIONS, DECISIONS		
Resolutions		
41-NQ/TW	Political Bureau's Resolution on environmental protection during the period of accelerated industrialization - modernization of Vietnam	15 November 2004
23/2006/ND-CP	Decree on the implementation of the Law on Forest Protection and Development	3 March 2006
186/2006/QD-TTg	PM's Decision promulgating the Regulation on Forest Management	14 August 2006
119/2006/ND-CP	Decree on organization and operation of the Forest Protection Service	16 October 2006
380/QD-TTg	PM's Decision on the pilot policy on payment for forest environment services	10 April 2008
99/2010/ND-CP	Decree on the policy on payment for forest environment services	24 September 2010
2284/QD-TTg	PM's Decision of the Prime Minister approving the scheme on implementation of Decree 99/2010/ND-CP	13 December 2010
117/2010/ND-CP	Decree on Organization and Management of the Special-Use Forest System	24 December 2010
24/2012/QD-TTg	PM's Decision on investment policy of special-use forest development stages 2011-2020	1 June 2012
Regarding marine protected areas and inland water protected areas		
57/2008/ND-CP	Decree on the promulgation of a regulation governing marine protected areas of national and international importance	2 May 2008
1479/QD-TTg	PM's Decision on approval of the master plan for an inland water protection area system by 2020	7 October 2008
Regarding species protection		
139/2004/ND-CP	Decree on the processing of administrative infringements on forest management and protection and forest product management	25 June 2004
32/2006/ND-CP	Decree on management of endangered, rare and precious forest animals and plants	30 March 2006

(continued)

Table 9.2 (continued)

Number	Description	Date
82/2006/ND-CP	Decree on management of export, import, re-export and introduction from the sea, transit, breeding, rearing and artificial propagation of rare, endangered and precious wild animals and plants (implementation of CITES)	10 August 2006
Regarding biodiversity conservation and wetland conservation		
109/2003/ND-CP	Decree on the Conservation and Sustainable Development of Wetlands	23 September 2003
65/2010/ND-CP	Decree detailing and guiding a number of articles of the biodiversity law	11 June 2010
69/2010/ND-CP	Decree on biosafety for genetically modified organisms, genetic specimens and products of genetically modified organisms	21 June 2010

Source USAID (2013)

of Agriculture and Rural Development, the Departments of Science, Technology and Environment, the Forest Protection Departments, the Fisheries Departments and the Departments of Culture and Information (ICEM, 2003). Table 9.3 outlines the main institutions involved in PA administration in Vietnam.

Since the late 1990s, the emphasis on national forest policy has shifted from production to protection, including the management of forests for conservation, along with local livelihoods and economic development. There has been significant reform of state-owned enterprises, a loosening of government controls in agriculture, and a decentralization of authority to the lowest level appropriate from 1998. This decentralization process has had important implications for PA management. Responsibility for all nature reserves and most national parks has been divested to Provincial People's Committees in accordance with Decision 08/2001/QD-TTg, 11 January 2001 (ICEM, 2003). With decentralization, provincial and local-level government institutions have assumed an important role in the conservation of PAs. Functions and responsibilities of the different institutions at national and provincial levels hint at a lack of clarity in mandates, the fragmentation of decision-making and overlap in responsibilities (USAID, 2013).

9.2.3 Management of Protected Areas

The greatest institutional change over the last 20 years in Vietnam was the devolution of management authority to local government under the *Doi Moi* reforms, which led to a shift in authority and responsibility from the center to the provinces, and by extension to the districts and communes (USAID, 2013). However, at the local level, provinces now have to compete with each other to attract jobs and investment and

Table 9.3 Institutions involved in tropical forest and biodiversity management

Institutions	Main functions and agencies
Ministry of Agriculture and Rural Development (MARD)	Governmental body performing state management functions in the fields of agriculture, forestry, salt production, fishery, irrigation/water services and rural development nationwide. Under MARD, there are a number of line agencies and institutions working on biodiversity conservation: Vietnam Forest Administration Directorate of Fisheries, and Forest Inventory and Planning Institute (FIPI). MARD implements its mandate through a number of national, provincial and district-level implementation entities such as Forest Protection Units.
Ministry of Natural Resources and Environment (MoNRE)	MoNRE is the state agency tasked with national-level administration of water resources, mineral resources, geology, environment, meteorology, hydrology; geodesy and cartography, as well as the management of marine waters and islands. MoNRE is also responsible for the management of natural ecosystems aside from forests, including development and management of natural conservation areas in wetlands and limestone ecosystems
Ministry of Planning and Investment (MPI)	Responsible for national-level planning and investment including that for environmental protection, biodiversity conservation and protected areas
Ministry of Finance	Responsible for overall financial management including the development of the state budget, taxes, fees and other revenues sources, coordinating monetary and fiscal policy, overseeing government accounts, etc The ministry therefore is responsible for the financial aspects of environmental protection, biodiversity conservation and protected areas, including financing of the Vietnam Conservation Fund and Vietnam Environment Fund
Ministry of Education and Training	Responsible for pre-school education, general education, professional education, higher education, and continuing education, including environmental education
Ministry of Culture, Sport and Tourism	Responsible for coordinating ecotourism activities, including those in PAs

(continued)

Table 9.3 (continued)

Institutions	Main functions and agencies
PPCs and provincial level line agencies	Provincial People's Committees are the state administrative organizations at the provincial level. The PPCs are responsible for provincial level budgeting, ensuring adherence to the constitution, national laws, and resolutions of the People's Council on the provincial level. To meet their mandates, the PPCs have management control of the provincial level analogues to national level line ministries, such as the Provincial Departments for Agriculture and Rural Development (DARDs), Departments of Natural Resources and Environment (DoNREs), and the Departments of Fisheries Resources Exploitation and Protection (DoFREPs)
District People's Committees	The analogue of PPCs at the District level. The review of literature did not reveal strong participation of DPCs in forest management
Commune People's Committees	The commune level can be considered a sub-district level. The CPC is the lowest hierarchical level of administration of the Provincial People's Committee (PPC). The commune usually consists of several villages
Commune, community groups citations	A commune or village level group/association may secure use rights over protected forests
Households	Households may obtain use rights over production and protection forests. The rights are usually granted by CPCs. It qualifies households to obtain bank loans
Management Boards	Management boards, usually integrated by technical ministry personnel, may be established to oversee management of protection and special-use forests. Their composition and mandate are flexible. Management boards may be established at different levels. They rely on line ministry personnel (MARD) to implement management

Source USAID (2013)

fill budget deficits. This competition, and the limited ability of central ministries to exert appropriate oversight and control, has resulted in the unregulated use of natural resources and concomitant degradation of natural habitats, increase in large-scale pollution, and species loss.

Responsibility for the management of a PA in Vietnam depends on whether it falls entirely within a single province or if it straddles provincial boundaries. MARD has management responsibility for those PAs that extend beyond more than one province while the PPCs are responsible for PAs contained entirely within one province. This arrangement has resulted in a highly decentralized governance system, with only six national parks directly managed by MARD. Most remaining SUFs are managed at the provincial level. In reality, while the PPCs typically nominate a management board and retain control over PA budgets, they delegate the day-to-day administration of PA operations and conservation activities to the Provincial Department of Agriculture and Rural Development or its Sub-Department of Forest Protection. When the territory of PAs crosses provincial boundaries, they can only be established by agreement among the PPCs from all provinces in which the PA is located (USAID, 2013).

PAs are zoned according to three resource-use categories: strict protection zone, ecological restoration zone, and service-administration zone. Furthermore, a buffer zone is established around the PA, although no formal standards for buffer zone management have been established. Strictly protected zones offer only limited access, while ecological restoration zones are closed to visitors. The last sub-zone, the service-administration zone, not only has provision for management specific facilities but also has envisaged research facilities, facilities for tourism recreation and entertainment. Areas outside the PA are managed as buffer zones to mitigate against the encroachment of human activities into the parks (Jones et al., 2016).

Most PAs have a small core budget that comes from the province while of the 33 national parks in Vietnam, six (Cuc Phuong, Cat Tien, Yordon, Ba Vi, Tam Dao, Bach Ma) receive their budgets directly from the national government (see Table 9.4). However, these funds are seldom enough to cover the PAs' full operations and maintenance costs (USAID, 2013). The Forest Protection Department has identified three components of this funding problem: an overall lack of funding for protected area management; varying annual budget allocations; and an imbalance in investment priorities for PAs with a tendency to attach special importance on infrastructure development while giving insufficient investment priority to conservation (USAID, 2013).

Government funding for SUFs comes from a wide range of funds to 'develop' national parks (both provincial and national). The 5 Million Hectare Reforestation Program is commonly used for forest protection contracts and tree planting in buffer zones, as well as 'ecological regeneration' areas within PA core zones. The 135 Program from the Infrastructure Development Fund administered by Ministry of Planning and Investment can finance local infrastructure, such as schools and irrigation systems, in buffer zones, and is often used to construct buildings and roads inside PAs. Management boards can also apply to the Department of Science, Technology and Product Quality of MARD for funding to support research, although capacity to review prepare proposals to access these funds is limited (ICEM, 2003). International donors, such as WWF, FFI, ILO are another source of funds for conservation management. Most donor support is for large, site-specific projects and is often used for investment and the conservation needs that the government budgets cannot meet. This source of funds is increasingly important and functions as a vehicle to attract foreign researchers and visitors to PAs (ICEM, 2003).

PA financing varies considerably between different SUFs and from province to province. Poorer provinces, where most PAs are situated, are less able to invest in SUF management compared to provinces with stronger economies and local revenue streams. For example, several poorer provinces are required to individually manage

Table 9.4 National Parks in Vietnam

Region	Name	Year established	Area (ha)	Location
Northwest	*Hoàng Liên*	2002	29,845	*Lào Cai, Lai Châu*
	Ba Bể	1992	7,610	*B'ắc Kạn*
	Bái Tử Long	2001	15,783	*Quảng Ninh*
	Xuân Sơn	2002	15,048	*Phú Thọ*
	Tam Đảo	1996	36,883	*Vĩnh Phúc, Thái Nguyên, Tuyên Quang*
	Du Già	2015	15,006	*Hà Giang*
	Phia O'ắc - Phia Đén	2018	10,593	*Cao B`ăng*
Red River Delta	*Ba Vì*	1991	10,815	*Hà Nội, Hòa Bình*
	Cát Bà	1986	15,200	*Hải Phòng*
	Cúc Phương	1962	22,200	*Ninh Bình, Thanh Hóa, Hòa Bình*
	Xuân Thủy	2003	7,100	*Nam Định*
North Central Coast	*Bến En*	1992	14,735	*Thanh Hóa*
	Pù Mát	2001	91,113	*Nghệ An*
	Vũ Quang	2002	55,029	*Hà Tĩnh*
	Phong Nha-Kẻ Bàng	2001	85,754	*Quảng Bình*
	Bạch Mã	1991	22,030	*Th`ừa Thiên–Huế, Quảng Nam*
South Central Coast	*Phư'ớc Bình*	2006	19,814	*Ninh Thuận*
	Núi Chúa	2003	29,865	*Ninh Thuận*
Central Highlands	*Chư Mom Ray*	2002	56,621	*Kon Tum*
	Kon Ka Kinh	2002	41,780	*Gia Lai*
	Yok Đôn	1991	115,545	*Đ'ăk Nông, Đ'ăk L'ắk*
	Chư Yang Sin	2002	58,947	*Đ'ăk L`ắk*
	Bidoup Núi Bà	2004	64,800	*Lâm Đồng*
	Tà Đùng	2018	20,937	*Đ'ăc Nông*
Southeast	*Cát Tiên*	1992	73,878	*Đồng Nai, Lâm Đồng, Bình Phư'ớc*
	Bù Gia Mập	2002	26,032	*Bình Phư'ớc*
	Côn Đảo	1993	15,043	*Bà Rịa–Vũng Tàu*

(continued)

Table 9.4 (continued)

Region	Name	Year established	Area (ha)	Location
	Lò Gò-Xa Mát	2002	18,765	Tây Ninh
Mekong Delta	Tràm Chim	1994	7,588	Đồng Tháp
	U Minh Thượng	2002	8,053	Kiên Giang
	Mũi Cà Mau	2003	41,862	Cà Mau
	U Minh Hạ	2006	8,286	Cà Mau
	Phú Quốc	2001	31,422	Kiên Giang

Source GIZ and VN Forest (2019) (Mountainous NPs are marked in *italics*)

four or more SUFs, but have a small budget for this purpose (IUCN, 2002). Analyzing the structure of PA financing, researchers (An et al., 2018) conclude that central and provincial budgets were crucial funding sources to cover the costs of PA management and conservation. Significant differences were found between the two groups of national parks with respect to central and provincial budgets, as expenditure of the state budget for MARD-managed national parks was higher than for the provincially-managed national parks (Emerton et al., 2011).

9.3 Nature-Based Tourism in the Hoàng Liên National Park (HLNP)

9.3.1 Nature-Based Tourism in NPs

The Vietnamese government has declared nature-based tourism (NBT), often termed ecotourism *(du lịch sinh thái)* in government documents, to be one of the country's key tourism products for development. The ecotourism concept used by the administrators is simply specified by the location in which the tourism activities take place, i.e. inside a PA setting. Therefore, ecotourism is synonymous with the nature-based tourism term used elsewhere in this book.

Nature-based tourism is expected to help offset some of the management costs of PAs, generate income for local populations and promote the acceptance of nature conservation as an indirect driver of economic impact (Buckley, 1999). The legal framework for PA tourism is contained in Article 16 of Decision 08/2001/QD-TTg. It states that PA management boards can organize, lease out or sub-contract the provision of tourism services and facilities to organizations, households and individuals, in compliance with existing financial management regulations and subject to a majority of earnings being reinvested in managing, protecting and developing the protected area (ICEM, 2003). The Vietnam Forestry Development Strategy 2006–2020 (Prime Minister Office, 2007) details the mechanism to utilize ecotourism for the environmental services of protected areas. The action plan to implement the strategy notes that "forestry development planning should be made available to the public, and to pilot and scale up the tendering and leasing … Special-use forest for ecotourism and recreation purposes" (Prime Minister's Office, 2007, p. 25).

Despite the existence of a legal framework for ecotourism inside PAs, there is a lack of parallel institutional arrangements. It is unclear whether the agency responsible for NBT should be its management board, Vietnam's National Administration of Tourism or a district/provincial agency. The private sector's involvement in NBT is also ambiguous (ICEM, 2003). Further clarifying the sectors involved in NBT, Decision 104/2007 QD-BNN dated 27/12/2007 of MARD on the management of ecotourism activities in national parks and nature reserves outlines three forms of businesses in the national parks and nature reserves: (a) businesses self-organized by the management board of the parks; (b) private sector investment in national parks, and (c) public-private partnerships, i.e. joint-ventures for tourism initiatives (Pham, 2016).

With the withdrawal of government subsidies as a result of decentralization, park facilities had to be re-invented as self-sustaining enterprises (Suntikul et al., 2010), or opened up for investment in infrastructure from the private sector. Following the decentralization of park management to provincial and municipal levels, national parks with relatively well-established infrastructure and proximity to tourist hubs put emphasis on economic benefits from entrance ticket sales by maximizing the number of visitors to the park. The majority of visits to Vietnam's PAs are made by domestic tourists (Pham, 2016). This is the case of Cuc Phuong and Ba Vi National Park (Ly & Nguyen, 2017). Another national park with UNESCO World Heritage designation, Phong Nha-Ke Bang has adopted a model of park management that involves collaboration between the public and private sector (Ly & Nguyen, 2017). However, this natural heritage site suffers from vandalism due to visitor overload (Pham, 2016) and a lack of carrying capacity planning tools to cope with growing tourist arrivals.

Decentralization in park management empowers local communities to participate in tourism activities in national parks. However, constraints hindering the development of NBT in Vietnam's national parks derive from the limited capacity of local communities, lack of a comprehensive policy, legal and institutional framework, as well as limited promotional efforts (Pham, 2016). Firstly, local people that reside in the national parks' core and buffer zones tend to be reliant on agricultural production inside the park for their livelihoods and have not yet considered or adapted to tourism as an alternative source of income. This is partly due to a low level of awareness of environmental protection among locals and the majority of domestic visitors. Secondly, many PAs lack a clear nature-based tourism development strategy or action plan. In addition, there has yet to be the establishment of strict standards and criteria to maintain quality of infrastructure, technical facilities and services to meet the stricter requirements of NBT.

Many of the tourists visiting PAs are students on school outings, and there is great potential to expand environmental education activities for them. There is a need, however, to balance the potential costs and benefits of mass domestic tourism and the attractions which tourists come to experience (ICEM, 2003). Group travel is the most popular form of travel arrangement. Group travel is often organized by schools, universities for education purposes or by public or private sector organizations for their workers and family as a part of an annual holiday offered as fringe benefits (Bui & Jolliffe, 2011). Fieldwork conducted for MARD in 2016 by consultant teams (Pham, 2016) reveals that the majority of the visits are day trips or within 2 days and 1 night (Cuc Phuong, Cat Tien, Phong Nha-Ke Bang). International tourists tend to stay in national parks and nature reserves longer than domestic visitors. For example, in 2015, international visitors stayed for an average of 3.1 days compared to just 1.5 days for their domestic counterparts. Visitors to national parks and nature reserves in Vietnam are also unevenly distributed spatially. National parks with easy access from major national roads, good connections to provincial roads and with adequate accommodations regularly receive a large number of visitors. These include Phong Nha-Ke Bang, Ba Vi, Cat Ba and Cat Tien National Parks. Though NBT activities are varied, the majority of activities are concentrated in areas with better infrastructure and convenient access. Despite great potential for more diversified activities, remote areas with greater biodiversity or unique landscapes still struggle to attract visitors, and persuade them to extend their stay in the respective national parks.

Overall, a conventional understanding of nature-based tourism is omnipresent in the current discussion of the nexus of tourism and national parks. Vietnam's NBT market has some unique characteristics, being overwhelmingly dependent on domestic visitors who perceive national parks and NBT experiences as a form of leisure and a getaway from city life. Thus, the value of natural resources for enhancing educational and learning experience is undervalued. A remote mountain area, as demonstrated in the case of Hoang Lien National Park (HLNP) below, has to transform itself to accommodate the trend of NBT for recreation and relaxation as preferred by the en-mass domestic market. Such a transformation must be backed by a resource recentralization policy through tourism investment.

9.3.2 Transformative Nature-Based Tourism Landscape in the Mountain Area of HLNP

The case of Hoang Lien National Park presented below illustrates the controversial effects of policy implementation resulting from policies for government control over mountain resources, diversification of local livelihoods through tourism, and justification for partnership investment in tourism facilities. Above all, numerous questions arise over a mega-scale tourism development project in relation to the protection of biodiversity, landscape and culture at a fragile mountain area.

The Hoang Lien National Park (HLNP) in Vietnam is located southeast of the Himalayas in the northwest of Vietnam within the two provinces of Lai Chau and Lao Cai. Comprising one of the last remnants of native forest in the northern Vietnamese highlands, it was upgraded from its former status of natural reserve to become the Hoang Lien National Park on July 12, 2002. Covering an area of 29,845 hectares, the HLNP comprises 11,875 hectares of strictly protected area, 17,900 hectares for forest regeneration and 70 hectares for service-administration services. The HLNP lies between the altitudes of 380 and 3,143 meters above sea level and covers large swathes of geological, topographical and climatic diversity. The diversity of HLNP's vegetation is characterized by high elevation due to its mountain terrain (Kieu & Nguyen, 2014).

The mountain areas of HLNP are predominantly populated by ethnic minorities that speak local languages and follow traditional lifestyles based primarily on subsistence farming. The livelihood of these local people strongly depends on the availability of natural resources, while the scarcity of arable land coupled with population growth has led to increasing pressure on the forest. Exploitation of forest resources by ethnic minorities, responding to socio-economic pressures, such as conversion of forest to farmland, is generally thought to be the leading cause of rapid forest degradation and deforestation on Southeast Asia (Hoang et al., 2014). However, deforestation is not necessarily associated with poverty. The creation of non-agricultural jobs related to the development of tourism activities in rural areas has been suggested as a viable means to offset pressures on forests (Nyaupane & Poudel, 2011). The shift to tourism-related jobs can generate additional income for households, lead to the abandonment of less-productive farmland and consequently, the spontaneous establishment of secondary forest on former agricultural plots (Gaughan et al., 2009). However, the nexus of tourism development, poverty eradication and conservation do not show evidence of a positive relationship. People living in the HLNP acknowledge the economic potential of the NBT sector in the park, but do not see themselves as actual beneficiaries (Puntscher et al., 2017).

Resources for development of nature-based tourism in HLNP include its natural landscape, biodiversity and diverse culture due to the ethnic minority groups inhabiting in the area. Three NBT strategies are approved by the HLNP administration under the three different forms of NBT mentioned earlier: (a) self-organization of business by the management board of the parks to the core conservation area for research and education; (b) public-private partnership, i.e. joint-venture between park management and tourism businesses establishing a connection from major tracking routes in the outskirts of Sapa to Fansipan; and (c) private sector investment in the national park under an environmental leasing contract for the Fansipan Sapa Cable-Tourist Services Co. Ltd (Lao Cai People Committee, 2019).

The most popular tourist destination in the Northwest region is Sapa district. Although the town is located adjacent to the buffer zone of the HLNP, Sapa is paramount for NBT analysis of HLNP as it is the access hub for HLNP visitors and the base for tourist services. The socio-cultural landscape of Sapa extends beyond that of the HLNP. Under this circumstance, physical zoning of a core-buffer zone for national parks fails to account for the interconnectedness of the natural and socio-cultural environments comprising the tourism landscape, as argued by Jones et al. (2020). Therefore, in this analysis of NBT in the mountainous region of HLNP, the authors constantly anchor the analysis to the tourism base in Sapa District.

Sapa district covers an area of 680 km^2, with a population of 55,900 according to the 2010 national population census (GSO, 2010). The district is home to ethnic minority groups of the H'mong, the Yao, the Tay, the Giay, the Xa Pho. The Tay occupy the fertile valleys and middle altitudes. The other ethnic groups such as H'mong and Yao have settled on steep forest slopes generally above 800 meters. Prior to the 1960s, only a few Kinh lowlanders (Kinh is a major ethnic group account for more than 80% of Vietnamese population) lived in Sapa Town, and more migrated to the town after the 1960s stimulated by the New Economic Zone Policy of the national government (Michaud & Turner, 2006). The Kinh have mainly been involved in administration, tourism, and education and have settled in the district's capital, while most of the other ethnic groups have practiced different types of subsistence agriculture mostly in the form of shifting cultivation (Tugault-Lafleur and Turner, 2009). Today, the ethnic groups cultivate rice on permanent terraced paddy fields; maze and other crops on upland fields, and many households cultivate cardamom under forest cover as a substitute cash crop, after the ban on opium in 1992 (Turner, 2012). The diversity and distribution of ethnic groups along the Hoang Lien Mountain range at different altitudes underpins their roles in the supply chain of mountain tourism. For example, the H'mong group provides porter and guide services for mountaineers, trekkers and adventure tours. The Tay group, with larger houses located in mid-altitude areas, offers homestay services for adventurers. The Yao and Giay ethnic groups often participate in tourism souvenir production, while the majority Kinh group dominates tourism infrastructure and services in the Sapa town lowland areas. Continuing their mountainous lifestyle traditions, ethnic minorities groups inhabit the core and buffer zones of the NPs, and earn their livings from forest products and resources.

Sapa district opened for international tourism in 1993. Tourism is now the most important economic activity in the area, and it generated 58% of the district's GDP in 2010, while the poverty rate decreased gradually from 36% in 2000 to 21% in 2009 (GSO, 2010). Tourism development, however, has changed the landscape of the HLNP. An analysis of land cover change for the period 1993–2014 detected possible associations between forest cover change and socio-economic, cultural and biophysical variables at the village level (Hoang et al., 2014). In detail, between 1993 and 2006, Sapa district experienced a net decrease of forest converted to arable land, although this trend was reversed in the period 2006-2014. Deforestation is lower in villages that have access to alternative income sources, either from cardamom cultivation under forest canopy or from tourism activities (ibid.).

The main challenges are to find a balance between the rapid development of tourism activities and the preservation of the authentic socio-cultural features of the very ethnic minorities that make the area so exotically attractive for tourists. On February 2nd, 2016, the corporate giant Sun Group impressed its Vietnamese audience by inaugurating the 'Fansipan Legend – Indochina Summit' cable car,

Table 9.5 Visitor number and revenue from tourism in Hoang Lien National Park (2015–2019)

Year	Visitor number (people)	Revenue (million dong)	Deposit to national budget (million dong)	Revenue retained for administration (million dong)
2014	68.269	3.081	2.465	616
2015	118.069	7.115	5.692	1.423
2016	107.245	7.240	5.792	1.448
2017	82.925	6.321	5.057	1.264
2018	67.414	5.240	4.192	1.048
2019 (expected)	62.000	5.200	4.160	1.040
Total	505.922	34.197	27.358	6.839

Source Lao Cai People Committee (2019). Report on ecotourism activities in Hoang Lien National Park (2014–2019)

starting at a height of 1500 meters on the outskirts of the town of Sapa, transporting tourists to the summit of Mount Fansipan, the highest peak in Vietnam. In the past, small numbers of Vietnamese and foreign tourists could access Mount Fansipan by a three-day hike. But with the 18-minute cable car ride landing just 143 m from Fansipan's summit, the former wilderness in the middle of Hoang Lien National Park is now replaced with modern structures of restaurants, and entertainment complexes for Vietnamese weekend tourists. The tourist town of Sapa now serves an extra function as the base camp area for the Fansipan Legend cable car. Table 9.5 shows the dramatic increase of visitation to the HLNP in 2015 (when the cable car began trial operation) and 2016 (after the official opening). The number declined in 2017–2018 owing to an increase in cable car fares, which resulted in a lower number of users.

The state government of Vietnam has embraced mass tourism targeted at the growing national middle class, with domestic tourists now more important for the industry than international visitors (Michaud & Turner, 2017). The engineering of modern Sapa can be read as the Vietnamese state's intent to bring these remote highlands under central political control by way of exhaustive 'distance demolishing technologies' with an emphasis on infrastructure projects. This in fact reflects a tendency of recentralization of resource management by diversification of partnerships in the circle of centralization–decentralization–re-centralization observed in Southeast Asian resource politics more broadly (Nevins & Peluso, 2008). As Michaud and Turner (2017) contend: "Sapa is now firmly bound to follow a modernist agenda and reach a new level of integration into state projects, in this case, blending scientific socialism and neoliberal principles on a grand scale. In such a project there is little room left for cultural distinction and local desires" (p. 47). The environmental resource leasing model, within which tourism development policy is directed by the state government, and tourism infrastructure results from investment of private groups such as Sun Group, while targeting the *en masse* domestic market, has proven its function as a mechanism of market-economy with socialist orientation in Vietnam.

9.4 Case Study: Mountain Tourism on Mount Fansipan

Mount Fansipan (Phan Xi Păng), the tallest peak in Indochina (3,143 m) is located some 250 km north of Hanoi in the Hoàng Liên Mountain Range, which marks the southeastern-most extent of the collision between Indian and Eurasian plates that formed the Himalayas. Mount Fansipan is a granite peak formed around 250–260 million years ago between the Permian and Triassic periods. As Fansipan is on the national border between Vietnam and China, climbers must carry personal identification documents at all times.

Mount Fansipan is a popular destination for international trekkers that visit Sapa. A visitor study in 2009 revealed that the purpose of 44% of the 223,045 visits to Hoang Lien National Park involved trekking. Of the trekkers, the number of foreigners, at 78,925, far exceeded the 18,126 domestic visitors. Sixty per cent of foreign visitors to Hoang Lien stated "conquering the Fansipan summit" to be their motivation, compared with only 17% of domestic visitors (Hoang, 2010). The gateway to the mountain is nearby Sapa (1500 m ASL), a district-level market town of Lào Cai Province in Northwest Vietnam. Since 2002, ecotourism projects for the conservation of Hoang Lien National Park were introduced (Manh, 2014). There are three available trails divided by the degree of difficulty as described in Table 9.6. Though Mount Fansipan is open all year round, the months suitable for climbing are from October to May. The period from June to September is rainy season, so conditions can be treacherous. Thus, trekkers are warned against climbing during rainy periods from June to September due to the slippery trail conditions.

Table 9.6 Mountain trekkers' season, trails and costs

Climb season	Open year-round, but the most suitable months are from October to May, especially April to May, when the weather is warm and flowers are in bloom. June to September is rainy season, so conditions can be treacherous
Climbing trails	*Easy*: Starting from Trạm Tôn (1900 meters), the trail is gradually sloped, which attracts the most climbers. The hike takes around 10-12 h to reach to the summit, and can be completed in one day *Moderate*: Starting from San Sả Hồ (1260 meters), the trail is more challenging with obstacles. It takes around 10 hours to reach the summit *Difficult*: Starting from Cát Cát village (1245 meters), this is the longest and most challenging trail, taking around 18 hours to reach to the summit
Costs (to hike)	National Park entrance fee: VND 150,000 per day Toilet fee: VND 5,000 per day Trekking fee: VND 30,000 per day Insurance fee: VND 5,000 per day Hiring porter: VND 200,000 per person per day Package fee from tour company of around VND 1,450,000 (based on 2 persons). This covers guide fees, park fees, and pick up/drop off transfer
Cable car	A one-way adult ticket costs VND 700,000

Source Authors' compilation

Tourism activities on Mount Fansipan have changed dramatically since 2016, with a cable car to the summit of Fansipan and the construction of a highway link from Hanoi to Sapa, which has likely accelerated the number of visitors to Sapa, the base area of Mount Fansipan (Michaud & Turner, 2017). The cable car construction was justified by its economic benefits for locals, including job creation and poverty alleviation, and was completed in 2016. Ascending 6,293 meters up the Muong Hoa Valley, the Fansipan Cable Car claims itself to be the world's longest three-rope, non-stop cable car. It also holds the record for the greatest elevation gain with a combined 1,410 meters for the three-rope cable cars. Users pay VND 700,000 (one-way adult fare) to alight near the summit and walk the final 600 stairs to the peak. The ride takes about 18 min, and operates daily from 7am to 6 pm. Not only does the cable car dramatically shorten the time to reach the summit (reduced from 3-days trekking to a 15-minute cable car ride), the cable car also makes the summit accessible to people of all ages who can afford to pay VND 700,000 for one way (or VND 1,400,000 for a return ticket). Obviously, poor local ethnic minorities who have been living on the mountain for generations are not target consumers and beneficiaries from this development initiative. The modernization of the mountain area with technology also calls for questioning the notion of nature-based tourism in the northern highlands of Vietnam.

9.5 Conclusion

This chapter offers a contemporary review of governance and management of protected areas in the context of NBT initiatives in Vietnam. Meta-analysis of the legal and institutional framework of PAs reveals issues related to overlapping laws and regulations, leading to an unclear division of responsibilities among institutions. In this context, the decentralization of PA management to provincial and municipal levels, combined with limited funding and further governance difficulties, pose formidable barriers to fulfilling the goal of biodiversity protection.

Despite great potential, nature-based tourism in Vietnam's PAs remains at an embryonic stage. While NBT is an alternative income source for local communities, most of the activities are rather basic and limited in scope and scale. The current legal and institutional framework for PA governance hinders the development of the ecotourism sector to its full potential. While opportunities come with the decentralization of governance and management of national parks, challenges also arise when local communities and authorities are not provided with adequate support, financially or technically, to establish a functioning management system, making ecotourism either unsustainable or unable to develop to its full potential.

Different scenarios for tourism development in PAs and national parks in Vietnam are discussed in this chapter. Firstly, tourism activities in the national parks flourish as a result of decentralization policies towards park management, allowing national park administrators to open resources to alternative income sources and investment from the private sector under state-directed policy. However, the top-down, centralized governance system and unclear boundary of responsibility towards tourism management have hindered the effectiveness of this policy. Secondly, tourism development in mountainous PAs such as Hoang Lien encompasses a range of issues concerning not only fragile natural environment but also diverse cultural contexts of the ethnic minorities who have long inhabited such areas. NBT development provides alternative livelihoods making locals less dependent on forest resources; however, it may widen the socio-cultural gaps between different ethnic groups participating in the supply chain of the tourism industry when it comes to benefit and resource sharing. Furthermore, centralized power and a unique market economy with socialist orientation may prioritize certain investment and infrastructure projects even in fragile mountainous ecosystems with little concern for the sustainability of natural and cultural resources. The discussion of the utilization of natural resources for NBT development, therefore, demands much further debate and increased transparency in governance and decision-making.

References

An, L. T., Markowski, J., & Bartos, M. (2018). The comparative analyses of selected aspects of conservation and management of Vietnam's national parks. *Nature Conservation, 25,* 1–30. https://doi.org/10.3897/natureconservation.25.19973.

Buckley, R. (1999). Planning for a national ecotourism strategy in Vietnam. *Workshop on Development of a National Ecotourism Strategy for Vietnam.* Hanoi, Vietnam.

Bui, H. T., & Jolliffe, L. (2011). Vietnamese domestic tourism: An investigation of travel motivation. *Austrian Journal of Southeast Asian Studies, 4*(1), 10–29.

Emerton, L., Bishop, J., & Thomas, L. (2006). *Sustainable financing of Protected Areas: A global review of challenges and options.* Gland, Switzerland and Cambridge, UK: IUCN.

Emerton, L., Pham, X. P., & Ha, T. M. (2011). *PA Financing mechanism in Vietnam: Lessons learned and future direction.* Hanoi: GIZ.

Gaughan, A. E., Binford, M. W., & Southworth, J. (2009). Tourism, forest conversion, and land transformations in the Angkor basin, Cambodia. *Applied Geography, 29*(2), 212–223.

GIZ & VN Forest (2019). *Report of Management of SUFs, PAs and the Solutions for Sustainable Development,* Hanoi December, 2019.

GSO. (2010). *Niên Giám Thống Kê – Tổng Điều Tra Dân Số 2010* [National Vietnamese census— Year book 2010]. Hanoi: General Statistics Office.

Hoang, T. T. (2010). *Bước đầu nghiên cứu hoạt động du lịch Trekking tại vườn quốc gia Hoàng Liên theo quan điểm du lịch sinh thái* [Exploratory study on trekking tourism in Hoang Lien National Park: An ecotourism view]. Retrieved from http://luanvan.net.vn/luan-van/de-tai-buoc-dau-nghien-cuu-hoat-dong-du-lich-trekking-tai-vuon-quoc-gia-hoang-lien-theo-quan-diem-du-lich-sinh-thai-1519/.

Hoang, H. T. T., Vanacker, V., Van Rompaey, A., Vu, K. C., & Nguyen, A. T. (2014). Changing human—Landscape interactions after development of tourism in the northern Vietnamese Highlands. *Anthropocene, 5,* 42–51.

ICEM. (2003). *Vietnam National Report on Protected Areas and Development. Review of Protected Areas and Development in the Lower Mekong River Region.* Indooroopilly, Queensland.

IUCN. (2002). *Financing mechanisms study phase II. Second mission report.* IUCN—The World Conservation Union unpublished report to FPD, MARD, UNOPS and UNDP, Hanoi.

Jones, T., Bui, H., & Ando, K. (2020). Zoning for World Heritage sites: Dual dilemmas in development and demographics. *Tourism Geographies* (in press). Retrieved from https://doi.org/10.1080/14616688.2020.1780631.

Jones, T., Dinh, N., & Taranov, I. (2016). Comparing approach to environmental governance: A review of protected areas and nature-based tourism in Kyrgyzstan and Vietnam. *Meiji Journal of Governance Studies, 3*, 43–57.

Kieu, Q. L., & Nguyen, T. T. (2014). Study on the distribution of characteristics of the vegetation in high elevations in Hoang Lien National Park of Vietnam. *Journal of Vietnamese Environment, 6*(2), 84–88.

Lao Cai People Committee. (2019). *Report on ecotourism activities in Hoang Lien National Park (2014–2019)*. Lao Cai: Hoang Lien National Park.

Ly, T. P., & Nguyen, T. H. H. (2017). Application of carrying capacity management in Vietnamese national parks. *Asia Pacific Journal of Tourism Research, 22*(10), 1005–1020.

Manh, N. V. (2014). *Hợp tác Lào Cai - Aquitaine - 5 dấu ấn nổi bật* [Lao Cai International Cooperation—Five Major Projects]. Retrieved from http://dulichlaocai.vn/1558/TinChiTiet/Hop-tac-Lao-Cai–Aquitaine–5-dau-an-noi-bat.pvd.

Michaud, J., & Turner, S. (2006). Contending visions of a hill-station in Vietnam. *Annals of Tourism Research, 33*(3), 785–808.

Michaud, J., & Turner, S. (2017). Reaching new heights. State legibility in Sa Pa, a Vietnam hill station. *Annals of Tourism Research, 66*, 37–48.

Mittermeier, R. A., Robles Gil, P., Hoffmann, M., Pilgrim, J. D., Brooks, T. M., Mittermeier, C. G., et al. (2004). *Hotspots revisited: Earth's biologically richest and most endangered ecoregions*. Mexico City: CEMEX.

Nevins, J., & Peluso, N. L. (2008). Introduction. In J. Nevins & N. L. Peluso (Eds.), *Taking Southeast Asia to market: Commodities, nature, and people in the neoliberal age* (pp. 1–26). Ithaca: Cornell University Press.

Nguyen, H. D. & Vu, V. D. (2007). Bảo tồn đa dạng sinh học ở Việt Nam- mối liên hệ v´ới Phát triển bền v˜ứng (SD) và biến đổi khí hậu [Biodiversity conservation in Vietnam in relation to sustainable development and climate change]. *International Symposium on Biodiversity and Climate Change*, May 22–23, Hanoi.

Nyaupane, G. P., & Poudel, S. (2011). Linkages among biodiversity, livelihood, and tourism. *Annals of Tourism Research, 38*(4), 1344–1366.

Pham, H. L. (2016). Sustainable ecotourism development in protected areas and national Parks in Vietnam. *International conference: Sustainable tourism development: The role of government, business and educational institution* (pp. 117–132). National Economic University Press.

Price Minister Office. (2007). *Strategy for forest development 2006–2020*. Vietnam.

Puntscher, S., Duc, T. H, Walde, J., Tappeiner, U., & Tappeiner, G (2017). *The acceptance of a protected area and the benefits of sustainable tourism: In search of the weak link in their relationship* (Working Papers in Economics and Statistics). University of Innsbruck. http://eeecon.uibk.ac.at/.

Stolton, S., Nguyen, D., & Dudley, N. (2004). *Categorizing protected areas in Vietnam. In Protected Areas Program—Parks, 14*(3). Protected Area Categories, IUCN.

Suntikul, W., Butler, R., & Airey, D. (2010). Implications of political change on national park operations: Doi Moi and tourism to Vietnam's national parks. *Journal of Ecotourism, 9*(3), 201–218.

Timmins, R. J., & Duckworth, J. W. (2001). Priorities for mammal conservation in the ROA. In M. C. Baltzer, Nguyen Thi Dao, and R.G. Shore (Eds.), *Towards a vision for biodiversity conservation in the forests of the lower Mekong Ecoregion Complex*. WWF Indochina/WWF US, Hanoi and Washington, DC.

Tugault-Lafleur, C., & Turner, S. (2009). The price of spice: Ethnic minority livelihoods and cardamom commodity chains in upland northern Vietnam. *Singapore Journal of Tropical Geography, 30*(3), 388–403.

Turner, S. (2012). Making a living the Hmong way: An actor-oriented livelihoods approach to everyday politics and resistance in upland Vietnam. *Annals of the Association of American Geographers, 102*(2), 403–422.

USAID. (2013, August). *Vietnam tropical forest and biodiversity assessment: US Foreign Assistance Act, Section 118–119—Report*.

Van, N. T. (2019, December). *Biodiversity conservation in special use forests in Vietnam—The efforts of WWF and its recommendations*. National Symposium on Management of SUFs, PAs and the Solutions for Sustainable Development, Hanoi.

Vo, Q. (1995). Conservation of Flora, Fauna and endangered species in Vietnam. *Tropical Forest Ecosystems*, vol. 55 BIOTROP Special Publication.

Huong T. Bui is Professor of Tourism and Hospitality cluster, the College of Asia Pacific Studies, Ritsumeikan Asia Pacific University (APU), Japan. Her research interests are Heritage Conservation; War and Disaster-related Tourism, Sustainability and Resilience of the Tourism Sector.

Long H. Pham is Associate Professor and Dean of the Faculty of Tourism Studies, University of Social Sciences and Humanities, Vietnam National University in Hanoi. His major research interests are ecotourism, community-based tourism, and sustainable tourism development. He is also a leading consultant of tourism development in protected areas in Vietnam for international organisations such as British Council Vietnam, KOICA, GIZ, USAID, ILO and JICA.

Thomas E. Jones is Associate Professor in the Environment & Development Cluster at Ritsumeikan APU in Kyushu, Japan. His research interests include Nature-Based Tourism, Protected Area Management and Sustainability. Tom completed his PhD at the University of Tokyo and has conducted visitor surveys on Mount Fuji and in the Japan Alps.

Chapter 10
Mountainous Protected Areas in Myanmar: Current Conditions and the Outlook for Nature-Based Tourism

Yana Wengel, Nandar Aye, Wut Yee Kyi Pyar, and Jennifer Kreisz

10.1 Introduction

Establishing protected areas (PAs) is indispensable for the conservation of global biodiversity (Brooks et al., 2004) and regarded as an essential mechanism for ecological protection (Locke & Dearden, 2005; McDonald & Boucher, 2011). With global inspiration from the UN Sustainable Development Goals 14 and 15 that highlight the importance of PAs (UN, 2019), there has been a widespread global understanding of the importance of protecting the natural ecosystem, biodiversity and endangered species to ensure ecological security and sustainable development due to increased exploitation of natural resources (Dearden et al., 2005; Leroux et al., 2010). Having a range of PA options can alleviate human pressures, providing enhanced protection of biodiversity and the sustainable use of resources (MacKinnon & Hatta, 1996). Poignantly, biodiversity conservation and the effectiveness of natural resource management are critical to ensure the livelihood of local people, where land degradation and poverty are fundamentally connected (Beffasti & Galanti, 2011). Since PAs in tropical regions are largely established on the traditional habitats of indigenous communities (Arriagada et al., 2016; Ferraro et al., 2011), the conservation predominantly depends on the social acceptability and economy of these habitually less affluent communities reliant on PAs resources (Adams et al., 2004; Godde et al., 2000; Voyer et al., 2015).

Y. Wengel (✉) · J. Kreisz
Hainan University—Arizona State University Joint International Tourism College, Haikou, China

N. Aye
National Kaohsiung University of Hospitality and Tourism, Kaohsiung, Taiwan

W. Y. K. Pyar
Mother Earth Tourism Specialists Myanmar, Yangon, Myanmar

LuxDev Aid & Development Agency, Yangon, Myanmar

Tourism is widely acknowledged as a way to stimulate local economic development for developing countries (Fayos-Solà, 2005), although environmental degradation resulting from tourism has been a critical factor for developing countries, mainly because of the constitution of nature-based tourism (NBT) (Leroux et al., 2010). Local communities in remote mountain and forest areas are habitually less affluent and live on a high dependency of natural resources. To gain more benefits from tourism to PAs recommend to include PAs into destination development and design of regional tourism destinations (Eagles & Hillel, 2008).

The UN Department of Economic and Social Affairs classified Myanmar as a Least Developed Country (DESA, 2018). Myanmar's government works not only towards the improvement of economic and medical systems, poverty alleviation and modernisation of education, but also towards sustainable natural resource and environmental management of the country's vast environmental resources situated within the natural habitats of the diverse cultural groups (World Bank, 2017). As one of the most biologically diverse regions in Asia, geographically, Myanmar occupies the Indo-Burma Biodiversity Hotspot with high levels of biodiversity and endemism for its irreversibility, irreplaceability and threatened species (Mittermeier et al., 2004; Myers et al., 2000). Furthermore, Myanmar has the highest forest coverage in Southeast Asia with more than 48% of forests stretching over the country's land area of 338,033 km^2 (Yang et al., 2019).

A historical forerunner to PAs was established around Buddhist monasteries from the eleventh century with the emergence of a conservation concept (Aung, 2007). Supporting rich biodiversity in its variety of habitats that range from snow-capped mountains to coral reefs, the PAs of Myanmar conserve spectacular natural, cultural and spiritual values, providing recreational and educational opportunities for the local communities (Beffasti & Galanti, 2011). Several vital ecosystems of global importance appear in Myanmar's 8 unique biodiversity corridors (Tordoff et al., 2005) which contain a great richness of species in mountain areas such as north of Kachin, Shan and Chin State (Khine & Schneider, 2020). High levels of biodiversity, compared to the relatively low population, is providing the country with an excellent opportunity for conserving a wide variety of species (Renner et al., 2015).

The remote areas of Myanmar's hill jungle, Chin, Kachin and Shan States, mostly untouched by humans, are represented by primary forest and are 'carte blanche' for most biodiversity still to be discovered (Renner et al., 2015). The PAs' unique combination of geography with the presence of high mountains, rivers, lakes, and seasonal rainfalls provide protection and habitat to roughly 250 mammal species, more than 1,000 birds, 370 reptiles, and 7,000 plants, including 39 species of mammals, 45 of birds, 21 of reptiles, and 38 of critically endangered plants (NCEA, 2009). However, its rich biodiversity continues to be threatened by increasing demographic and socio-economic pressures (Myers et al., 2000; NCEA, 2009; Zhang et al., 2018).

The main challenges are linked to deforestation through legal and illegal teak trade, which has resulted in habitat destruction and the decline of wildlife population (Bhagwat et al., 2017; Leimgruber et al., 2011). Between 2000 and 2015, Myanmar was amongst the global top three endangered countries with an annual net loss of 5460 km^2 forest area, the equivalent to a loss of 1.8% per year (MacDicken et al., 2016). Furthermore, the implementation of conservation actions in Myanmar is challenged by a lack of comprehensive management plans, budget, and quality and quantity of staff (Beffasti & Galanti, 2011). Furthermore, wildlife is threatened by excessive hunting caused by poverty and population growth (Rao et al., 2002; Wen et al., 2017). Smiley Evans et al. (2020) highlight that the establishment of a strong understanding of human-wildlife livelihood relationships is necessary for sustainable conservation management policies, which can ensure that humans and wildlife coexist symbiotically.

This chapter provides an overview of the conservation system, governance and management policy, and practice in Myanmar's PAs. By looking at areas over 1000 metres we describe mountain sites potentially suitable for NBT activities. We provide insights into integrating NBT in the conservation of PAs with community encouragement and involvement, and describe the case of Myanmar's most popular and accessible peak, Mount Natma Taung.

10.2 Myanmar's National System of Protected Areas

10.2.1 Overview of Myanmar's Systems of Protected Areas

Myanmar's PA system is guided by a new law since 2018 (MONREC, 2018). One distinct change stipulated in the Conservation of Biodiversity and Protected Areas Law is the possibility to establish 'buffer zones' in cooperation with the local community to maintain a balance between sustainable socio-economic development of communities and the conservation of biodiversity. Furthermore, the law adds a new category of 'community protected areas'.

As of 2020, Myanmar has 46 designated PAs and 21 proposed PAs with 41,156.07 km^2 in total land and marine area covering 6.1% of the entire country's area (Table 10.1). As a signatory to the UN Convention on Biological Diversity, Myanmar aims to impove conservation efforts and to expand PAs to 17% of terrestrial and inland water areas and 10% of coastal and marine areas by 2020 (Forest Department, 2015). However, as of June 2020, the target had not been met.

Myanmar's PAs have seven categories:

1. Scientific nature reserve
2. National park
3. Marine national park
4. Nature reserve
5. Wildlife sanctuary
6. Geo-physically significant reserve area
7. Community protected area.

Table 10.1 Myanmar's protected areas

ID	Site name	Designation	Year	Area (km^2)	Altitude in metres	Key protected species
Kachin State						
1	Pidaung	Wildlife Sanctuary	1927	122.07	155–665	Barking deer, Wildboar, Reptiles
2	Hkakaborazi*	National Park	1998	3812.46	900–5,900	Takin, Musked deer, Red panda
3	Hponkanrazi*	Wildlife Sanctuary	2003	2703.95	295–5,165	Gibbon, Wild dogs, Mongooses
4	Indawgyi	Wildlife Sanctuary	2004	814.99	105–1,400	Sambar, Serow, Goral
5	Hukaung Valley*	Wildlife Sanctuary	2004	6,371.37	185–3,435	Tiger, Sambar, Serow
6	Bumhpabum*	Wildlife Sanctuary	2004	1,854.43	140–3,435	Elephant, Leopard, Jackal
7	Inkhaingbum*	National Park	2017	300.52	No data	Hog deer, Pangolin, Leaf deer, Red goral
8	Imawbum*	National Park	2020	1,562.8	No data	Bears, Leopard, Red Goral, Jungle cats
Sagaing Region						
9	Chatthin	Wildlife Sanctuary	1941	269.36	165–260	Eld's deer, Sambar, Barking deer
10	Alaungdaw Kathapa*	National Park	1989	1,402.79	135–1,335	Tiger, Leopard, Elephant
11	Minwuntaung	Wildlife Sanctuary	1972	205.88	75–305	Barking deer, Hog deer,
12	Htamanthi*	Wildlife Sanctuary	1974	2150.73	105–2,465	Tiger, Elephant, Sambar, Macaque
13	Hukaung Valley* (extension)	Wildlife Sanctuary	2010	11,002.19	125–3,255	Bear, Wildboar, Serow
Chin State						
14	Natmataung*	National Park	2010	713.15	740–3,070	Gaur, Serow, White-browed Nuthatch
15	Kyauk Pan Taung*	Wildlife Sanctuary	2013	130.61	25–1,310	Wild cats, Barking deer, Wild boar

(continued)

Table 10.1 (continued)

ID	Site name	Designation	Year	Area (km^2)	Altitude in metres	Key protected species
16	Bwe Par Taung	National Park	2019	176.16	No data	
Shan State						
17	Taunggyi*	Bird Sanctuary	1920	7.228	1,045–1,750	Avifauna
18	Inlay Lake*	Wildlife Sanctuary	1985	533.72	830–1,270	Waterbirds, Migratory birds, Crane
19	Loimwe	Protected Area	1996	41.36	925–1,920	Bear, Pangolin, Avifauna
20	Parsar*	Protected Area	1996	76.51	370–1,105	Jungle fowl, Chinese pangolin, Avifauna
21	Panlaung-Pyadalin Cave*	Wildlife Sanctuary	2002	333.8	150–1,555	Clouded leopard, Serow, Gibbon, Avifauna
Mandalay Region						
22	Shwe-U-Daung*	Wildlife Sanctuary	1927	175.96	180–1,845	Banteng, Sambar, Macaque
23	Pyin-Oo-Lwin*	Bird Sanctuary	1927	127.35	975–1,210	Barking deer, Avifauna
24	Popa Taung*	Mountain Park	1989	128.54	285 – 1,490	Barking deer, Wildboar, Dusk leaf monkey
25	Lawkananda	Wildlife Sanctuary	1995	0.47	45–70	Myanmar star tortoise, Eld's deer, Avifauna
26	Minsontaung	Wildlife Sanctuary	2001	22.56	195–375	Barking deer, Myanmar star tortoise, Jackal, Wild cat
Rakhine State						
27	Rakine Yoma*	Elephant Reserve	2002	1,755.7	20–1,270	Elephant, Gaur, Leopard
Kayin State						
28	Kahilu	Wildlife Sanctuary	1928	160.55	20–260	Serow, Mouse deer, Hog deer
29	Mulayit*	Wildlife Sanctuary	1935	138.54	80–2010	Barking deer, Wildboar, Macaque

(continued)

Table 10.1 (continued)

ID	Site name	Designation	Year	Area (km^2)	Altitude in metres	Key protected species
30	Htuangwi Taung	Geo-features Significant area	2018	427.35	No data	Barking deer, Bear, Macaque
31	Se Taung	Wildlife Sanctuary	2018	233.82	No data	Macaque, Wild boar, Jungle cat
32	Essathaya Cave	Geo-features Significant area	2019	0.303	No data	Porcupine, Pangolin, Bat, Cobra
Bago Region						
33	Moeyungyi Wetland	Wildlife Sanctuary	1988	103.6	0–30	Migratory birds
34	North Zamrari	Wildlife Sanctuary	2014	983.13	No data	Clouded leopard, Gaur, Bear, Banteng
Magwe Region						
35	Wethtikan	Wildlife Sanctuary	1939	3.73	0–35	Waterbirds
36	Shwesettaw	Wildlife Sanctuary	1940	464.09	56–555	Eld's deer, Sambar, Wild dog
37	Chungponkan	Wildlife Sanctuary	2013	1.5	No data	Myanmar golden deer, Wildcat, Jackal
Yangon Region						
38	Hlawga	Nature Park	1989	6.24	No data	Sambar, Barking deer, Hog deer
Ayeyerwaddy Region						
39	Thamihla Kyun	Wildlife Sanctuary	1970	136.7	0–35	Marine turtle, Waterbirds
40	Meinmahla Kyun	Wildlife Sanctuary	1994	137.58	0–30	Crocodiles, Sea birds
Thanintharyi Region						
41	Moscos Island	Wildlife Sanctuary	1927	49.2	No data	Barking deer, Sambar, Waterbirds
42	Lanpi Island	National Park	1996	204.48	No data	Elephant, Pangolin, Marine biotics
43	Tanintharyi	Nature Reserve	2005	1,699.99	No data	Tapir, Gurney's Pitta, Bear,

(continued)

Table 10.1 (continued)

ID	Site name	Designation	Year	Area (km²)	Altitude in metres	Key protected species
Mon State						
44	Kelatha Wildlife Sanctuary	Wildlife Sanctuary	2002	22.46	0–355	Sambar, Barking deer, Wild boar
45	Kyaikhtiyo Wildlife Sanctuary*	Wildlife Sanctuary	2001	156.2	50–1,090	Goral, Gaur, Sambar
46	Hpabaubg Taung	Nature Reserve	2018	1,891	No data	Bat, Black-crowned night heron

Source Authors' elaboration based on data from MONREC

Notably, alternative designation exists as certain PAs do not fall within the seven categories proposed by the new law and have a 'descriptive' designation instead, for example, 'bird sanctuary.'

The majority of PAs in Myanmar are wildlife sanctuaries (under the International Union for Conservation of Nature Category IV); there are seven ASEAN Heritage Parks, two UNESCO-MAB Biosphere Reserves, and four Ramsar sites (wetlands of international importance). Based on available information, in total, 21 PAs reach an altitude of over 1,000 m above sea level (marked with '*' in Table 10.1) and could be suitable for the development of mountain NBT. However, many of these areas are off-limits to international, and, in some cases, domestic tourists, due to socio-political challenges and lack of accessibility (MLIP, 2020).

10.2.2 Governance and Management of Protected Areas

Myanmar's first conservation history originated in the eleventh century, while first systemised environmental protection was captured in 19 legal documents during the colonial period (Aung, 2007). Despite the economic and socio-political struggle of the post-colonial period, the Forest Department expanded the PA system between 1981-1984 as a part of the Nature Conservation and National Park Program (Beffasti & Galanti, 2011; MRBC, 2018). Nowadays, the Ministry of Natural Resources and Environment Conservation (MONREC) is responsible for environmental policy and management on national, regional, and local levels. The management of PAs is embodied by the new Biodiversity and Conservation of Protected Areas Law (2018). Furthermore, the law is enforced by the targets of the Myanmar Sustainable Development Plan (2018–2030), which focuses on policy development to improve protection and management of the natural environment and ecosystems through increased conservation efforts, improved infrastructure, and increased enforcement against illegal natural resource-related practices, for example, poaching, illegal wildlife trade and logging (Ministry of Planning and Finance, 2018).

MONREC is responsible for the overall management and PAs administration. Specifically, Forest Department is responsible for monitoring natural conservation of ecosystems and species, biodiversity and natural habitats and its operational unit, Nature and Wildlife Conservation Division (founded in 1985) overseas all PAs (Aung, 2007; Forest Department, 2020; MONREC, 2018). In terms of international cooperation, since recently, Myanmar initiated conservation cooperation with international development organisations, UN agencies and international non-governmental organisations (NGO) and as of July 2019 there are 33 ongoing conservation projects (Kyaw, 2020). Additionally, private sector businesses contribute to conservation efforts. For example, Moatama Gas Transportation Company (Total S.A.), the Tanintharyi Pipeline Company (Petronas) and PTT Exploration and Production provide funding for Tanintharyi Nature Reserve, while the Htoo Group of Companies supports Hlawga Park, Pyin Oo Lwin Botanical Garden, Nay Pyi Taw, and Yadanabon and Yangon Zoological Gardens (Myanmar Centre for Responsible Business, 2018).

Management and conservation of PAs are complex processes which require participation from a range of different stakeholders (Dearden et al., 2005). In Myanmar, private-public partnerships for conservation are limited as historically the PAs are managed by the divisions of MONREC, but in 2010 the Hlawga Wildlife Park site management changed to joint management between government and private companies (Beffasti & Galanti, 2011). Myanmar Ecotourism Policy and Management Strategy, released in 2015, prioritises the development of ecotourism to provide environmental, social and economic benefits for destination community and explores the opportunities of public-private partnerships in tourism (MONREC & MOHT, 2015). Furthermore, community forestry initiatives have been recently introduced, and community groups are participating in the management of some PAs (Tint et al., 2011, 2014).

Involvement of the communities can help to gain a deeper understanding of the area, local resources, and relevant issues. Community participation and interactions with conservation stakeholders facilitate the creation of enabling the environment for the community to have a voice in decision-making over aspects of which affect their livelihoods. For instance, Myanmar's new conservation law and National Biodiversity Strategy and Action Plans highlight the importance of community participation in effective conservation of PAs (Kyaw, 2015; MONREC, 2018) and stipulate an emphasis on the role of the local community in their participation. The law has placed Community Protected Areas under the category of PAs and the Forest Department is responsible for the technical support and guidance of the Community Protected Areas management. To enable more active involvement of the community, several international organisations have provided capacity-building initiatives aimed at the development of effective conservation management (Hockings et al., 2018; Khine & Schneider, 2020).

Overall, the Forest Department has implemented several conservation initiatives designed to promote community participation. Some initiatives include designated 'buffer zones' for the local communities' socio-economic activities (Myanmar Centre for Responsible Business, 2018). For example, local communities from Alaungdaw Kathapa and Natma Taung National Parks joined biodiversity conservation training, participated in conservation activities and provided their feedback towards the draft of the conservation plan (Win, 2018). However, despite the engagement of local communities in some PAs, conflicts of interest have arisen. Due to the lack of resources and limited opportunities for sustainable living, communities depending on the natural resources of protected areas have been involved in illegal deforestation and poaching to sustain their livelihoods (N. Aye, personal communication, May 17, 2020). Furthermore, political instability and civil conflict, especially in mountainous regions, are some of the main obstacles to effective community involvement, subsequently resulting in no current Community Protected Area in the existing PAs.

Previous research suggests the positive effect of community involvement and states that community-managed forests have similar deforestation rates to state-managed PAs (Hayes, 2006; Porter-Bolland et al., 2012). Oldekop et al. (2016) claim that national parks' conservation management is more successful when communities are involved in the process. Hence, regardless of the legal status and definition of PAs, the conservation rules are understood and recognised by the PA users, including local communities and tourists, are crucial for effective conservation. Allendorf et al. (2018) emphasised the need for the Government of Myanmar and civil actors to give precedence to policies and programs promoting the institutionalisation of the involvement of local communities. Thus, to ensure the livelihood of vulnerable communities and their effective conservation management, a potential solution to the problem could be the establishment of clearly defined and well-managed buffer zones that meet the requirements of neighbouring communities as well as involve the residents in decision-making and co-management of PAs (J. Kimengsi et al., 2019a).

10.2.3 Tourism in Myanmar's Protected Areas

The Ministry of Hotels and Tourism (MOHT) supports tourism to protected areas by working with MONREC to advance tourism to PAs and open up new ecotourism destinations which presently are off-limits to tourists (MOHT, 2013). In 2015, Myanmar released an ecotourism policy and management strategy. One distinguishing feature of tourism policy in Myanmar is the differentiation between the nature-based and ecotourism. According to MOHT, "nature-based tourism is leisure travel is undertaken largely or solely to enjoy natural attractions and engage in a variety of outdoor activities (e.g., bird watching, hiking, fishing, and beach-combing)" (MOHT, 2013, p. 68) while ecotourism refers to "tourism-related activity in and around its protected areas, that focuses upon management tools, systems and processes to deliver three elements:

1. biodiversity and ecosystem conservation;

2. education and learning to enable hosts and visitors to understand and engage with management approaches to protect and conserve the natural and cultural assets of these areas; and,
3. economic and social benefits to communities in and around protected areas that (a) reduce their demand for the natural assets of these areas, and (b) engage them in collaborative approaches to protected area management" (MONREC & MOHT, 2015, p. 20).The government has selected 21 pilot ecotourism sites, and aims to establish management plans for ten priority PAs by 2020. However, as of June 2020, the preparation and adoption of this policy has a consultative approach whereby the stakeholders are engaged in information collection and policy formulation processes. Additionally, the government has been considering partnering with investors to establish sustainable projects in PAs, which can contribute to the conservation of biodiversity and sustainable community development. One such example of a co-managed project is an eco-resort in Lampi Island Marine National Park that aims to support social welfare and conservation considerations simultaneously by providing NBT products (TLF, 2020).

Considering Myanmar's unique environmental resources, NBT can potentially contribute to social and economic development in remote areas. Introducing responsible tourism in remote PAs can not only positively strengthen conservation efforts but also provide an alternative source of income, thus minimising illegal activities, such as deforestation and poaching (Brockington et al., 2006; Buckley et al., 2016; Smiley Evans et al., 2020). The MOHT advocates the development of the tourism sector for Marine Protected Areas in the 2013–2020 Myanmar Tourism Master Plan, focusing on ecotourism by using market-led solutions to support PAs with rare wildlife, beautiful landscapes, seascapes, diverse ethnicities and cultures (MOHT, 2013). However, the implementation of tourism plans and policies depends entirely on proper PA management supported by resources and political decisions at the national and local level.

10.3 Myanmar's Mountainous Protected Areas

Myanmar's diverse geological landscape spreads from plains in the south to high mountains east of the Himalayas. The Himalayan arc stretches over 2,400 km from west-northwest to the east and is positioned between the Western and Eastern Syntaxis bends; from western and eastern bends, the Himalayas terminate at Nanga Parbat (8,125 m) and Namche Barwa (7,755 m) respectively (Wadia, 1931). From north to south, the Himalayas are 150-250 km wide and form a natural barrier between the Tibetan Plateau and the alluvial plains of the Indian subcontinent (Le Fort, 1975; Valdiya, 1998).

Different researchers determined the borders of the Himalayas based on different criteria. While some authors argue that the Himalayan range reaches as far as Thailand, others state that the Himalayan border is the Brahmaputra river (Apollo, 2017). According to Singh (1971), three meso-physiographic regions of the Himalayas (Western, Central, and Eastern) are subdivided into eight regions with the northern mountains of Myanmar (the Kachin Hills, Kumon range, and Patkai and Naga hills) as being part of the Eastern Himalayan Purvanchal range. However, other scholars exclude the Purvanchal range from the Himalayan arc (Apollo, 2017; Burrard, 1934; Chatterjee, 1964).

Neighbouring the eastern Himalayas, Myanmar's northern mountain ranges reach nearly 6000 metres in altitude. In terms of topography, the country's landscape is dominated by central lowlands surrounded by steep, rocky highlands (Fig. 10.1).

Myanmar's mountain ranges run mostly in a north to south direction and form a natural border in the north, north-west, east, and south-east. Table 10.2 summarises the most significant mountain ranges of the country. The systematic mapping of the northern Myanmar area is restricted due to limited infrastructure and armed insurgencies. Although recent political involvement has resulted in reconciliations with some ethnic groups, safe access to large swathes of Kachin, Shan, Kayin, and the Kayah States remains limited. Thus, significant areas of the country remain inadequately explored (Barber et al., 2017).

Myanmar's highest mountains are located in the northern hill jungle between the Eastern Himalayas and Kumon range. The country's highest peak, Hkakabo Razi (5,881 m) lies in the northernmost part of the country and is part of the Kachin hills, a mountain highland neighbouring the south-eastern slopes of the Himalayas (Renner et al., 2015). On different maps, Hkakabo Razi's altitude ranges from 5,691 m to 5,887 m, with 5,881 m being the most frequently used altitude (Long, 2013). However, because cartographic inaccuracies are identified by progress in technology and improved measuring standards (Apollo et al., 2020), the designation of Hkakabo Razi as South-East Asia's highest peak has recently been disputed. In 2013, Mt Gamlang Razi was re-measured (using Archer Field PC® with a Hemisphere GPS XF101 receiver) at 5,870 metres ($\pm$ 2 meters), and the use of modern GPS technology suggests its true altitude may be more precise (Brown, 2014).

Along the western border with India and Bangladesh Myanmar's mountains run south in a belt composed of many ranges, including the Patkai, Naga hills, the Mangin, and the Chin Hills, which continue southward to the extreme southwestern corner of the country. The Arakan (Rakhine) Mountains extend southeastward along the coast. Notable peaks in the west include Mt Saramati (3,860 m), located precisely on the border of Myanmar and India, and Mt Natma Taung (3,053 m), known as Mount Victoria among international tourists. Further south in central Myanmar is the Pegu Range, which hosts Mount Popa (1,518 m), another peak popular among domestic and international tourists.

Eastern Myanmar is dominated by Shan hills made up of mountain ranges, which are separated by valleys and basins (Mitchell, 2017) at an average elevation of 900 m (Hadden, 2008). In the southeast of Myanmar, the Dawna and Bilauktaung ranges with narrow valleys mark the border with Thailand on the Malay Peninsula.

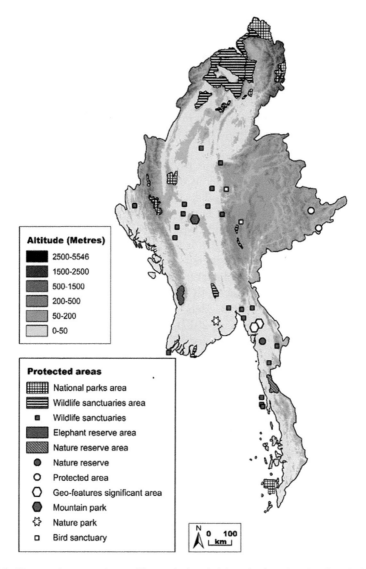

Fig. 10.1 Myanmar's protected areas (*Source* Authors' elaboration based on data from MONREC)

10.4 NBT Potential in Myanmar's Mountainous Protected Areas

In this section, we focus on the tourism potential of the 21 PAs located at an altitude of over 1,000 m above sea level. However, the emphasis is on 'potential' rather

Table 10.2 Myanmar's mountain ranges

	Range names	Location	Highest point
	Kachin hills	The most northern mountain range of Myanmar lies east of the Himalayas. The range is bordered on the north-west by the Arunachal Pradesh state of India, on the north by the Tibet autonomous region of China, and on the east by Yunnan province of China	Mt Hkakaborazi (5,881 m)
Indu-Burman Ranges	Kumon Range	This range is located south of Kachin hills and is believed to be the south-eastern extension of the Himalayan orogenic belt	Highest peak (1,263 m)
	Gaolingonshan or Nu Shan	The range is in north-east of Myanmar and is an extension of the Gaoligong Mountains, (a sub-range of the southern Hengduan Mountain Range) extending into Myanmar from Chinse bordering province Yunan	Mt Wona (3,916 m) peak located in China
	Patkai Bum range and Naga Hills	The range is on the north-eastern border of Myanmar (Sagaing region) and India (Nagaland state). Naga hills lie south of Patkai range along the Indian border and are the northern extension of the Arakan Mountains. According to some sources, Patkai Bum and Naga Hills are part of the Purvanchal range (extending from India)	Mount Saramati (3,826 m)

(continued)

Table 10.2 (continued)

	Range names	Location	Highest point
	Mangin Range	The range lies south of Kumon Range and is in Sagaing region. The Mangin Range extends 17.2 km in a southeast-northwest direction	Peak Kanbat Taung (1,249 m) level.
	Chin Hills	Chin Hills are located in the eastern part of the Patkai Range in north-western Myanmar on India border	Mt Natma Taung (Khaw-nu-soum or Mount Victoria, 3,053 m)
	Arakan Mountains, Rakhine Mountains, Rakhine Yoma,	The range is the southern extension of Chin Hills and is located between the coast of Rakhine State and the Central Myanmar Basin	No data available
	Pegu Range, Pegu Yoma or Bago Yoma	Pegu Range is located between the river Irrawaddy and the Sittaung River in central Myanmar	Mount Popa (Taung Ma-Gyi) (1,518 m)
Sino-Burman Range	Shan Hills, Shan Yoma, Shan Highland, Shan plateau	Shan hills are in central-eastern Myanmar on the border of China, Laos and Thailand. They include Daen Lao Range and Karen Hills (Kayah-Karen Mountains)	Peak Loi Leng (2,673 m)
	Dawna Range, Dawna Hills, Dawna Taungdan	Its northern end is in Kayah State where it meets the Daen Lao Range (a subrange of the Shan Hills). The range runs southwards along with Kayin State as a natural border with Mon State. In the west, it forms parallel ranges to the northern end of the Tenasserim Hills further south and south-east	Mt Mela Taung (Mulayit Taung) (2,080 m)

(continued)

Table 10.2 (continued)

Range names	Location	Highest point
Tenasserim Hills, Tenasserim Range	Located in Burma on the border with Thailand, Bilauktaung is a subrange of Tenasserim hills which is in south-east Myanmar on the border with Thailand. It extends from the Dawna Range for about 400 km along the frontier area to the Kra Isthmus	Peak Myinmoletkat Taung (2,072 m)

Source Authors' elaboration

than actual tourism to these areas as mountain NBT remains largely 'off-the-beaten-track' (even for local tourists). NBT infrastructure thus remains scant and market development sporadic as it is neither systematically managed nor promoted.

Some research claims the northern mountains of Myanmar as part of the Greater Himalayas (Manish & Pandit, 2018). Although they are not generally included in definitions of the Himalayas, they are part of the greater Hindu Kush Himalayan river system (Wester et al., 2019). Myanmar's highest and most famous mountains span across the Chin, Kachin and Shan states, with Mt Hkakaborazi (5,881 m) being the highest mountain in Southeast Asia. Northern Myanmar is rich in diverse and globally threatened wildlife, including tigers, red pandas, black musk deers, white-bellied heron, and numerous other birds roaming the dense jungle (Than et al., 1997). The Northern Mountain Forest Complex (NMFC) includes one national park (Hkakabo Razi) and three wildlife sanctuaries (Hponkan Razi, Huahung valley and Bumhpabum). Supported by NGOs and Norway Heritage Foundation, Myanmar Forest Department is collaborating with UNESCO to prepare a nomination of the Hkakabo Razi Landscape as Myanmar's first Natural World Heritage Site (Forest Department, 2017).

The Kachin mountains, and especially areas around Putao, are perhaps the most popular for domestic and international trekkers and mountain climbers. Flood-plains near Putao, dense jungles, and fast-flowing streams of mountain gorges, lead the way to the most accessible summit of Mount Phongun Razi (3,635 m). For mountaineers, the region offers the more technical peaks of Phrangran Razi (4,328 m) and Phonyin Razi (4,297 m). However, due to the intensified long-running conflict in this area, Putao and other areas of Kachin have been closed to international tourists since 2017.

Mountains south-west of Kachin State are part of the Indo-Burman Range, and some of the peaks are popular with tourists. One of Myanmar's highest and perhaps most-climbed mountains, Mt Natma Taung (Khaw-nu-soum or Mount Victoria), is located in little-visited Natma Taung National Park which holds the ASEAN Heritage Park status and is designated as having Outstanding Universal Value by UNESCO

(UNESCO, 2014). The national park is home to a wide range of rare flora and fauna and ethnic tribes known for their tattoo-faced women folk. Mt Natma Taung summit can be reached on foot. Although Mt Natma Taung is not a difficult peak, domestic tourists visit the summit predominantly by motorbike. Local nature protection organisations have warned about the impacts of such activity. Experts have lobbied for a proper management conservation plan for Mt Natma Tung and highlight that mountain access by motorbike or car should be banned. The second highest peak in Chin State is Kennedy Peak (2,703 m).

One of the few 'snow peaks' located outside of Kachin state is Mount Saramati (3,826 m). Mt Saramati rises above the neighbouring peaks of the mountainous border of Nagaland state (India) and the Sagaing Region (Myanmar), but the peak is usually reached from the Indian side (Sayers, 2017).

Mountain ranges along the Indian and the Bangladesh border, Rakhine Yoma, are generally lower in altitude. However, Rakhine Yoma Elephant Range (a protected area which is home to wild elephants, sun bears, gibbons and rare turtles) is increasingly being promoted by nature and wildlife tourists interested in hiking. Many other mountain areas in Rakhine state are difficult to reach and off-limits due to political unrest in the region.

A former hill station in Mandalay region, Pyin-Oo-Lwin (Maymyo) sits over 1,000 m above sea level. The city is well known among domestic and international tourists for its botanical gardens and colonial-style architectural houses. Pine trees, eucalyptus, and silver-oak abound the town. Recently, trekking around Pyin Oo Lwin to the jungle and waterfalls has grown in popularity among domestic and international tourists since the area offers visitors an authentic countryside trekking experience, retaining its cultural and ecological diversity.

In Shan State, the eastern border with Chin State, the Shan Highlands are clearly separated from central Myanmar by a meridian fault, which is expressed in the relief as a ledge 600 m high. The surface of the highlands is strongly dissected by river valleys. The average altitude in the highlands is approximately 900 m above sea level. However, there are several mountain ranges with peaks up to 1,800–2,600 m. Several locations in Shan state are popular with domestic and international tourists. As such, areas between Kalaw hills and Inle Lake are famous for trekking. In recent years, domestic NBT has been on the rise, and Shan state has become a popular destination for outdoor enthusiasts. Shan cities are surrounded by hills, valleys, and mountains, such as Shae-myin-nout-myin Taung (East-and-west-side-viewing Mountain), the highest peak in southern Shan at 2,362 m.

Further to the northeast, many tourists visit Taungyi, Hsipaw and Lashio by train or by car along the Burma Road leading to China. The hills of Mal Nal region near Taunggyi offer lower altitude trekking and a unique cultural experience of the Pa-O culture at an elevation of about 1,500 m. Lashio is emerging as a new trekking and waterfalls destination. Tachileik, located on the Myanmar-Thai border (eastern Shan State), serves as a gateway to the heart of the Golden Triangle (the area where Thailand, Laos and Myanmar meet at the confluence of the Ruak and Mekong rivers). Surrounded by a picturesque landscape, the town offers several hiking trails. On the west of the town, other activities, including trekking, rock climbing, camping, and enjoying the riverside, are available only to domestic tourists, where local authorities have discouraged foreign tourists from visiting. Another potential mountain NBT destination with unique flora and fauna, Parsar PA, is located on the eastern side of the Tachileik district. However, the development of off-the-beaten-track mountain destinations in Shan state is problematic due to the region's challenging socio-political situation.

South of the Shan highlands, Kayah State is home to pristine forests, spectacular mountain views, gorgeous rivers, and abundant waterfalls. Hoya region has beautiful low altitude mountains with fascinating scenery, while Mt Loin Nan Pha (1,527 m), which is located in Dimawhso Township, is getting increasing attention from international and domestic visitors. Although the Kayah State Government has recognised the potential of Kayah and prioritised the development of the hotel and tourism sector, the development has been delayed as parts of the region are governed by Kayan Ethnic Armed Group, requiring a special permit to visit certain areas, such as White Elephant Mountain (1,250 m) on the Thai border (Min, 2019). Further south, in Kayin State, a popular peak, Mt Zwegabin (722 m), offers stunning views of nearby hills. The highest point of Dawna Range, Mt Mela Taung (or Mulayit), is popular among hikers and campers and receives visitors from Myanmar and Thai sides of the border.

The central part of Myanmar is divided into two unequal parts by the Bago Yoma Ranges, the larger Ayeyarwaddy Valley and the smaller Sittaung Valley. Bago Yoma mountains (also known as the Pegu mountains) extend 435 km north-south between the Irrawaddy and Sittang rivers. The average altitude of the range is 600 metres with an extinct volcano Mount Popa (1,518 m) at its highest point.

Another popular mountain tourism destination is Mount Popa, which lies close to Bagan. Also called Taung Ma-Gyi, the mountain is a composite volcano in the northernmost part of the Pegu mountain range lying within Popa Mountain Park. Most tourists climb steps on a steep volcanic plug called Popa Taung Kalat, topped with a picturesque temple at 737 m. However, only a few tourists visit the true summit of Mt Popa at 1,518 m.

10.5 The Case of Natma Taung National Park and Mt Natma Taung

Natma Taung National park is located in the southern Chin State and is one of the largest PAs in Myanmar, covering about 713.46 km^2 (Aung et al., 2015). About 6,000 ethnic Chin people live around the national park with an estimated 100 people residing inside the park (World Heritage Centre, 2020). Mt Natma Taung (or Mt Victoria as known among international tourists) reaches the altitude of 3,053 m. This mountain is among the highest mountains in Myanmar and perhaps the most popular one due to its accessibility. The summit is most accessible during the dry season

Fig. 10.2 Road to the summit of Mt Natma Taung (*Source* Photo credits: Htet Eaindray and Thiha Lu Lin)

(October–January) via one of two main routes: the first easier option via Kanpetlet township and the second route via Mindat township.

There is no public transportion within the national park except four-wheel-drive vehicles and motorbikes rentable around the foot of the mountain. Until 2019, it was possible to reach the summit only by four-wheel-drive vehicles (see the road in Fig. 10.2). But recently, only vehicles with a special permit have been allowed.

Natma Taung National Park is a popular destination for domestic tourists and an emerging destination for international tourists, although Chin State only recently opened to foreign tourists in 2013 (Mon, 2018). In 2019, over 30,000 domestic and 3,762 international tourists travelled to Chin State, of which 8,029 domestic and 769 international tourists summited Mt Natma Taung by paying the 10,000 MMK (~ 10 USD) Zone Fee Entry to the national park (MOHT, 2020).

Trekking from the 'base camp carpark' to the summit usually takes about three hours. While access by 4×4 wheel drive vehicles was banned in 2019, it is still possible to summit the mountain by motorcycles within 30 minutes.

Local nature protection activists, guides and representatives from Yangon's Mountaineering Club reported that mismanagement of tourism in the national park has led to significant negative impacts on the unique biosphere of Mt Natma Taung. Many tourists still access the peak by motorbike, which leads to erosion, land degradation, and noise pollution, all of which affect wildlife and authentic tourists' experience in the national park. Another direct impact of mismanaged tourism activity is pollution from littering and sanitation facilities. Additional threats impacting the biodiversity are road construction, sand and rock mining for road construction, hunting and firewood extraction by local Chin communities. Furthermore, the tea plantations within the boundaries of the park accelerate forest degradation in the area (Beffasti & Galanti, 2011).

Although Natma Taung National Park has a high potential for NBT, towing to its natural landscape with mountains, unique biodiversity, and local culture, a proper integrated management plan needs to be put in place to preserve this destination for the future. To date, there is no formal management plan for the national park (World Heritage Centre, 2020), but two pilot buffer zones are functioning in the north and south, and Chin people practice shifting agriculture in the specified 'local use zone.'

The current decline in wildlife and negative impacts from tourism have enabled discussion toward establishing a participatory management plan with Chin villages, which has the potential to improve resource management outcomes in the national park (Kimengsi et al., 2019b). Furthermore, implementing a carrying capacity plan within the national park, and especially on the summit of Mt Natma Taung, could help to mitigate negative impacts caused by tourists and extensive use of motorised transport. Other alternatives, for example, the introduction of a cable car as a quick and accessible way to the summit, could provide a more sustainable model (Erfurt-Cooper, 2014). Improved education and capacity development for local villagers can also support the integrated co-management approach. Despite the national-level strategies and action plans aimed to sustainably manage PAs, protect biodiversity, and promote ecotourism while improving the livelihood of local villages through employment in tourism, the example of Natma Taung National Park demonstrates that current implementation of such strategies and plans is still ineffective.

10.6 Conclusion

The chapter provides a comprehensive review of Myanmar's PAs, its management and the potential NBT in mountain areas. However, due to current economic and socio-political challenges, the capacity of country's to introduce tourism to mountain areas remains in doubt. The argument draws attention towards an integrated management approach to the development of mountain NBT in Myanmar by involving not only the governmental authorities supported by international NGOs and private business, but also the local stakeholders depending on the PAs resources. As such, involving unique ethnic communities living in and around PAs could provide a better-balanced, more sustainable management style focused on conservation, protection of natural and cultural resources by providing employment opportunities (in tourism and conservation) and building the capacity of the native population. Creating employment opportunities for local ethnic communities living in and around PAs can provide a more sustainable development system for the conservation and protection of natural resources.

Among the Southeast Asian countries, Myanmar has vast biodiversity resources and many potential mountainous NBT destinations. However, despite large areas of forest cover, high levels of deforestation, hunting and destructive agricultural practices have resulted in rapid loss of natural habitats and species. Over the last decade, Myanmar has expanded its network and increased efficiency of the management of PAs. As of 2020, the country has 46 protected areas split into seven categories, each of which serves diverse functions according to its legislation and objectives. Understanding that Myanmar's current tourism offer is mostly centred around cultural and urban heritage, MOHT and MONREC focus on diversification of the sector by establishing ecotourism. The government and the International Centre for Integrated

Mountain Development are currently working on ecotourism policy and management strategies which will not only support conservation but also promote sustainable NBT products and services in Myanmar's PAs. The government has designated 21 ecotourism sites and works towards the establishment of ecotourism products for the domestic and international markets.

To ensure that ecotourism supports biodiversity conservation, community-based income generation and strengthens the management of PAs, we recommend adopting an integrated management approach. The integrated approach is needed to co-construct the ecotourism products, which can help to protect the environment, address the needs of vulnerable local communities, and promote sustainable NBT in Myanmar's PAs. As discussed in Sect. 10.5, uncontrolled tourist activity, such as at Mt Natma Taung, can have considerable negative impacts and can lead to rapid biodiversity loss. Hence, the integration of carrying capacity plans is a necessary step in the development of NBT in PAs. Many mountainous areas located along the borders are characterised as zones of instability and conflict dominated by shadow economies, including human trafficking and illegal trade of timber, wildlife, precious stones and drugs (Wester et al., 2019). These hold back regional development and undermine biodiversity conservation and NBT development. Hence, we recommend focussing not only on the conservation of natural resources but also on creating sustainable livelihood opportunities for the local population by engaging them more intensively in conservation actions and decision-making processes. Furthermore, strengthening educational efforts and capacity-building of the local population living in and around the PAs can provide human resources with the necessary skills to serve as the staff needed to manage PAs, community forests, and buffer zones (MONREC, 2018). Considering current socio-political challenges, the development of NBT in mountain areas faces broad challenges related to safety, but future research can map out potential sites ready for NBT.

References

Adams, W., Aveling, R., Brockington, D., Dickson, B., Elliott, J., Hutton, J., Roe, D., Vira, B., & Wolmer, W. (2004). Biodiversity conservation and the eradication of poverty. *Science (New York, N.Y.), 306*, 1146–1149. https://doi.org/10.1126/science.1097920.

Allendorf, T. D., Swe, K. K., Aung, M., & Thorsen, A. (2018). Community use and perceptions of a biodiversity corridor in Myanmar's threatened southern forests. *Global Ecology and Conservation, 15*. https://doi.org/10.1016/j.gecco.2018.e00409.

Apollo, M. (2017). The population of Himalayan regions—By the numbers: Past, present and future. In R. Efe & M. Öztürk (Eds.), *Contemporary studies in environment and tourism* (pp. 145–159). Cambridge Scholars Publishing.

Apollo, M., Mostowska, J., Maciuk, K., Wengel, Y., Jones, T., & Cheer, J. M. (2020). Peak bagging and cartographic misrepresentations. *Current Issues in Tourism*. https://doi.org/10.1080/13683500.2020.1812541.

Arriagada, R. A., Echeverria, C. M., & Moya, D. E. (2016). Creating protected areas on public lands: Is there room for additional conservation? *PLoS ONE, 11*(2), https://doi.org/10.1371/journal.pone.0148094.

Aung, M. U. (2007). Policy and practice in Myanmar's protected area system. *Journal of Environmental Management, 84*(2), 188–203. https://doi.org/10.1016/j.jenvman.2006.05.016.

Aung, P. S., Adam, Y. O., Pretzsch, J., & Peters, R. (2015). Distribution of forest income among rural households: a case study from Natma Taung national park, Myanmar. *Forests, Trees and Livelihoods, 24*(3), 190–201. https://doi.org/10.1080/14728028.2014.976597.

Barber, A. J., Zaw, K., & Crow, M. J. (2017). *Myanmar: Geology, resources and tectonics.* Geological Society of London.

Beffasti, L., & Galanti, V. (2011). *Myanmar Protected Areas: Context, current status and challenges.*

Bhagwat, T., Hess, A., Horning, N., Khaing, T., Thein, Z. M., Aung, K. M., et al. (2017). Losing a jewel—Rapid declines in Myanmar's intact forests from 2002–2014. *PLoS ONE, 12*(5), 1–22. https://doi.org/10.1371/journal.pone.0176364.

Brockington, D. A. N., Igoe, J. I. M., & Schmidt-Soltau, K. A. I. (2006). Conservation, human rights, and poverty reduction. *Conservation Biology, 20*(1), 250–252. https://doi.org/10.1111/j.1523-1739.2006.00335.x.

Brooks, T. M., Bakarr, M. I., Boucher, T., Da Fonseca, G. A. B., Hilton-Taylor, C., Hoekstra, J. M., Moritz, T., Olivieri, S., Parrish, J., Pressey, R. L., Rodrigues, A. S. L., Sechrest, W., Stattersfield, A., Strahm, W., & Stuart, S. N. (2004). Coverage provided by the global protected-area system: Is It Enough? *BioScience, 54*(12), 1081–1091. https://doi.org/10.1641/0006-3568(2004)054%5b1081:Cpbtgp%5d2.0.Co;2.

Brown, T. (2014, January 29 2014). *Gamlang Razi—Setting the elevation straight.* http://blog.junipersys.com/gamlang-razi-elevation/.

Buckley, R. C., Morrison, C., & Castley, J. G. (2016). Net effects of ecotourism on threatened species survival. *PLoS ONE, 11*(2), https://doi.org/10.1371/journal.pone.0147988.

Burrard, G. (1934). The place of Mount Everest in history. *Survey Review, 2*(14), 450–464.

Chatterjee, P. S. (1964). *Fifty years of science in India–progresss of geography.* Indian Science Congress Association.

Dearden, P., Bennett, M., & Johnston, J. (2005). Trends in global protected area governance, 1992–2002. *Environmental Management, 36*(1), 89–100. https://doi.org/10.1007/s00267-004-0131-9.

DESA. (2018). *Least developed country category: Myanmar profile.* UN Department of Economic and Social Affairs. Retrieved 6 July 2020 from https://www.un.org/development/desa/dpad/least-developed-country-category-myanmar.html.

Eagles, P., & Hillel, O. (2008, February 11–15). *Improving protected area finance through tourism* 2nd meeting of the ad hoc open-ended working Group on Protected Areas (WGPA-2), Rome, Italy.

Erfurt-Cooper, P. (2014). *Volcanic tourist destinations.* Berlin, Springer.

Fayos-Solà, E. (2005, 18–20 October 2004). Tourism's potential as a sustainable development strategy. WTO tourism policy forum, Washington, DC.

Ferraro, P., Hanauer, M., & Sims, K. (2011). Conditions associated with protected area success in conservation and poverty reduction. *Proceedings of the National Academy of Sciences of the United States of America, 108,* 13913–13918. https://doi.org/10.1073/pnas.1011529108.

Forest Department. (2015). *National biodiversity strategy and action plan (2015–2020).* MONREC. https://www.forestdepartment.gov.mm/eng/content/preparation-process-myanmar%E2%80%99s-first-natural-world-heritage-site-hkakabo-razi-landscape.

Forest Department. (2017). *Preparation process for Myanmar's first Natural World Heritage Site: Hkakabo Razi Landscape.* MONREC. https://www.forestdepartment.gov.mm/eng/content/preparation-process-myanmar%E2%80%99s-first-natural-world-heritage-site-hkakabo-razi-landscape..

Forest Department. (2020). *Organizational structure of forest department.* MONREC. https://www.forestdepartment.gov.mm/node/63.

Godde, P., Godde, P. M., Price, M. F., & Zimmerman, F. M. (2000). *Tourism and development in mountain regions.* Wallingford, CABI.

Hadden, R. L. (2008). *The geology of Burma (Myanmar): An annotated bibliography of Burma's geology, geography and earth science.*

Hayes, T. M. (2006). Parks, people, and forest protection: An institutional assessment of the effectiveness of protected areas. *World Development, 34*(12), 2064–2075. https://doi.org/10.1016/j.worlddev.2006.03.002.

Hockings, M., Stolton, S., Dudley, N., & Deguignet, M. (2018). *Protected area management effectiveness.*

Khine, P. K., & Schneider, H. (2020). First assessment of pteridophytes' composition and conservation status in Myanmar. *Global Ecology and Conservation, 22*, 1–9. https://doi.org/10.1016/j.gecco.2020.e00995.

Kimengsi, J., Aung, P., Pretzsch, J., Haller, T., & Auch, E. (2019). Constitutionality and the co-management of protected areas: Reflections from Cameroon and Myanmar. *International Journal of the Commons, 13*. https://doi.org/10.5334/ijc.934.

Kimengsi, J. N., Aung, P. S., Pretzsch, J., Haller, T., & Auch, E. (2019b). Constitutionality and the co-management of protected areas: Reflections from Cameroon and Myanmar. *International Journal of the Commons, 2*(13), 1003–1020.

Kyaw, N. N. (2015). *National biodiversity strategy and action plan 2015–2020.*

Kyaw, N. N. (2020). *Forestry in Myanmar 2019–2020.*

Le Fort, P. (1975). Himalayas: The collided range. Present knowledge of the continental arc. *American Journal Science, 275*(A), 1–44.

Leimgruber, P., Zaw Min, O. O., Aung, M., Kelly, D. S., Wemmer, C., Senior, B., & Songer, M. (2011). Current status of Asian Elephants in Myanmar. *Gajah, 35,* 76–86.

Leroux, S. J., Krawchuk, M. A., Schmiegelow, F., Cumming, S. G., Lisgo, K., Anderson, L. G., & Petkova, M. (2010). Global protected areas and IUCN designations: Do the categories match the conditions? *Biological Conservation, 143*(3), 609–616. https://doi.org/10.1016/j.biocon.2009.11.018.

Locke, H., & Dearden, P. (2005). Rethinking protected area categories and the "new paradigm". *Environmental Conservation, 32*(1), 1–10.

Long, k. (2013, September 19). Gamlang Razi expedition reaches summit. *Myanmar Times.* https://www.mmtimes.com/national-news/8200-gamlang-razi-expedition-reaches-summit.html.

MacDicken, K., Jonsson, Ö., Piña, L., Marklund, L., Maulo, S., Contessa, V., Adikari, Y., Garzuglia, M., Lindquist, E., Reams, G., & D'Annunzio, R. (2016). *The global forest resources assessment.*

MacKinnon, K., & Hatta, G. (1996). *Ecology of Kalimantan.* Periplus.

Manish, K., & Pandit, M. K. (2018). Geophysical upheavals and evolutionary diversification of plant species in the Himalaya. *PeerJ, 6,*. https://doi.org/10.7717/peerj.5919.

McDonald, R., & Boucher, T. (2011). Global development and the future of protected area strategy. *Biological Conservation, 144,* 383–392. https://doi.org/10.1016/j.biocon.2010.09.016.

Min, S. (2019). *Kyah State: Striving with developmental opportunities.* Ministry of Information. https://www.moi.gov.mm/moi:eng/?q=news/26/04/2019/id-17468.

Ministry of Planning and Finance. (2018). *Myanmar sustainable development plan (2018–2030).*

Mitchell, A. (2017). *Geological belts, plate boundaries, and mineral deposits in Myanmar.* Elsevier.

Mittermeier, R. A., Gill, P. R., Hoffman, M., Pilgrim, J., Brooks, T., Mittermeier, C. G., Lamoreux, J., & Fonseca, G. (2004). Hotspots revisited: Earths biologically richest and most endangered terrestrial ecoregions. *Cemex.*

MLIP. (2020). *Regarding foreigners and tourist travelling in the Country, Restricted Areas are as follow.* Ministry of Labour, Immigration and Population. http://www.mip.gov.mm/restricted-areas-for-foreigners-tourist-travelling-in-the-country/?fbclid=IwAR2CaNz92HlnwIUY7cQ16YKNezeI1jvhkVyuWiyKDsnXRweMr94Mtmj_JHU.

MOHT. (2013). *Tourism master plan 2013–2020.*

MOHT. (2020). *Attractions entry fees.* Ministry of Hotels and Tourism. Retrieved 22 June 2020 from https://tourism.gov.mm/attractions-entry-fee/.

Mon, M. S. (2018, May 11). Chin State opens up. *Frontier Myanmar.* https://frontiermyanmar.net/en/chin-state-opens-up?fbclid=IwAR0yXrkAPCXLs3ScO9ruQgsZArIBzqmFhGZV30wf-ozGpd1vp3wDWTmfZ8g#:~:text=A%20sharp%20increase%20in%20visitors,who%20visited%20Chin%20during%202017.

MONREC. (2018). *Protection of biodiversity and protected area law 2018*.
MONREC, & MOHT. (2015). *Myanmar ecotourism policy and management strategy for protected areas*.
Myanmar Centre for Responsible Business. (2018). *Biodiversity in Myanmar, including protected areas and key biodiversity areas*.
Myers, N., Mittermeier, R., Mittermeier, C., Fonseca, G., & Kent, J. (2000). Biodiversity hotspot for conservation priorities. *Nature, 403*, 853–858. https://doi.org/10.1038/35002501.
NCEA. (2009). *Fourth National Report to the United Nations Convention of Biological Diversity*.
Oldekop, J. A., Holmes, G., Harris, W. E., & Evans, K. L. (2016). A global assessment of the social and conservation outcomes of protected areas. *Conservation Biology, 30*(1), 133–141. https://doi.org/10.1111/cobi.12568.
Porter-Bolland, L., Ellis, E. A., Guariguata, M. R., Ruiz-Mallén, I., Negrete-Yankelevich, S., & Reyes-García, V. (2012). Community managed forests and forest protected areas: An assessment of their conservation effectiveness across the tropics. *Forest Ecology and Management, 268*, 6–17. https://doi.org/10.1016/j.foreco.2011.05.034.
Rao, M., Rabinowitz, A., & Khaing, S. T. (2002). Status review of the protected-area system in Myanmar, with recommendations for conservation planning. *Conservation Biology, 16*(2), 360–368. https://doi.org/10.1046/j.1523-1739.2002.00219.x.
Renner, S., Rappole, J. H., Milensky, C., Aung, M., Shwe, N., & Aung, T. (2015). Avifauna of the Southeastern Himalayan Mountains and neighboring Myanmar hill country. *Bonn Zoological Bulletin—Supplementum, 62*, 1–75.
Sayers, D. (2017). *Climbing Mt Saramati: From Myanmar and India with travel on the Chindwin*. Troubador Publishing Limited.
Singh, R. L. (1971). *India: A regional geography*. National Geographical Society of India.
Smiley Evans, T., Myat, T. W., Aung, P., Oo, Z. M., Maw, M. T., Toe, A. T., Aung, T. H., Hom, N. S., Shein, K. T., Thant, K. Z., Win, Y. T., Thein, W. Z., Gilardi, K., Thu, H. M., & Johnson, C. K. (2020). Bushmeat hunting and trade in Myanmar's central teak forests: Threats to biodiversity and human livelihoods. *Global Ecology and Conservation, 22*, e00889. https://doi.org/10.1016/j.gecco.2019.e00889.
Than, U. T., Moe, T. A., & Wikramanayake, E. (1997). *Southeastern Asia: Northern Myanmar*. WWF. https://www.worldwildlife.org/ecoregions/im0140
The Lampi Foundation (TLF). (2020). *Turtle Conservation Project*. Retrieved on 3 September 2020 at https://www.lampifoundation.org/projects.
Tint, K., Baginski, S. O., & Gyi, M. K. K. (2011). *Community forestry in Myanmar: Progress and potentials*.
Tint, K., Springate-Baginski, O., Macqueen, D. J., & Gyi, M. K. K. (2014). *Unleashing the potential of community forest enterprises in Myanmar*.
Tordoff, A. W., Eames, J., Eberhardt, K., Baltzer, M., Davidson, P., Leimgruber, P., Uga, U., & Than, U. A. (2005). *Myanmar: Investment opportunities in biodiversity conservation*. BirdLife International.
UN. (2019). *The Sustainable Development Goals Report*.
UNESCO. (2014). *Natma Taung National Park*. Retrieved 31 August from https://whc.unesco.org/en/tentativelists/5873/.
Valdiya, K. S. (1998). *Dynamic Himalaya*. Universities Press.
Voyer, M., Gladstone, W., & Goodall, H. (2015). Obtaining a social licence for MPAs—Influences on social acceptability. *Marine Policy, 51*, 260–266. https://doi.org/10.1016/j.marpol.2014.09.004.
Wadia, D. N. (1931). The syntaxis of the northwest Himalaya: Its rocks, tectonics and orogeny. *Record Geology Survey of India, 65*(2), 189–220.
Wen, Y., Htun, T., & Ko, A. C. (2017). Assessment of forest resources dependency for local livelihood around protected area: A case study in Popa Mountain Park, Central Myanmar. *International Journal of Sciences, 3*, 34–43. https://doi.org/10.18483/ijSci.1176

Wester, P., Mishra, A., Mukherji, A., & Shrestha, A. B. (Eds.). (2019). *The Hindu Kush Himalaya assessment: Mountains, climate change, sustainability and people*. Springer. https://doi.org/10.1007/978-3-319-92288-1.

Win, S. (2018). *Steps towards establishment of protected area management plans in Alaungdaw Kathapa and Natmataung*. Wildlife Conservation Society. https://myanmar.wcs.org/News/articleType/ArticleView/articleId/11293.aspx.

World Bank. (2017). *Myanmar performance learning review of the country partnership framework*. https://www.worldbank.org/en/country/myanmar/overview#2.

World Heritage Centre. (2020). *Natma Taung National Park*. UNESCO. https://whc.unesco.org/en/tentativelists/5873/.

Yang, R. L., Y.; Yang, K.; Hong, L.; Zhou, X.. (2019). Analysis of forest deforestation and its driving factors in Myanmar from 1988 to 2017. 2019, 11, 3047. *Sustainability, 11*(11), 1–15. https://doi.org/10.3390/su11113047.

Zhang, Y., Prescott, G. W., Tay, R. E., Dickens, B. L., Webb, E. L., Htun, S., et al. (2018). Dramatic cropland expansion in Myanmar following political reforms threatens biodiversity. *Scientific Reports, 8*(1), 16558. https://doi.org/10.1038/s41598-018-34974-8.

Yana Wengel is Associate Professor at the Hainan University—Arizona State University Joint International Tourism College in Haikou (China). She takes a critical approach to tourism studies, and her interests include volunteer tourism, tourism in developing economies and creative methodologies. Yana's current project examines issues of mountaineering tourism, focusing on the '7 Summits Challenge'. She has an interest in creative qualitative tools for data collection and stakeholder engagement.

Nandar Aye is a Master's student at the National Kaohsiung University of Hospitality and Tourism in Taiwan and she holds a bachelor's degree in Tourism from Mandalar University (Myanmar). Her interests lie in the coexistence of nature, tourism, and local communities, along with socio-ecological issues, community-based tourism, nature-based tourism, and wildlife tourism. She has worked as a tour guide and is involved in wildlife tourism and conservation projects.

Wut Yee Kyi Pyar is a founder of Mother Earth Tourism Specialists Myanmar and a Research Trainer at Luxembourg Development Cooperation Agency. Wut Yee has a special interest in slow tourism and conducted research on creating a sustainable livelihoods framework through Food Tourism. She is also a Futurepreneur for the ASEAN tourism sector, nominated by Young Southeast Asian Leaders Initiative Program, United States.

Jennifer Kreisz is English Language instructor at Hainan University—Arizona State University Joint International Tourism College in Haikou (China). She holds a Master's degree in English Education with a focus on linguistics and interlanguage pragmatics from Korea University and is a certified Canadian teacher. Jennifer presents in English Education and Cross-Cultural Linguistic conferences. She volunteers in local Educational and Sustainable Development initiatives.

Part IV
South Asia

Chapter 11
Indo-Himalayan Protected Areas: Peak-Hunters, Pilgrims and Mountain Tourism

Michal Apollo ⓘ, Viacheslav Andreychouk ⓘ, Joanna Mostowska ⓘ, and Karun Rawat

11.1 Introduction

There are over 100,000 protected areas (PAs) covering 11.7% of the Earth's surface, from less than 1 million km^2 in 1970 to 18.8 million km^2 in 2003 (Phillips, 2004; Sheppard, 2004). Two thirds of these PAs are located in less developed countries (Zimmerer, 2006). Rooted in the Eurocentric concept of human and nature as separate, conservation of PAs adheres to the principle of creating 'people-free' zones in the natural areas where the natural process can play out without any human interference. As a result, human displacement practice has been implemented to achieve his land use objective of conservation as the most extreme version of the conservation paradigm (Kabra, 2009).

In a different context of developing Asia, India holds the recognition of being a 'ho-spot' for PAs due to two important biodiverse regions covering the Himalayas including the north-eastern hills along the northern border, and the Western Ghats in peninsular India (Ninan et al., 2000). Due to the different level of economic development of most parts of the Indian Himalayas, currently there is no possibility of introducing a comprehensive, rational, and balanced approach to the natural environment in the region (Sachs, 2015). However, there are ongoing attempts to selectively preserve areas characterised by rare flora and fauna, although the PAs created in this

---

M. Apollo (✉)
Department of Tourism and Regional Studies, Institute of Geography, Pedagogical University of Krakow, Krakow, Poland
e-mail: michal.apollo@up.krakow.pl

V. Andreychouk · J. Mostowska
Faculty of Geography and Regional Studies, University of Warsaw, Warsaw, Poland

K. Rawat
Department of Tourism, University of Otago, Dunedin, New Zealand
e-mail: karun.rawat@postgrad.otago.ac.nz

© The Author(s), under exclusive license to Springer Nature Switzerland AG 2021
T. E. Jones et al. (eds.), *Nature-Based Tourism in Asia's Mountainous Protected Areas*, Geographies of Tourism and Global Change,
https://doi.org/10.1007/978-3-030-76833-1_11

way put nature protection on a par with commercial goals, like tourism. In recent decades, each part of the Indian Himalayas has delineated a variety of areas with natural protection status. Among these, national parks definitely dominate, although the conservation model is often highly pressured by commercial demands. In this chapter we will focus on the PAs of the Indian Himalayas, emphasising the challenges and opportunities standing before all stakeholders connected with these PAs.

11.2 Insights from Geography, Conservation, History and Development

11.2.1 Geography and Biodiversity Status

The Himalayas (Sanskrit: *Deana-gari*) form the highest mountain system on Earth, located between the Tibetan Plateau and the Indus, Ganges and Brahmaputra lowlands and across the two subcontinents—South Asian (Hindustan) and Central Asian. The Himalayas are an important orographic, climate, landscape and ecological frontier, separating two geographical worlds from each other—lifted to the height of the endless cold mountain wastes of Tibet from the hot humid, vibrant, landscapes of the Ganges lowlands and the plains of Assam (see e.g., Andrejczuk, 2016a; Yin & Harrison, 2000).

The Himalayas stretch almost latitudinally from north-west to south-east, and further east over a length of about 2400 km (Fig. 11.1). The width of this mountain system ranges from 350 km in the western part (Kashmir) to 150 km in the east (Arunachal Pradesh). The range has the shape of a gentle arch, open to the northeast. In the north, the valleys of the upper Indus and the Brahmaputra separate the Himalayas from the Trans-Himalayas, which are on the edge of the Tibetan Plateau (Andrejczuk, 2016a). The area of the Himalayas is about 550,000 km^2 (Apollo, 2017b). The Himalayan mountain system forms a separate physio-geographical land. It neighbours such lands and natural regions as: the Tibetan Highlands in the north, the Burmese Mountains in the east, the Hindustan Plain in the south, the Kashmir Valley and the Hindu Kush range in the west (Andrejczuk, 2016a; Apollo, 2017b).

On a global scale, the Himalayan environment is characterised by the largest ecological contrasts. There is huge topographic prominence, varied insolation and contrasts in north–south exposures of slopes (Singh, 1971). The masses of warm summer oceanic monsoon, as well as dry icy anticyclone, are blocked, which paralyses Central Asia in winter. It is a land not only of contrasts, but also records. As many as ten out of the 14 eight-thousanders in the world are located in the Himalayas. One hundred Himalayan peaks reach a height of over 7200 m and mountain passes here have an average altitude of 5000 m.

Within this mountainous system there exists unprecedented *landscape diversity* resulting from a strong fragmentation into longitudinal mountain zones and ridges, transverse valleys and gorges, separate massifs and extensive valleys. Vertical sections of the Himalayas contain the most complete spectrum of climate and vegetation zones on Earth—from tropical rainforests, with extremely rich biodiversity, to ice-capped and lifeless mountain peaks and valleys (Fig. 11.2). The richest biomes occur in parts of the eastern Himalayas (Andrejczuk, 2016b). According to the World Wildlife Fund, there are over 4000 vascular plant species on 10,000 km^2. About 30% of forest plants are exclusively Himalayan species (oaks, rhododendrons, pines). In the eastern Himalayas alone over 10,000 species of plants, around 300 species of mammals, 176 of reptiles, 105 of amphibians and 269 species of freshwater fish have been described to date. In the last ten years, the number of newly discovered species has exceeded 300 (WWF Report, 2009). The number of new species not yet documented seems sure to grow.

Due to the variety of physiography and climatic conditions, the Himalayas are very diverse. If the main natural asset of the eastern parts of the Himalayas is exceptional biodiversity, especially botanical, then in the case of the north-western part of the Indian Himalayas (Punjab Himalayas), fauna is a special value—rare animals, primarily mammals, inhabiting the higher (over 2000 m above sea level) parts of the mountains. If, in the north-western Himalayas the biogeographic background consists of palearctic species, then in the eastern Himalayas—tropical. If the botanical showcase of the north-western Himalayas is coniferous forests: pine, cedar, spruce and juniper, then in the eastern Himalayas—evergreen tropical forests with bamboo, and in the higher parts of the mountains—with magnolias and rhododendrons (Andrejczuk, 2016b; Conservation International, 2012; Negi, 1992, 1993, 2002).

Unfortunately, due to anthropogenic impacts, especially within the lower floors of the mountains (cutting down mountain forests for tea plantations, teak trees, for fuel, etc.), many species of plants and animals are endangered (Negi, 1992, 1993). According to data from the *Center for Applied Biodiversity Sciences* of the international organisation *Conservation International*, 3160 endemic plant species, eight endemic bird species, four mammal and four amphibian are at risk in the Himalayas. Other sources cite far greater numbers of endangered species, owing to the destruction of mountain forests, which are a shelter and home to most species of Himalayan animals. Due to the aforementioned threats, this area of the Himalayas has been classified in the *Hotspot Regions of the World* list, as it is very valuable in nature, with progressive environmental degradation and disappearing biodiversity. Since Himalayan forests and meadows are the source of over 600 species of medicinal plants, used for thousands of years in the so-called ethnomedicine, i.e., folk medicine, the threat is enormous also for pharmaceutical reasons (Andrejczuk, 2016b).

India is an excellent example of a Himalayan country with a unique richness of biodiversity due to diversity of physiography and climatic conditions. Overall, a significant proportion (72.3%) of the Himalayan mountain system is located within India. It consists of three parts insulated from each other: west, central and eastern (see Fig. 11.1). These parts differ significantly in their environmental characteristics (Sect. 11.2.1) and nature-based tourism development (Sect. 11.3). That is why a PA has been established. As of 2019, the PAs in the Indian Himalayas cover 34,766 km^2, roughly 9% of the total surface area (see Table 11.1; Fig. 11.3). In terms of the

Table 11.1 Percentage of the total area under protection in the Indian Himalaya regions

Region	Area (km^2)[a]	National Parks Area[b] (No. of NPs)	Other areas under protection[b] (No. of areas)	PAs: total area (No. of PAs)	Percentage of the total area under protection
1. Kashmir Himalayas	222,236	3916 (3)	10,245 (15)	14,161 (18)	6.37
2. Punjab Himalayas	64,017	141,484 (3)	6165 (35)	7649 (38)	11.95
3. Kumaun Himalayas	41,762	4915 (6)	2675 (7)	7590 (13)	18.17
4. Sikkim Himalayas	7096	1784 (1)	399 (7)	2183 (8)	30.76
5. Darjeeling Himalayas	6193	318 (3)	207 (4)	525 (7)	8.48
6. Arunachal Himalayas	55,101	483 (1)	2175 (7)	2658 (8)	4.82
Total (Indo Himalayas)	396,405	12,900 (17)	21,866 (75)	34,766 (92)	8.77

[a]Apollo (2017b) and [b]National Wildlife (2019)

percentage of PAs (on the scale of their entire area), the Sikkim Himalayas are unrivalled and over 30% of the region is protected. The Kashmir Himalayas have the smallest PA—just over 6%.

11.2.2 History and Development of Protected Areas

Nature conservation has a long tradition in Himalayan history (Zurick & Pacheco, 2006). Wise use of natural resources was a prerequisite for many hunter-gatherer societies of the Himalayas, which date back to at least 6000 BC (Kothari et al., 1989). After India gained independence in 1947, many of these reserves were subsequently declared as national parks or sanctuaries. Wildlife, together with forestry, has traditionally been managed under a single administrative organisation within the forest departments of each state or union territory, with the role of central government being mainly advisory. In 1970, the Indian Board for Wildlife drafted a national wildlife policy (Negi, 1992). This policy identified the causes of wildlife depletion and made specific recommendations for wildlife conservation in the country. The major threats to wildlife species and habitats identified were: habitat changes, use of pesticides, lack of legislative support, commercial exploitation, introduction of exotics, poaching, biotic interference, use of crop protection guns, lack of organisation and guidelines for management. The policy recommended that establishment of a central organisation to maintain territorial integrity of wildlife areas and suggested that 4% of total land area be managed as national parks by a central organisation. This led to the enactment of the Wildlife (Protection) Act in 1972, which provides for three categories of PA: national parks, sanctuaries and closed areas. However, levels of protection afforded in each category differ, as do the degrees of restriction on human activities. The adoption of a National Policy for Wildlife Conservation in 1970 and the enactment of the Wildlife (Protection) Act in 1972 led to a significant growth in the PA network in India.

Overall, national parks are given the highest level of protection, with no grazing and no private landholding or rights permitted within them. Sanctuaries are given a lower level of protection, and certain activities may be permitted within them for better protection of wildlife or for any other good and sufficient reason. The state government may declare an area closed to the hunting of wild animals for a specified period; other activities are permitted to continue.

11.3 Indian Himalayan Tourism

Tourism is one of the most dynamic sectors of the economy of many developing Asian countries. As an instrument of regional policy, it activates local communities and builds solid foundations for economic development (Apollo, 2015; Apollo & Rettinger, 2019; Nepal, 2000, 2003, 2008). Mountain tourism is a multifaceted phenomenon (Apollo, 2017a; Mieczkowski, 1995), which depends on the increase in the economic status of countries and communities living in mountain regions (Messerli & Ives, 1997).

Since the mid-twentieth century, tourism has become the main element shaping and influencing the development of Himalayan regions (Bisht, 2008). The Indian Himalaya region was visited by almost 57.5 million tourists in 2012 alone, an increase of 2.5% from the previous year (Table 11.2). On this basis, the number of high-mountain tourists in 2012 was estimated at almost 57.5 million, including 56.7 million domestic tourists and almost 770,000 foreign tourists. The Himalayan regions of Kumaun (26.9 million) and Punjab (17.1 million) received the greatest share of visitor arrivals. The structure of tourist traffic in all regions is dominated by domestic (regional) tourists, whose number is systematically increasing. The majority of visitors to the mountain region is of pilgrims practicing Buddhist or Hindu religious or cultural rites (Gawlik et al., 2021). Among domestic tourists, pilgrimage tourism dominates, while foreigners choose adventure tourism, especially high-altitude trekking and, increasingly common, high-altitude climbing (Apollo, 2016; Bisht, 2008; Gawlik et al., 2021).

These are most popular Indian Himalaya areas for mountain hiking, trekking or climbing. The classification below was made in terms of individual physiogeographical mesoregions (see Fig. 11.1), divided into eight smaller parts corresponding to administrative units (for more details see Apollo, 2017b).

The **Kashmir Himalayas** are defined as the part of the Himalayan range in the state of Jammu and Kashmir (Fig. 11.1). They are inhabited by 12.5 million people

Table 11.2 Domestic and foreign visitor trends to the Himalayas (2011–2012) and rate of change

Himalayas regions	2011		2012		Rate of change (%)	
	Domestic	Foreign	Domestic	Foreign	Domestic	Foreign
Kashmir Himalaya	13,071,531	71,593	12,427,122	78,802	−4.93	10.07
Punjab Himalaya	15,514,792	521,699	16,570,637	534,414	6.81	2.44
Kumaun Himalaya	25,946,254	124,653	26,827,329	124,555	3.4	−0.08
Sikkim Himalaya	552,453	23,602	558,538	26,489	1.1	12.23
Darjeeling Himalaya	NA	NA	NA	NA	-	-
Arunachal Himalaya	233,227	4,753	317,243	5,135	36.02	8.04
Total	55,318,257	746,300	56,700,869	769,395	2.5	3.09
	56,064,557		57,470,264		2.51	

Source ITS (2013)

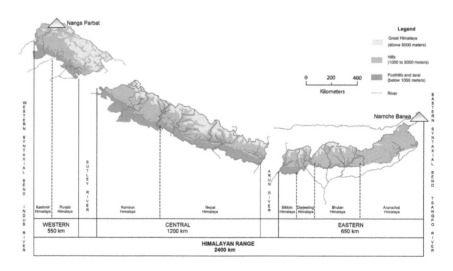

Fig. 11.1 Division of the Himalayas (Apollo, 2017b)

(Apollo, 2017b). The ratio of residents to visitors is almost one to one with 12.5 million tourists in 2012, among whom domestic tourists dominate (see Table 11.2). Kashmir's Himalayas owe their extraordinary tourist popularity to the pilgrims of three religions (Hinduism, Buddhism and Islam). Hiking routes to places of worship are often at high altitudes. Trekking and mountain climbing are a fraction of the total tourist traffic of the Kashmir Himalayas, although the percentage of those who

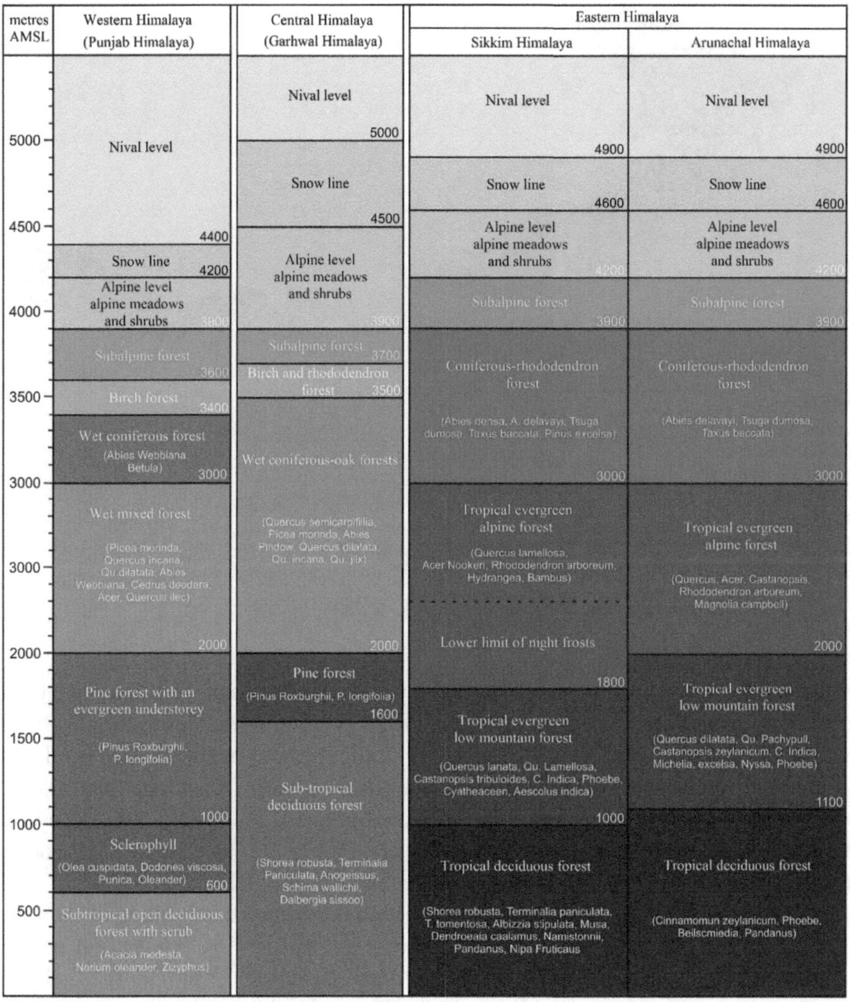

metres AMSL	Western Himalaya (Punjab Himalaya)	Central Himalaya (Garhwal Himalaya)	Eastern Himalaya - Sikkim Himalaya	Eastern Himalaya - Arunachal Himalaya
5000	Nival level	Nival level (5000)	Nival level (4900)	Nival level (4900)
4500	Snow line (4400)	Snow line (4500)	Snow line (4600)	Snow line (4600)
4000	Alpine level alpine meadows and shrubs (4200)	Alpine level alpine meadows and shrubs (4200)	Alpine level alpine meadows and shrubs (4200)	Alpine level alpine meadows and shrubs (4200)
	Subalpine forest (3600)	Subalpine forest (3700)	Subalpine forest (3900)	Subalpine forest (3900)
3500	Birch forest (3400)	Birch and rhododendron forest (3500)	Coniferous-rhododendron forest	Coniferous-rhododendron forest
3000	Wet coniferous forest (Abies Webbiana, Betula) (3000)	Wet coniferous-oak forests	(Abies densa, A. delavayi, Tsuga dumosa, Taxus baccata, Pinus excelsa) (3000)	(Abies delavayi, Tsuga dumosa, Taxus baccata) (3000)
	Wet mixed forest (Picea morinda, Quercus incana, Qu. dilatata, Abies Webbiana, Cedrus deodara, Acer, Quercus ilex)	(Quercus semicarpifolia, Picea morinda, Abies Pindow, Quercus dilatata, Qu. incana, Qu. ilX)	Tropical evergreen alpine forest (Quercus lamellosa, Acer Nookeri, Rhododendron arboreum, Hydrangea, Bambus)	Tropical evergreen alpine forest (Quercus, Acer, Castanopsis, Rhododendron arboreum, Magnolia campbeli)
2000	(2000)	(2000)	Lower limit of night frosts (1800)	(2000)
1500	Pine forest with an evergreen understorey (Pinus Roxburghii, P. longifolia)	Pine forest (Pinus Roxburghii, P. longifolia) (1600)	Tropical evergreen low mountain forest (Quercus lanata, Qu. Lamellosa, Castanopsis tribuloides, C. Indica, Phoebe, Cyatheaceen, Aescolus indica)	Tropical evergreen low mountain forest (Quercus dilatata, Qu. Pachypuli, Castanopsis zeylanicum, C. Indica, Michelia, excelsa, Nyssa, Phoebe) (1100)
1000	Sclerophyll (Olea cuspidata, Dodonea viscosa, Punica, Oleander) (1000, 600)	Sub-tropical deciduous forest (Shorea robusta, Terminalia Paniculata, Anogeissus, Schima wallichii, Dalbergia sissoo)	(1000)	
500	Subtropical open deciduous forest with scrub (Acacia modesta, Nerium oleander, Zizyphus)		Tropical deciduous forest (Shorea robusta, Terminalia paniculata, T. tomentosa, Albizzia sipulata, Musa, Dendroeala caalamus, Namistonnii, Pandanus, Nipa Fruticaus)	Tropical deciduous forest (Cinnamomun zeylanicum, Phoebe, Beilscmiedia, Pandanus)

Fig. 11.2 Altitudinal zones of the Himalayas (2016b). Adapted from Andrejczuk

practice this alpine activity is growing dynamically. The most popular regions include (1) Ladakh and (2) the Kashmiri Valley. These diametrically opposed areas account for almost all the tourist traffic in the region.

The **Punjab Himalayas** are the western part of the Himalayan chain covering the area of four Indian states, i.e., Himachal Pradesh, Punjab, Chandigarh and Haryana. It is inhabited by over 13 million people (Apollo, 2017b) and annually visited by over 16 million visitors with domestic tourists dominate (see Table 11.2). To international visitors, the Punjab Himalayas comes second in popularity with 500,000 tourists in 2012 are right behind the Nepalese (800,000) which is the most-visited part of the range by foreign tourists. The dominant part of the Punjab Himalayas is the state of

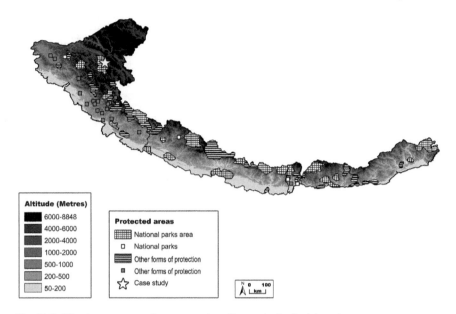

Fig. 11.3 Himalayan areas under conservation. *Source* Author's elaboration

Himachal Pradesh. It attracts tourists to its spectacular mountain scenery, hospitable climate, diverse flora and fauna, a varied culture, pilgrimage centres and numerous opportunities for active leisure: climbing, trekking, paragliding, fishing, river rafting, skiing, etc. (Jreat, 2004). For domestic tourists, the main reason for their visit is rest and recreation (72%), only 14% indicated pilgrimage (TSHP, 2012). Though pilgrims do not dominate the tourist traffic, numerous temples located in Himachal are crowded (Apollo et al., 2020b; Gawlik et al., 2021; TSHP, 2012).

The **Kumaun Himalayas** partly overlap with the administrative area of the state of Uttarakhand (formerly Uttaranchal). They are inhabited by six million people (Apollo, 2017b), and each year is visited by over 26 million people (tourists), dominated by domestic tourists (see Table 11.2). The region is a popular destination for Hindu pilgrims who have been coming to these areas for centuries. Hiking routes to places of worship, although they are often at high altitudes, tend to be well-prepared in terms of tourist infrastructure (e.g., in situ service). With millions of pilgrims, practicing (adventure) trekking or mountain climbing constitutes a negligible percentage of total tourist traffic (Gawlik et al., 2021). The research of Gupta and Bhatt (2012) carried out in the Kumaun Himalayas area showed that up to 82% of respondents consider pilgrimage to be the main reason for the visit. The experience of adventure is declared by only 12% of respondents, although this number is gradually increasing.

The **Sikkim Himalayas** are the part of the Himalayan chain that lies in the Indian state of Sikkim. Slightly over 600,000 people live there (Apollo, 2017b), and another nearly 600,000 people (tourists) visit annually. Domestic tourists dominate (see Table 11.2). A large number of domestic tourists are the pilgrims of two religions (Hinduism and Buddhism), who visit numerous temples located mainly in the southern district (Sikkim South). Trekking and climbing are a part of the total tourist traffic of the Sikkim Himalayas. However, an upward trend is noticeable. The most popular trekking area of the third highest mountain in the world—Kangchenjunga (8,589 m) (Reynolds, 2013; Weare, 2009). Located slightly south of the main Himalayan ridge—Kangchenjunga receives the main force of summer monsoons. This results in significant snowfall and leads to frequent avalanches, for which the mountain is famous. Similar snow characteristics are preserved by the southern 6,000- and 7,000-m mountains. The region is considered very dangerous for mountaineers.

The **Darjeeling Himalayas** lie in the northern part of Indian West Bengal (state). With an area of only 6,193 km^2, the area is inhabited by more than 5.7 million people. This makes it the most densely populated (924 people per km^2) area of the Himalayas (Apollo, 2017b). The Darjeeling region has been famous for its picturesque hill stations since the mid-eighteenth century, which served as bases for the pioneers of mountain hiking (Zurick & Pacheco, 2006). Located at a much lower altitude than the Sikkim Himalayas located in the north, they offer great opportunities for hiking and trekking. One of the best routes runs through the Singalila Mountains separating India from Nepal, in the area of which is the national park (Singalila).

The **Arunachal Himalayas** lie in the easternmost part of the Himalayan chain. They are administratively in the Indian state of Arunachal Pradesh, which has nearly 900,000 inhabitants (Apollo, 2017b). Every year, the region is visited by over 320,000 people (tourists), among whom domestic tourists dominate (see Table 11.2). Arunachal Pradesh is the wildest and the least-known state in India. It is often referred to as the *Final Frontier*, paraphrasing the nickname of the 49th state (Alaska) of the United States of America (*Last Frontier*). A unique culture (26 ethnic groups) and the unspoiled nature of Arunachal are attractive to tourists (Megu, 2007). The most popular hiking and trekking area is the Tawang Valley sandwiched between Bhutan and China. The temples there (mainly Buddhist) are frequently visited by pilgrims (mainly domestic tourists). Foreign tourists often head towards the Bailey Trail (named after F. M. Bailey) for environmental and cultural experiences.

11.4 Challenges and Opportunities for Mountain Tourism in the Himalayan PAs

11.4.1 Challenges: What to Protect?

Given this wide geographical coverage (see Sect. 11.2.1), the potential to promote mountain tourism activities in India is enormous. For instance, the diverse physical and geological variations of the Indian Himalayas offer unique flora and fauna, invasive wildlife and species, and a home to millions of mountain communities who have been residing in these landscape for centuries. Encountering these various features of the Indian Himalayas while undertaking mountain activities can enhance the tourism experiences in many ways. However, the utilisation of these features for tourism purposes is challenging and might impose threats to the fragile ecosystem (Apollo & Andreychouk, 2020).

Promoting or implementing mountain tourism in PAs often coincides with the multiple motives of different stakeholders. The stakeholders who are directly and indirectly involved with the PAs in mountainous areas of India are local mountain communities, state officials, non-governmental organisations, outside private tourism and hospitality investors, and the tourists (Negi & Nautiyal, 2003). The lack of clearly defined boundaries for each of the stakeholders with regards to the conservation efforts and tourism development is among main challenges for the management to implement sustainable PAs in India. Unfortunately, the management is still not certain what to protect due to the lack of guidelines and instrumental measures (Sekhar, 2003).

The establishment of PAs implies some restricted rules for the use of its natural resources, however, the cost of the establishment of PAs is paid by the local communities because their livelihoods are dependent on forest resources which are mostly shared within PA boundaries (Arriagada et al., 2016; Badola et al., 2018; Bhattarai et al., 2017). These communities are vulnerable because they bear the losses directly (e.g., restrictions to access to forest fuel resources) and indirectly (e.g., crops and livestock raided by wild animals dispersed from PAs) (Sekhar, 1998). Along this line, management also undermines local communities' participation in PA policy and its implementation and as well as unequal distribution of tangible economic benefits retained from tourism sharing or displacing the locals who are (not all but in most cases) displaced by the establishment of PAs or tourism development (Badola et al., 2018). It is quite ironic that when the government claims that they clearly and strictly follow the guidelines of Articles 10 and 11 which encourage national governments to adopt sound measures to the objectives of PAs, which should provide maximum incentives to the local people who support officials to undertake conservation practices and tourism development. In return, officials should equally support and provide offset benefits such as employment and engagement in conservation practices to the communities due to the land and livelihood displacement. In practice, the limited efforts of the management in providing or considering local communities benefits who are immediately affected by a PA in India is in stark contrast with Convention on Biological Diversity (CBD) goals (Sekhar, 2003).

These practices had led to having negative effects on local attitudes towards conservation, conservation practices and wildlife species who depredate their crops and livestock. Due to these reasons, many local communities in India turn hostile and confront the authorities while attempting to implement conservation policies and tourism developments plans (Apollo, 2015; Apollo et al., 2020a; Negi & Nautiyal, 2003; Sekhar, 2003). However, once tourism is introduced in the park area, most tourism benefits were channelled to outsiders' tourism private investors and hoteliers (Negi & Nautiyal, 2003; Sekhar, 2003).

Broadly speaking, this is a worldwide phenomenon, where the private tourism and hospitality sector quickly moves *in* and *around* PAs and outsiders benefit most from wildlife or nature tourism. This usually occurs when management does not have clear enforced guidelines for private investors around PAs, ranging from low-scale accommodation to the high-scale exotic resort and hotel service providers. For example, in the case of Nagarhole National Park, if the government had strictly followed the CBD guidelines on sustainable development, the management policies should have first been sensitive towards the local community's needs and interests (Ioannides, 1995). On the contrary, state officials were ineffective in coming up with such measures in support of this paradigm which favours and benefits the local communities. There are two major reasons. First, there are no enforced measures that guarantee a certain percentage of jobs for local communities when outside private investors operate their business in the shared land of local communities and PAs (Barrett et al., 2001). Besides, the policy does not monitor the environmental impacts of these high- and low-scale business operators who generate pollution through their tourism establishments. Sekhar (2003) commented that external tourism operators do not place their interest and effort within a larger context of local community development and the environment because their temporary objective is just to make profits. Therefore, due to the lack of capability in local communities' capacity to invest, the outside private investors have an opportunity to maximise their business turnover and can be an obstacle to biodiversity conservation (Barrett et al., 2001; Weinberg et al., 2002).

Second, the unfairness and mismanagement of conservation and tourism development in India is also affected by the corruption and bribery from local to state level and is also the other main reason for the dominance of private tourism agencies in PA regions. For example, in Keoladeo National Park, Ranthambore Tiger Reserve or in Sariska Tiger Reserve in Rajasthan, India, are cases to illustrate the issues highlighted over the inadequate measures of tourism policy. As a result, local villagers were unhappy and protested about tourism development and conversation efforts on several occasions, due to the state officials' self-regulated policies.

To tackle these problems, Apollo (2015) suggested five proposals for local communities (LCs): (1) LCs must look more critically upon the quality of the natural environment, because today's tourists wish to escape from urban pollution, noise, crime and other related stress to the relative calm of a mountain environment (Godde et al., 1999); (2) LCs must constructively combine the new with the old and not lose their tradition and culture which is as important as the natural environment and the landscape for tourists; (3) LCs have to develop new or redevelop old in situ services (i.e., accommodation, catering, transport) to ensure tourists have a worthwhile stay; (4) LCs must maintain control of the local tourism market, because when locals see no material benefits from tourism, they may develop open hostility towards visitors; (5) LCs must keep the division of social roles, mostly to control the diversity of employment. In a time of crisis or natural disaster (e.g., the earthquake in Nepal in 2015, heavy monsoon rains in Garhwal in 2013), a lack of tourists may result in a humanitarian crisis in places where people employed in the tourism industry dominate the employment structure.

11.4.2 Opportunities: How to Protect?

A series of answers to the question on how to protect tourism centres on finding a sustainable solution. To some extent, there is a potential to resolve major issues and ongoing challenges officials in India are now facing to implement sustainable conversation to promote tourism in PAs. There is a pressing need to design an appropriate mountain tourism policy which equally meets the needs of those who are directly and indirectly involved in mountain areas of and around PAs. First and foremost, designing appropriate policies to ensure equal local community participation, shared economic and labour benefits from tourism with the locals by reaching out to more rural households within the communities, and respectfully compensating the local values, places and culture in exchange for establishment of PAs in their neighbourhood (Zhong et al., 2011). Second, through capacity building, raising long-term awareness amongst the local communities about the impact of failed PAs in local environmental ecology, the traditional economy, livelihoods and well-being (UNDP, 2010). Third, devising tourism policy with restricted measures, for example, in India much mountain tourism in PAs is at an early stage, tourism can also impact the local culture, values and traditions. Regulating the tourist growth pattern and activity from the early stages could lead to minimal impact on local communities as well as wildlife species (migratory or feeding patterns), and endangered plants (Chandola, 2012). If management sees these as vantage points, the initiative towards the development of mountain tourism without compromising the conservation practices is the best step forward. Therefore, to achieve a sustainable goal to promote mountain tourism and thereby to generate support for conservation from the local communities, the management should exercise and prioritise management initiatives and policies for the future (Muradian & Rival, 2012).

Previous research has demonstrated that in planning and developing tourism in PAs, local communities' attitudes, influence and adequate involvement in tourism development and conservation practices play a crucial role (Apollo, 2015; Apollo et al., 2020a; Badola, 1998; Badola et al., 2018). If locals are given an equal leadership role in decision making and designing tourism development plans, in return support to the management towards conservation and tourism development have shown some promising outcomes. Therefore, if seen from the conservation perspective, there is an advantage to sustainably promote mountain tourism in PAs. For instance, to most local mountain communities the mountains are respected as the home of local deities.

Accessibility to the mountain areas is one of the major needs, not only seen from the tourism perspective but also from providing basic access and rights (Apollo, 2017a). In many ways, accessibility to mountain tourism and PAs can have a profound impact and if monitored can have a positive outcome for the associated stakeholders. First, to mitigate negative *cultural impact*, before designing and implementing mountain tourism activity in PA shared areas with local mountain communities, consideration should be given to the communities and villages which are closest to the PA zones because they are the ones who will be affected (e.g., environmentally) if tourist activity is increased in their local areas. To reduce such impact, in the initial stages regulated and limited access to a different location should be enforced through policies and laws. For example, in 1994 the Bhutan government closed the 40th highest unclimbed mountain in the world. The reason Ganghkar Puensum (6000 m) mountain was closed to climbers out of respect for local spiritual beliefs. All of these recent successful examples illustrate how management prioritised the local ecology over the tourism economy and valued local mountain communities' beliefs over tourism development objectives in PAs and established a stable social relationship of trust and reciprocity (Badola et al., 2012).

Second, implementation of mountain tourism can be beneficial without compromising the conservation efforts and biological diversity of local communities, when at the outset of designing a policy plan the *ecological impact* of mountain tourism on the local environment is taken into consideration. For instance, mountain tourism activities such as hiking, tramping and trekking should not coincide with the traditional occupation of communities (e.g., harvesting season), the ecosystem of wildlife (e.g., breeding or mating season), and the growing season of important grasslands, herbs and crops. Even though the seasonal demand of these activities clashes with the local or wildlife routine way of life, respect and soft measures should be taken into account. One practical example to illustrate this point can be seen from the initiative by the Forest Department (FD) in India. According to the FD policy, land and wildlife species and plants which are shared between the local farmers and the mountain activity, a message to tourists about their roles and responsibilities while undertaking such activates is clearly outlined. Besides, while hiking or trekking, various signposts are put up in places by the officials in each area of the route, detailing the mating or breeding season of sheep (e.g., a reproductive season of sheep) (Apollo et al., 2018). These are minor initiatives, but have a profoundly positive impact on tourist experiences as they create awareness. To the locals, management consideration is seen as a sign of respect, and in return they support the conservation efforts and tourism development initiates, which is the reason why many trekking and hiking trails pass through the private property of local farmers or owners. Hence, there is no conflict in most cases between local farmers or communities with the regional government.

Although there is no denial that accessibility to mountain areas is crucial (Apollo, 2017a), accessibility without enforced measures to minimise the environmental pollution generated through mountain tourism can have a serious impact on the *ecology* of the mountain, mountain communities, species and plants which live in the shared environment. Therefore, implementing some environment pollution measures could be an efficient step to control the adverse impact of mountain tourism in PAs, the sooner the better. There are many pieces of evidence available in other parts of the world to highlight this issue due to the lack of proper and strict enforced management measures, whether that be the handling of littering resulting from food and human waste of mountaineers (Apollo, 2016, 2017c). A well-known example that might support this argument is the most popular mountain destination in the world—Sagarmatha/Mount Everest. The government of Nepal has consistently lagged behind in controlling the littering issues on the mountain as well as in the base camp areas, or observed by Kaseva (2009) the case of the most polluted mountain on Earth—Kilimanjaro. However, in recent years, some changes and precautionary measures have been undertaken by the officials, but the issues of managing littering remain unresolved and out of control, due to the lack enforced measures.

Third, mountain tourism can have a positive impact on the local economy, but officials need to change the traditional contradictory view which discourages tourism development and encourages conservation. As observed by Sekhar (2003) in India, rather than encouraging policy tools such as 'participatory governance structure' that has the potential to conserve the physical land and local environment, and empower local people by providing job options for sustainable livelihoods, officials are mostly influenced by a conservation view leading to restrictions on tourism development. This perception needs some flexible approaches which can be done by using locals and their ways of life as a potential tool to develop mountain tourism and sustainably promote PAs. For instance, tourists desire to travel to the tranquil landscape, learn about mountain communities' cultures, beliefs and way of life, and are attracted to wildlife and the environment. These are the competitive advantage of PAs and an important reason for tourists to undertake mountain tourism activities. Mountains hold a central role, being symbols of local beliefs and religions, and it is a cultural representation based on the personal meaning attached to them. For instance, local high-altitude mountain guides of Nepal perform climbing rituals such as prayers at the basecamp for team success and safety before the actual expeditions climb (Ortner, 1999). These are micro aspects of expedition experience but culturally hold a significant value to the locals and if seen from a destination attraction perspective, these practices add meaningful experiences to the tourist or client-climbers who are not from the host destination.

Management should initiate job opportunities, particularly targeting locals (Badola et al., 2018). Tour guiding is one of the jobs that can support the management. Depending upon which conservation efforts are required, management should train and employ local people in this profession. As noted by Weiler and Black (2015), in niche tourism areas such as ecotourism, sustainable, adventure, and nature-based tourism, tour guides have a major role in educating and influencing tourist attitudes and behaviour towards practising sustainability, whether it be before the tour trip, ongoing or after the guided trips. In this respect, these mountain locals whose lives have been shaped by living around the mountains, their traditional knowledge and experiences could be a greater use and insightful to the management while designing the policies. Therefore, inviting and educating locals (e.g., free training courses) to be a tour guide in niche tourism areas has several advantages for mountain tourism policy planners.

Nature-based tour guides are seen as the frontline service provider because they contribute to creating tourist experiences (Pond, 1993) and also provide a nature interpretation service to tourists to enhance the core value of ecotourism, i.e., education about nature. In the meantime, they perform other encouraging roles to represent wider national characteristics of the destination to the tourist which they may or may not come into contact with. Tour guides play a role of 'mediator' and assist tourists to understand the various elements present in the destination, informing tourists about local cultures, raising awareness about the littering practices and including conservation efforts to save flora and fauna and wildlife species (Cohen, 1985). In this perspective, tour guides can be a valuable source of manpower to the tourism industry planner complimented by their local operational knowledge and traditional ground-level engagement. These aspects can also be used as a portion of significant information to evaluate destination status and conditions by the tourism planners (Albrecht, 2021). Besides, tour guides' day-to-day contact with the tourist can be the fastest way to obtain national-level data such as tourist demographics, changing demands, the purpose of travel, and choices of preferred accommodation (Kozak, 2004). Further, tour guides' everyday working conditions with the tourist can be useful information to understand the issues and challenges faced by the tourist at the destination (Prakash & Chowdhary, 2010). Therefore, in this sense, local guides can act as a 'tourism analyst' who can suggest and help to tackle the tourism industry officials and planners and complement their existing data, and can assist to promote tourism development sustainably and conservation efforts in many ways. However, the questions remain to what extent management sees these micro aspects of mountain tourist experience from mountain tourism activity as a potential attractive component of the destination, and in making sure that conservation efforts are not compromised.

Above, several considerations to deal with the management in respect of PAs and tourism development are outlined. The local communities' attitudes, influence and adequate involvement in tourism development and conservation practices play a crucial role in planning and developing tourism in PAs (Apollo, 2015, 2017a; Apollo et al., 2020a; Badola et al., 2018). If locals are given an equal leadership role in decision making and designing tourism development plans and benefit significantly from tourism activities, in return support to the management for conservation and tourism development have shown some promising outcomes.

An example of the challenges and opportunities described above are presented below in light of the case study of Stok Kangri.

11.5 Case Study: Mountain Tourism at Stok Kangri in Hemis National Park

Stok Kangri is the highest peak located in the Stok Range of the Ladakh in the Himalayas at 6154 m (6137 m according to other measurements). The peak is located in Ladakh, Hemis National Park, 12 km southwest of Stok village and around 15 km from Leh—the capital of Ladakh. Due to its non-technical ascent, Stok Kangri has become tremendously popular amongst trekkers and mountaineers (Table 11.3).

The mountain is located in Ladakh, which is also called 'Lesser Tibet'. It is the highest plateau in India which is located in the state of Jammu and Kashmir. Ladakh is situated between the northern and southern ranges of the Karakoram Himalayas, and most of its territory is above 3000 m. Due to its location, Ladakh's ecology is characterised by harsh terrain, limited accessibility, and extreme climate conditions,

Table 11.3 Overview of Stok Kangri case study site

Mountain range	Himalayas
Popular climbing season	Late July and August
Common routes to the top	The shortest and most popular route to the peak is along the Stok valley. Trekkers can walk from Leh to Stok village which takes approximately three hours, or if choosing to drive from Leh to Stok village, on an average it takes around 30 min
The average duration of the climb	It takes usually two to four days to reach base camp, 4.980 m from the village of Stok
The average cost to climb	Cost varies depending upon the tour package services

Source Authors' compilation

which in turn provides a cultural, physical and spiritual diversity that acts as a unique selling point for tourism (Goering, 1990). Ladakh officially started out welcoming 527 tourists in 1974, since then the number of tourists has grown exponentially (Menon, 2011). The Government of Jammu & Kashmir estimated that the total number of tourists grew from about 20,000 in 2001 to 150,000 in 2011 on account of offering major tourism products in Ladakh including nature, wildlife, adventure and religious tourism (Rajashekariah & Chandan, 2013).

Despite tourism being the major backbone of the economy, the primary occupation of most local communities is sustained through livestock grazing and subsistence agriculture (Michaud, 1991). Along with the rapid growth of tourism development, Ladakh has experienced drastic climate change that has changed the regional ecosystems (Barthwal-Chandola & Barthwal, 2014). For instance, there is a lack of tourism policies responding to the overcrowding issues and poorly managed infrastructure that threatens the environmental quality of the region, including water pollution, garbage (like gas bottles for cookers, food leftovers) and human waste. Research shows that the glaciers in the Ladakh region are shrinking at a rapid rate, and within two decades about 35% of them will disappear (Menon, 2011). Hence, sustainable ecological development has become one of the goals in the Ladakh region. The authorities have noticed the problem and there are more funds for sustainability initiatives such as renewable energy, particularly solar energy. Despite this, the number of tourists in Ladakh has risen and its impact is difficult to control.

The increasing popularity of the Stok Kangri peak has resulted in the delicate ecosystem of the region being over exploited. That is why, in December 2019, the 'All Ladakh Tour Operators Association' (ALTOA) in support of Stok village community (also known as Stok Nambardar Village Committee) announced a trail closure to trekkers from 2020 for three years (2020–2022). Water pollution problems are among the most significant impacts of over tourism on local communities. As tourism grew, so did the consumption of water; locals were facing a shortage of water on a daily basis and their drinking water was getting polluted by growing tourism. Locals in these areas solely depend upon the glacier water for their source of livelihood and as well as for the irrigation (Menon, 2011). Poor sustainability measures, lack of proper tourism infrastructure to supply the growing demand of tourists as well improper management of waste, have damaged the pristine landscape and spirit of the Stok Valley (Rajashekariah & Chandan, 2013).

The environment of the Hemis National Park is also badly affected by tourism (Geneletti & Dawa, 2009). This resonates with the authors' personal observation and data gained from using GIS modelling and remote imaging sensing showed that the park is heavily affected, mostly on account of tourism and climate change. For instance, Hemis National Park is easily accessible due to close proximity to Ladakh—the main tourist hub of the region. Second, the PA lacks a sufficient management plan to implement spatial zoning schemes or other forms of management policies (e.g., limiting tourist flows from 2020 until 2022) that can help to prevent further environmental degradation.

11.6 Conclusion

Establishment of PAs in India has been one of the most widely accepted means of biodiversity conservation so far. India, encouragingly, follows the sustainable goals discussed in Articles 10 and 11 of The Convention on Biological Diversity (CBD) (Porter et al., 1998). Tourism development in PAs should provide incentives to the local people who share the area with PAs or adjacent to it, meanwhile, such measures should also bear conservation practices. However, in practice such measures and success stories of following the CBD in India are few. Unlike successful stories of participatory conservation efforts of management and the local communities in Nepal (e.g., Annapurna Conservation Area) the integrated concept of mountain tourism with PAs in India is still young and an institutional framework to promote mountain tourism in PAs is still lacking. Although according to the 1997 National Tourism Policy Act of India, the policy suggests addressing the environmental, social and economic impacts of tourism for the local communities while undertaking conservation practices, the guidelines of legal or institutional framework to initiate sustainable measures when tourism is introduced in PAs areas is not outlined. Therefore, there is still a lack of clarity in policies and guidelines that could assist tourism planners when tourism is introduced in the conservation areas, and what to protect first and how to protect those, including humans and non-humans who are directly and indirectly involved within the shared premises.

References

Albrecht, J. (Ed.) (2021). *Managing visitor experiences in nature-based tourisms*. Cabi. https://www.cabi.org/bookshop/book/9781789245714/.

Andrejczuk, W. (2016a). Himalaje: szkic fizyczno-geograficzny – przyroda nieożywiona. *Acta Geographica Silesiana, 24*, 5–28.

Andrejczuk, W. (2016b). Himalaje: szkic fizyczno-geograficzny – biota, piętra roślinne i krajobrazy. *Acta Geographica Silesiana, 24*, 29–49.

Apollo, M. (2015). The clash-social, environmental and economical changes in tourism destination areas caused by tourism the case of Himalayan villages (India and Nepal). *Current Issues of Tourism Research, 5*(1), 6–19.

Apollo, M. (2016). Mountaineer's waste: Past, present and future. *Annals of Valahia University of Targoviste: Geographical Series, 16*(2), 13–32.

Apollo, M. (2017a). The true accessibility of mountaineering: The case of the High Himalaya. *Journal of Outdoor Recreation and Tourism, 17*, 29–43.

Apollo, M. (2017b). The population of Himalayan regions—By the numbers: Past, present, and future. In R. Efe & M. Ozturk (Eds.), *Contemporary studies in environment and tourism* (pp. 145–160). Cambridge Scholars Publishing.

Apollo, M. (2017c). The good, the bad and the ugly-three approaches to management of human waste in a high-mountain environment. *International Journal of Environmental Studies, 74*(1), 129–158.

Apollo, M., & Andreychouk, V. (2020). Mountaineering and the natural environment in developing countries: An insight to a comprehensive approach. *International Journal of Environmental Studies, 77*(6), 942–953.

Apollo, M., Andreychouk, V., & Bhattarai, S. S. (2018). Short-term impacts of livestock grazing on vegetation and track formation in a high mountain environment: A case study from the Himalayan Miyar Valley (India). *Sustainability, 10*(4), 951.

Apollo, M., Andreychouk, V., Moolio, P., Wengel, Y., & Myga-Piątek, U. (2020a). Does the altitude of habitat influence residents' attitudes to guests? A new dimension in the residents' attitudes to tourism. *Journal of Outdoor Recreation and Tourism, 31*, 100312.

Apollo, M., & Rettinger, R. (2019). Mountaineering in Cuba: Improvement of true accessibility as an opportunity for regional development of communities outside the tourism enclaves. *Current Issues in Tourism, 22*(15), 1797–1804.

Apollo, M., Wengel, Y., Schänzel, H., & Musa, G. (2020b). Hinduism, ecological conservation, and public health: What are the health hazards for religious tourists at Hindu temples? *Religions, 11*(8), 416.

Arriagada, R. A., Echeverria, C. M., & Moya, D. E. (2016). Creating protected areas on public lands: Is there room for additional conservation? *PLoS ONE, 11*(2), e0148094.

Badola, R. (1998). Attitudes of local people towards conservation and alternatives to forest resources: A case study from the lower Himalayas. *Biodiversity and Conservation, 7*, 1245–1259.

Badola, R., Barthwal, S., & Hussain, S. A. (2012). Attitudes of local communities towards conservation of mangrove forests: A case study from the east coast of India. *Estuarine, Coastal and Shelf Science, 96*, 188–196.

Badola, R., Hussain, S. A., Dobriyal, P., Manral, U., Barthwal, S., Rastogi, A., & Gill, A. K. (2018). Institutional arrangements for managing tourism in the Indian Himalayan protected areas. *Tourism Management, 66*, 1–12.

Barrett, C. B., Brandon, K., Gibson, C., & Gjertsen, H. (2001). Conserving tropical biodiversity amid weak institutions. *BioScience, 51*(6), 497–502.

Barthwal-Chandola, S., & Barthwal, S. (2014). *Marketing ecotourism in fragile ecosystem: Challenges and strategies in context of Ladakh*. Paper presented at the Annual Conference of the Emerging Markets Conference Board. Noida Campus: Centre for Marketing in Emerging Economies. IIM Lucknow, 9–11 January 2014.

Bhattarai, B. R., Wright, W., Poudel, B. S., Aryal, A., Yadav, B. P., & Wagle, R. (2017). Shifting paradigms for Nepal's protected areas: History, challenges and relationships. *Journal of Mountain Science, 14*(5), 964–979.

Bisht, R. C. (2008). *International encyclopaedia of Himalayas*. Mittal Publications.

Chandola, S. (2012). *An assessment of human-wildlife interaction in the Indus valley, Ladakh, Trans-Himalaya* (Doctoral dissertation). Saurashtra University.

Cohen, E. (1985). The tourist guide: The origins, structure and dynamics of a role. *Annals of Tourism Research, 12,* 5–29.
Conservation International. (2012). *Report.* https://www.conservation.org/.
Gawlik, A., Apollo, M., Andreychouk, V., & Wengel, Y. (2021). Pilgrimage tourism to sacred places of the high Himalaya. In D. Timothy & G. Nyaupane (Eds.), *Tourism and sustainable development in the Himalayas: Social, environmental, and economic encounters.* Routledge.
Geneletti, D., & Dawa, D. (2009). Environmental impact assessment of mountain tourism in developing regions: A study in Ladakh, Indian Himalaya. *Environmental Impact Assessment Review, 29*(4), 229–242.
Godde, P. M., Price, M. F., & Zimmermann, F. M. (1999). Tourism and development in mountain regions: Moving forward into the new millennium. In P. M. Godde, M. P. Price & F. M. Zimmermann (Eds.), *Tourism and development in mountain regions* (pp. 1–25). CABI Publishing.
Goering, P. G. (1990). The response to tourism in Ladakh. *Cultural Survival Quarterly, 14*(1), 20–25.
Gupta, S. K., & Bhatt, V. P. (2012). Changing expectations of traditional pilgrims. An analysis of expectations and motivations of tourists visiting Badri-Kedar tourism zone. In P. S. Manhas (Eds.), *Sustainable and responsible tourism: Trends, practices and cases* (pp. 54–65). Phi Learning Pvt. Ltd.
Ioannides, D. (1995). Planning for international tourism in less developed countries: Towards sustainability? *Journal of Planning Literature, 3*(9), 235–254.
ITS. (2013). *India Tourism Statistics 2012. Government of India. Ministry of Tourism.* www.tourism.gov.in.
Jreat, M. (2004). *Tourism in Himachal Pradesh.* Indus Publishing.
Kabra, A. (2009). Conservation-induced displacement: A comparative study of two Indian protected areas. *Conservation and Society, 7*(4), 249–267.
Kaseva, M. E. (2009). Problems of solid waste management on Mount Kilimanjaro—Challenge to tourism. *Waste Management Research, 28*(8), 695–704.
Kothari, A., Pande, P., Singh, S., & Variava, D. (1989). *Management of national parks and sanctuaries in India: A status report.* Indian Institute of Public Administration.
Kozak, M. (2004). *Destination benchmarking: Concepts, practices and operations.* CABI Publishing.
Megu, K. (2007). Prospects of cultural tourism in Arunachal Pradesh. In J. U. Ahmed (Ed.), *Industrialisation in North-Eastern region* (pp. 61–72). Mittal Publications.
Menon, S. (2011). *Two sides to Ladakh tourism.* The Hindu Business Line.
Messerli, B., & Ives, J. D. (1997). *Mountains of the world: A global priority.* Parthenon Publishing Group.
Michaud, J. (1991). A social anthropology of tourism in Ladakh, India. *Annals of Tourism Research, 18*(4), 605–621.
Mieczkowski, Z. (1995). *Environmental issues of tourism and recreation.* University Press of America.
Muradian, R., & Rival, L. (2012). Between markets and hierarchies: The challenge of governing ecosystem services. *Ecosystem Services, 1*(1), 93–100.
National Wildlife. (2019). *ENVIS centre on wildlife & protected areas.* http://www.wiienvis.nic.in/Database/Protected_Area_854.aspx.
Negi, C. S., & Nautiyal, S. (2003). Indigenous peoples, biological diversity and protected area management—Policy framework towards resolving conflicts. *International Journal of Sustainable Development and World Ecology, 10,* 169–179.
Negi, S. S. (1992). *Himalayan wildlife: Habitat and conservation.* Indus Publishing Company.
Negi, S. S. (1993). *Himalayan forests and forestry.* Indus Publishing Company.
Negi, S. S. (2002). *Discovering the Himalaya.* Indus Publishing Company.
Nepal, S. K. (2000). Tourism in protected areas: The Nepalese Himalaya. *Annals of Tourism Research, 27*(3), 661–681.

Nepal, S. K. (2003). *Tourism and the environment. Perspectives from the Nepal Himalaya*. Himal Books.

Nepal, S. K. (2008). Tourism-induced rural energy consumption in the Annapurna region of Nepal. *Tourism Management, 29*(1), 89–100.

Ninan, K. N., Jyothis, S., Babu, P., & Ramakrishnan, V. (2000). Economic analysis of biodiversity conservation: The case of tropical forests in the Western Ghats. *Bangalore: India-Environmental Management Capacity Building Technical Assistance Project, ISEC*.

Ortner, S. (1999). *Life and death on Mount Everest: Sherpas and Himalayan mountaineering*. Princeton University Press.

Phillips, A. (2004). History of the international system of protected area management categories. *Parks, 14*(3), 4–14.

Pond, K. (1993). *The professional guide: Dynamics of tour guiding*. Van Nostrand Reinhold.

Porter, G., Clemencon, R., Ofosu-Amaah, W., & Philips, M. (1998). *Study of GEFs overall performance*. World Bank.

Prakash, M., & Chowdhary, N. (2010). Tour guides: Roles, challenges and desired competences: A review of literature. *International Journal of Hospitality & Tourism Systems, 3*(1), 1–12.

Rajashekariah, K., & Chandan, P. (2013). *Value chain mapping of tourism in Ladakh* (Technical Report). https://doi.org/10.13140/2.1.1807.1046

Reynolds, K. (2013). *Trekking in the Himalaya*. Cicerone Press Ltd.

Sachs, J. (2015). *The age of sustainable development*. Columbia University Press.

Sekhar, N. U. (2003). Local people's attitudes towards conservation and wildlife tourism around Sariska Tiger Reserve, India. *Journal of Environmental Management, 69*(4), 339–347.

Sekhar, U. N. (1998). Crop and livestock depredation caused by wild animals in protected areas: The case of Sariska Tiger Reserve, Rajasthan, India. *Environmental Conservation, 25*(2), 160–171.

Sheppard, D. (2004). Editorial. *Parks, 14*(2), 1–5.

Singh, R.L. (1971). *India: A regional geography*. National Geographical Society of India.

TSHP. (2012). *Tourism survey for the state of Himachal Pradesh*. www.tourism.gov.in.

UNDP. (2010). United Nations Development Programme: *Capacity development: Stories of institutions. United Nations Development Programme*. Bureau for Development Policy Capacity Development Group.

Weare, G. (2009). *Trekking in the Indian Himalaya*. Lonely Planet Publications Pty Ltd.

Weiler, B., & Black, R. (2015). *Tour guiding research: Insights, issues and implications*. Aspects of Tourism 62. Channel View Publication.

Weinberg, A., Bellows, S., & Ekster, D. (2002). Sustaining ecotourism: Insights and implications from two successful case studies. *Society & Natural Resources, 15*(4), 371–380.

WWF Report. (2009). *The eastern Himalayas: Where worlds collide—New species discoveries*. www.wwf.se.

Yin, A., & Harrison, M. (2000). Geologic evolution of the Himalayantibetan orogen. *Annual Review of Earth and Planetary Sciences, 28*, 211–280.

Zhong, L., Deng, J., Song, Z., & Ding, P. (2011). Research on environmental impacts of tourism in China: Progress and prospect. *Journal of Environmental Management, 92*(11), 2972–2983.

Zimmerer, K. (2006). Cultural ecology: At the interface with political ecology—The new geographies of environmental conservation and globalization. *Progress in Human Geography, 30*(1), 63–78.

Zurick, D., & Pacheco, J. (2006). *Illustrated Atlas of the Himalaya*. The University Press of Kentucky.

Michal Apollo is an Assistant Professor at the Pedagogical University of Krakow, Institute of Geography, Department of Tourism and Regional Studies, and a Fellow of Yale University's Global Justice Program, New Haven, USA. Michal's areas of expertise are tourism management, consumer behaviours as well as environmental and socio-economical issues. Currently he is working on a concept of sustainable use of environmental and human resources. He is also an enthusiastic, traveller, diver, mountaineer,ultra-runner, photographer, and science populariser.

Viacheslav Andreychouk is a Full Professor and the head of the Geoecology Department at the Faculty of Geography and Regional Studies, University of Warsaw. Currently, his scientific research is concentrated on holistic concepts of current geography, which have been applied to optimize interactions in the man-environment system. Viacheslav has also been recognised as a speleologist, researcher and explorer of numerous caves.

Joanna Mostowska holds a Master in Geography and Regional Studies at University of Warsaw. Joanna is currently working on her PhD where she focuses on ultramarathons as a new branch in tourism. Her areas of interest are tourism management, especially in high mountain and polar regions as well as environmental and socio-economic issues.

Karun Rawat holds a PhD from the Department of Tourism, University of Otago. His doctoral research focused on the social role of emotions in commercial high-altitude mountaineering guides who have guided above the death zone mountains. His research interest includes mountaineering tourism, history of mountaineering and guides, commercial high-altitude mountaineering guides, and outdoor leadership.

Chapter 12
Nepal's Network of Protected Areas and Nature-Based Tourism

Rajiv Dahal

12.1 Introduction

Nepal is a landlocked Himalayan nation that lies between China and India with a population of 28.6 million in 2019 (CBS, 2019; World Bank, 2020). From the northern mountains to the southern plains, it is a hotspot for climatic and biological diversity. It has an area of 147,516 km^2 and its rectangular proportions measure on average 885 km in length and 193 km in breadth. Administratively, the country has seven provinces and divided across three broad physiographic regions: (i) the southern Terai plains, covers 23.1% of the total area, (ii) the Mid-Hills, cover 41.7%, and (iii) the northern Mountains, cover 35.2% respectively (CBS, 2017). Rich in natural and cultural diversity, varied landscape, rich cultural heritage, diversity of flora and fauna, majestic mountains, glaciers, lakes, and rivers, it has attracted travelers from around the world since 1951, when foreigners were first allowed to visit (Shrestha, 2016). Before that, Nepal was ruled by de facto rulers the 'Rana'. They had ruled Nepal for a hundred years with brutality and disparity. The change in regime was the first step in several political and economic transformations in Nepal, travel, and mobility for tourism and others were one such change (Shrestha, 2016).

Tourism has long been identified in Nepal as a powerful means for socio-economic transformation (Joshi & Dahal, 2019). However, the sector retains a relatively minor role in Nepalese development planning that has revolved around increasing agriculture productivity, infrastructure development, and hydropower production (Joshi & Dahal, 2019; Stevens, 1988). Despite insufficient consideration of the tourism sector, its increasing importance as a source of foreign exchange and employment and its continuing steady growth in an otherwise stagnant economy has brought growing attention in national economic planning (Gautam, 2011; Shrestha & Shrestha, 2012;

R. Dahal (✉)
Department of Travel and Tourism Management, Faculty of Social Sciences and Humanities, Lumbini Buddhist University, Lumbini, Nepal

© The Author(s), under exclusive license to Springer Nature Switzerland AG 2021
T. E. Jones et al. (eds.), *Nature-Based Tourism in Asia's Mountainous Protected Areas*, Geographies of Tourism and Global Change,
https://doi.org/10.1007/978-3-030-76833-1_12

Stevens, 1988). The country has experienced over six decades of tourism development. No international arrivals were allowed in Nepal until 1950. People with special status, were allowed to travel to Nepal, but were allowed within Kathmandu (the capital city) only. Thomas Cook and Sons, in 1955, was the first to start organizing a tour to Nepal, targeting westerners (Joshi & Dahal, 2019; Nepal, 2002; Shrestha, 2016). However, the initiation and popularization of organized trekking in the 1960s positioned Nepal as an exotic tourism destination in the world (Joshi & Dahal, 2019; Nepal, 2002; Shrestha, 2016). Up until the 1970s, Nepal was perceived as an extraordinary 'Shangri-la' destination, but this image has gradually eroded into a low-budget, unplanned, and unmanaged destination popular mainly for low spending tourists and the government's quest for volume-based tourism (Dahal, 2010; Joshi & Dahal, 2019; Nepal, 2002).

Currently, tourism contributes 2.2% to its GDP and it earns around US$ 703 million from one million tourists annually. The travel and tourism sector is a primary source of revenue, foreign exchange, and employment for the country, contributing 7.9% to the total GDP and 6.6% to the total employment (generating some 1.02 million jobs) as of 2018 (WTTC, 2019). Nepal's travel and tourism sector brought USD 2,051.4 million into the economy and supported 1.03 million jobs representing 6.9% of total employment in 2019 (WTTC, 2020). The National Tourism Strategy 2016–2025 envisages a fivefold increase in annual arrivals by 2025 (MoCTCA, 2016). The government has set a long-term target of a 25% contribution from the tourism sector (MoCTCA, 2018). Given the sector's direct positive impact on Nepal's economic growth, the government has been stepping up efforts to promote travel and tourism through arrays of activities—amending policies to make them more market-friendly, enhancing targeted marketing campaigns, and drawing investment programs and activities. One such effort was the Visit Nepal 2020 Initiative, to attract two million tourists and generate one million job opportunities in the sector by 2020. However, the COVID-19 pandemic affected such plans and the government had shelved the initiative as of August 2018.

Nature-based tourism is a significant driver of overall demand, with visiting Protected Areas (PAs) for trekking, mountaineering, jungle safari, canyoning, kayaking, etc. among the major activities. Nepal's PAs contain some of the highest mountains (above the sea) in the world, including Mt. Sagarmatha (Mt. Everest), Kanchenjunga, Lhotse, Makalu, Cho Oyu, Dhaulagiri I and Annapurna I. Similarly, the PAs on the plains (Terai) have dense forests that offer natural experiences to tourists. From mountaineering on top of the world to rowing down the wild rivers or navigating the dense jungle undergrowth for one-horned Rhinoceros and the Royal Bengal Tiger, Nepal's PAs offer unmatched opportunities for nature-based tourism. Besides more adventurous activities, the PAs are equally important for cultural/pilgrimage sites, such as Balmiki Ashram in Chitwan NP, Khaptad Baba ashram in Khaptad NP, Tengboche Monastery in Sagarmatha NP and Shey Monastery in Shey-Phokshundo NP. The PAs have been the center of attraction for many sages/gurus since the Vedic ages. Most of the spiritual/cultural sages traveled to mountains in search of the eternal soul, which had very character of mysticism, calmness, and spirituality (Chauhan, 2004; Shrestha, 1995). Many modern-day pilgrims are following the path albeit in collaboration with modern-day touristic activities such as yoga trekking and alternative lifestyle systems.

Sensing an increasingly vulnerable resource-base, Nepal has committed to placing more areas under protection by 2020 in line with the Convention of Biodiversity's Aichi Targets. Nepal's PA system is one of the most diverse and unique systems in the world, thus posing challenges for managing such areas to meet conservation needs. The PAs in Nepal has seen a gradual transformation in the management of such parks, from 'top-down' to a more collaborative, community-led approach. This has helped fuel nature-based tourism in the PAs, but there has been growing discontent among community and tourism service providers other stakeholders towards the policy, style, and practices of the authorities, at times demanding the handover of the PA's control to the community. There has been some progress in that area, particularly in cases where the community has been empowered to oversee PAs, such as Kanchenjunga Conservation Area. Although other PAs also advocate for community involvement and participation, more needs to be done to pacify growing resentment amongst the community residing in or around the PAs.

12.2 Nature-Based Tourism in Nepal: An Overview

The predominant form of tourism in Nepal is nature-based, thanks to Nepal's ecological diversity and cultural richness (Zurick, 1992). Nepal has 8 out of 14 eight-thousanders, which is a pull factor to attract tourists from all around the world. Consequently, Nepal has experienced unprecedented growth in the past sixty years, increasing from approximately 6000 tourist arrivals in 1962 to almost 1 million tourist arrivals in 2018 (MoCTCA, 2020). This is extraordinary, as Nepal allowed only a few outsiders until 1951.

Before 1951, the country was ruled by de facto rulers—the 'Rana', who allowed only a few foreigners for mountaineering exploration and expedition activities to enter/stay in Nepal (Shrestha, 2016). They had seized power from the Royal family and ruled for almost a century, seeking to gain control over the socio-political and economical system. The Ranas, influenced by the British-Indian diplomacy, pursued a 'closed-door policy' limiting the entry of high-ranking British Officers and aristocrats, mainly for big-game hunting. For others, Nepal remained largely isolated until the regime change in 1951. This led to the relaxation of borders to foreigners, and eventually witnessed a gradual increase of outsiders, mainly tourists (Shrestha, 2016).

Regarding the rise of tourism in Nepal, the initial impetus came from mountaineering as a climbing expedition in 1949 was attributed to the country's popularization by outsiders (UNESCAP, 1991). Mountaineering at that time required a long trek to unexplored and unmapped areas, and with the construction of an airport in 1954 (the current Tribhuvan International Airport), other types of tourists soon followed in the footsteps of the explorers and mountaineers (Gurung, 2007; Kunwar, 1999; Shrestha, 2016; UNESCAP, 1991). Tourism began to gain momentum after

1950, with the first ascent of an 8000 m peak recorded by Maurice Herzog on Mt. Annapurna I (8091 m). Another milestone involved an ascent of the world's highest peak, Mt. Sagarmatha (Mt. Everest), by Edmund Hillary and Tenzing Norgay Sherpa in 1953. This extraordinary feat received worldwide publicity and drew the eyes of the world to Nepal's mountains. However, the number of tourists visiting Nepal was still limited to 6000 in 1962 (MoTCA, 2009). Thereafter the number grew almost nine-fold in a decade (1962–1972, not including Indian tourists). Ever since the trend has been one of almost unbroken growth.

Tourism in Nepal has grown exponentially over the years, however with few dips (Fig. 12.1). The major events that took place between 1999 and 2005, both the internal as well as external issues, were to blame for such decline. Tourism growth trend had dipped due to international and national incidents. Examples of the former include the 2011 terror attacks in the US; and the 2003 SARS outbreak in the Association of Southeast Asian Nations (ASEAN) region. Domestic examples such as the Indian Airlines hijacking (1999); the Hrithik Roshan scandal (2000); and the Royal Massacre (2001) all contributed to downturns in the tourism industry. But, the main reason for the decline in tourist visitation (especially between 2000 and 2006) was the effect of 'the unrest' in the rural areas led by the Communist Party of Nepal (Maoist) (CPM). With the CPM clinching the majority of seats in the Constitution Assembly Polls of 2008 and subsequently forming and leading the then coalition government, the country has shown some positive signs towards national reconciliation. The long-awaited peace process has not only given hope to the people of Nepal but has equally enthused overseas travellers to visit the country (Dahal, 2012).

The tourist arrivals increased from 0.38 million in 2004 to 1.19 million in 2019 (MoCTCA, 2020). The increase in the number of tourists could be attributed to the political stability that the nation has been witnessing for the past few years. Tourism has been an important economic contributor to Nepal's economy. The recent

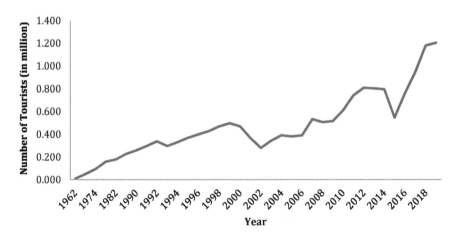

Fig. 12.1 International tourist arrivals in Nepal. *Source* MoCTCA (2020; various)

National Tourism Statistics 2019 shows that gross foreign exchange earnings from tourism rose from USD 158.7 million in 2000/2001 to around USD 670.6 million in 2018/2019 (MoCTCA, 2020). Due to the increase in international visitors, it is paramount to develop this sector to generate more revenue, employment, and other benefits, considering the current low level of tourism development. The planning activities in Nepal have revolved around increasing agriculture productivity, infrastructure development and hydropower production, but less on tourism (Joshi & Dahal, 2019). Regarding the much-needed foreign exchange, the government's tourism philosophy is to attract tourists and afterwards to hope to generate more income, employment and tax revenues. But tourism development depends upon the improvement of basic infrastructure, information, facilities, access, transportation options, safety and security, which are all needed in the case of Nepal.

The concept of conservation-friendly tourism-development has long been debated (Garrod & Fyall, 2000; Robinson, 2004; Wight, 1993). The increased tourism development at places of tourist interests, particularly in PAs, has far-reaching consequences, not only for tourism, but also for the ecosystem and the resources which tourism and other economic activities depend on. PAs, globally and in Nepal, have been used as an environmental conservation tool in maintaining representative samples of unpolluted and unaltered species and ecosystems for the future, and equally to limit the potential for environmental degradation through human management of resources (Grant et al., 1998). PAs are a key to tourism growth and development in Nepal. With increased tourism growth and development in Nepal, the demand for PA resources for nature-based tourism will increase.

12.3 Nepal's PAs System

12.3.1 Overview

Nepal is blessed with varied geographical conditions, from the plains in the south to the high Himalayas to the north. This results in natural hotspots for diverse ecosystems and flora and fauna, to the extent that Nepal accounts for 3.2% of global flora and 1.1% of fauna species (DNPWC, 2020a). Considering the abundance of natural resources and the subsequent need for conservation, Nepal has established an extensive network of PAs (see Fig. 12.2). As of 2020, there are 12 National Parks, 1 Wildlife Reserve, 1 Hunting Reserve, 6 Conservation Areas, and 13 Buffer Zones in Nepal (DNPWC, 2020a). Although Nepal occupies less than 1% of the global landmass, it has committed to the designation of 23.39% of the country for PA purposes, well above the Aichi Target 11 benchmark of 17%. The PAs of Nepal hosts popular spots not only for tourists but also equally among conservationists around the world (DNPWC, 2020a).

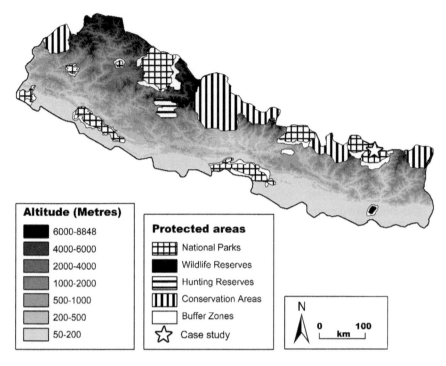

Fig. 12.2 Nepal's protected area system. *Source* Adapted from DNPWC (2019)

Nepal's network of PAs extends over an area of 34,419 km², which has helped in conserving forest genetic resources (MoFSC, 2016; NTNC, 2020b). Bhattarai et al. (2017) stated that "Nepal has been in the global forefront for nature conservation and has its commitments towards global conservation agendas" (p. 85).

12.3.2 Governance and Control

The term 'governance' gained recognition at the 5th IUCN World Parks Congress in Durban in 2003 that identified it as "central to the conservation of protected areas throughout the world" (WCPA, 2003, p. 33). The WCPA congress emphasized that governance is about leadership, power-sharing, vision, and commitments. Governance with PAs has to do with *policy* (stated intentions backed up by authority) converging with *practice* (the direct acts of humans affecting nature) (Borrini-Feyerabend & Hill, 2015). Also, it is about understanding, communicating and allocating power and resources in tandem with policy and practice (Borrini-Feyerabend & Hill, 2015, p. 171). Governance is to balance the goals of conservation with that of human and economic aspirations and developmental needs. In this sense, governance is the key to successful PA management and effectiveness (Dearden et al., 2005).

Effective PA management depends on the institutional set-up and frameworks. WCPA (2003) stated that the performance of such institutions against the mandates they are responsible for is the key to such effectiveness. Equally, it is about building relationships with stakeholders, which includes communities and community-based organizations, local government and society as a whole. In Nepal, there is a clear mandate, institutional set-up and framework to govern the PAs. From the early days, the main goal of establishing Nepal's PAs was to protect wildlife, especially rare and endangered species. Over time, PA governance has evolved from 'strict protection' of specific sites to 'ecosystem conservation', and then later to a 'landscape approach' model. The 'landscape approach' model incorporates human activities and the conservation of natural, historic, scenic, and cultural values (MoFSC, 2016).

PAs in Nepal are administered by the Federal Government under the jurisdiction of the Ministry of Forest and Soil Conservation (MoFSC). A separate department under the MOFSC looks after the governance and control of the PAs. The details about the authority and their framework are discussed in the section below.

12.3.3 Institutional Framework and Authority

The Department of National Parks and Wildlife Conservation (hereafter referred to as DNPWC) established in 1980 AD is the entrusted authority to conserve and manage wildlife and biodiversity in Nepal. The main goal of the DNPWC is to conserve wildlife and outstanding landscapes of ecological importance for the well-being of the people (DNPWC, 2020a). The overriding mandate is to conserve the country's major representative ecosystems, unique natural and cultural heritage, provide protection to valuable and endangered wildlife species, and encourage scientific research for the conservation of genetic diversity. The department implements various regular programs in and around the major PAs, such as habitat management, species conservation, anti-poaching operations, conservation education and ecotourism promotion.

Through its various divisions and sections (Fig. 12.3), the DNPWC oversees the governance of PAs. All PAs, except conservation areas, are regulated by the DNPWC in collaboration with Nepal Army and without the involvement of local people in management decisions (Bhattarai et al., 2017; DNPWC, 2020c). To support its conservation effort, the Government of Nepal has mandated National Trust for Nature Conservation (NTNC) to manage three PAs—Annapurna, Manaslu, and Gaurishankar Conservation Areas. NTNC also closely works with the government and other stakeholders in building conservation capacities across PA in the Terai plains region (NTNC, 2020b). The DNPWC and MOFSC oversee the management plan and other relevant jurisdiction about PAs delegated to the NTNC.

To manage PAs in Nepal, the DNPWC has enshrined five themes and strategies, under which all management plans are directed to guide/support the conservation needs and goals, namely: (i) Protection and Conservation of Biodiversity; (ii) Habitat management; (iii) Research, Monitoring, and Capacity Building; (iv) Species Conservation Special Program; and (v) Eco-tourism and Interpretation.

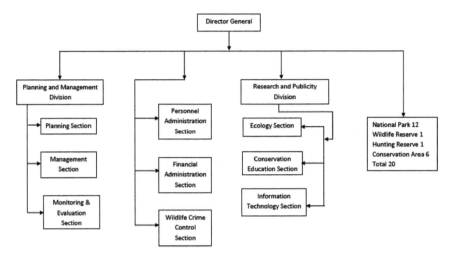

Fig. 12.3 DNPWC organizational structure. *Source* Unofficial translation from DNPWC (2020b)

12.3.4 Management Tools, Policies, and Priorities

Several laws and regulations are guiding Nepal's PAs and their management. Over the past five decades, there has been the promulgation of adequate policies on biodiversity conservation in Nepal. Such policies have evolved over various periods and conditions. Several legislations have been promulgated by the government to conserve the nation's biological diversity while harnessing the sustainable use of such resources.

The earliest legally binding document related to PAs and their governance was the Wildlife Conservation Act (1958), promulgated for wildlife patrols, specifically to protect rhinos in the Chitwan Valley (now the Chitwan National Park). The document did not deliberate on the enactment or establishment of PAs but was instrumental in the establishment of the authority (DNPWC) and the subsequent acts and regulations to govern the PAs. The National Park and Wildlife Conservation Act (NPWCA) (1973) was the first such document to provide a legal basis for the establishment of PAs in Nepal. With its 4th amendment in 1994, the NPWCA provided a legal framework to institutionalize Nepal's network of PAs (NTNC, 2020a). PA management is regulated by the NPWC Act (1973) and its regulations (1974). Clause 3 (1 Ka) of the Fifth Amendment of the act made 'management plans' mandatory for national parks, reserves, and conservation areas. Each plan must be approved by the DNPWC.

Nepal is a signatory to the Convention on International Trade in Endangered Species of Wild Fauna and Flora (CITES) since 1975. Since the early days, Nepal has been implementing CITES through various initiatives for the conservation and management of CITES-listed species (Joshi et al., 2017). CITES also guides the NPWC Act and Regulation viz-a-viz the conservation of biodiversity in PAs. The CITES Act 2073 has been enacted for the conservation of various species of the endangered fauna and flora. This Act has authorized the Chief Conservation Officer

Table 12.1 Policy and statutory instruments related to wildlife and PA management

Act/regulation/guidelines/directives	Policy/strategy/plan
• National Parks and Wildlife Conservation Act, 1973 • National Parks and Wildlife Conservation Regulation, 1974 • Chitwan National Park Regulation, 1974 • Forest Act, 1993 • Forest Regulations, 1995 • Wildlife Reserve Regulations, 1977 • Himalayan National Parks Regulation, 1979 • Conservation Area Management Guidelines, 1999 • Conservation Area Management Regulation, 1996 • Conservation Area Government Management Regulation, 2000 • Buffer Zone Management Guidelines, 1999 • Buffer Zone Management Regulation, 1996 • Environmental Protection Act, 2019 • Environmental Protection Regulation, 1997 • Control of International Trade of Endangered Wild Fauna and Flora Act, 2017 • Control of International Trade of Endangered Wild Fauna and Flora Regulation, 2019 • Protected Area Program Implementation Directive, 2016 • Local Self-Governance Act, 1999 • Elephant Management Regulations, 1966 • Wildlife Damage Relief Assistance Guidelines, 2013 (Third Amendment, 2018) • Yarsagumba Management (Collection and Extension) Directive, 2017	• National Conservation Strategy for Nepal, 1988 • Master Plan for the Forestry Sector, 1989 • Nepal Environmental Policy and Action Plan I, 1993 • Nepal Environmental Policy and Action Plan II, 1998 • Sustainable Development Agenda for Nepal, 2003 • Three Years Interim Plan (2007/2008–2009/2010) • National Wetland Policy, 2012 • National Forest Policy, 2015 • Nepal Biodiversity Strategy Implementation Plan, 2006 • Nepal Biodiversity Strategy, 2002 • Herbs and Non-Timber Forest Products Development Policy, 2004 • Working Policy on Construction and Operation of Development Projects in Protected Areas, 2008 • Working Policy on Wild Animal Farming, Breeding and Research, 2003 • Domesticated Elephant Management Policy, 2003 • Forest Sector Strategy 2016–2025

Note Except for these, each PA is governed by their separate act and regulation formulated and endorsed by the government of Nepal
Source Various

or PA officer assigned by him/her to work as an Investigation Officer in an illegal wildlife trade case and to file a case in District Court as per Clause 23. Relevant acts, policies and statutory instruments that have a direct or indirect bearing on PA management are presented in Table 12.1.

12.3.5 Stakeholders

Stakeholders include any individual or group that has a certain level of interest in any projects or development activities, including conservation (Golder & Gawler, 2005; Hashim et al., 2017; Kothari, 2008; Lockwood, 2010; Varvasovszky & Brugha, 2000).

They affect or could affect the conservation activities of any given area. These are the ones who have to compromise their resource benefits for the greater goal of conservation. Stakeholders' participation and inclusion in any PAs planning and management could greatly improve the success and effectiveness of PAs (Ivanić et al., 2017; Zafra-Calvo & Geldmann, 2020). Successful examples of stakeholder's involvement and participation could be attributed to almost all PAs of Nepal (Bhattarai et al., 2017; Poudyal et al., 2020; Shrestha et al., 2019). However, there are issues of stakeholder's rights and growing discontent. As of now, the park authorities have been successfully able to garner its support towards the logical distribution of benefits whilst ensuring conservation goals.

Federal Government is the sole authority overlooking the governance and financing of PAs in Nepal. There are several partner ministries, government bodies, community-based organizations, community users groups, forest users group, INGOs and NGOs, development aid agencies such as World Bank, International Finance Corporation (IFC), International Conservation Organizations (such as WWF, IUCN, UNESCO, etc.). Besides these, park staff, local people, herders, youth clubs, women groups, etc. are other key stakeholders. However, most of Nepal's PAs, except conservation areas, are mostly managed with insufficient stakeholder consultation.

12.4 Community and Protected Areas

The success of PA management lies in their relationship with the communities living in and around them (Allendorf et al., 2019). A community's positive attitude towards conservation efforts, their involvement, and participation in planning and management are keys to effective PA management (Andrade & Rhodes, 2012; Oldekop et al., 2016; Struhsaker et al., 2005; West & Brechin, 1991). PAs bring enormous socioeconomic impacts for local people (Braber et al., 2018). Designation of PAs aims to ensure the conservation of ecosystem services, which could limit communities' traditional social and economic systems, even pushing them into poverty (Brockington & Wilkie, 2015; Fisher & Christopher, 2007; Sherbinin, 2008). However, PAs globally has opened up opportunities for new income sources, such as nature-based tourism (Ferraro et al., 2011). With regards to PAs in Nepal, tourism has brought significant economic respite to the communities. This has equally helped them gain new knowledge, skills and awareness regarding tourism and conservation needs.

Previous research states that except for Nepal's conservation areas, most PAs are governed and managed with little or no consultation with the community (Bhattarai et al., 2017; Ghimire, 1994; Mishra, 1984). The local community greatly benefits PAs in their conservation goals too. PAs, where the community has been assured and shared of PA benefits, have shared great success. Communities have been instrumental in minimizing undue impacts of development and economic activities (including tourism) in and around PAs. The management plans for different national parks have emphasized the need for community participation. One example, the Koshi Tappu Wildlife Reserve Management Plan, 2018, acknowledged the need for local participation, along with the need for promotional activities and periodic monitoring to minimize impacts of tourism in the area. Most management plans now acknowledge the community's role and right but there is a gap in practice that has fuelled the resentment against park staff and officials.

Nepal has reported many cases of conflict between PAs and the community, including noted issues, such as (i) communities demanding customary rights (mainly, access to natural and cultural resources); (ii) compensation for damage and loss of life and/or properties by wildlife; (iii) increased migration of people from outside to the periphery of PAs due to better economic opportunities causing loss of opportunities for the indigenous and traditional landowners; (iv) translocation of communities from the PAs, hindering the subsistence and social-fabric; and (v) use of excessive force by patrol-team to the communities, leading to mistrust and resentment with the PAs authorities, etc. (Bhattarai et al., 2017; Budhathoki, 2004; DNPWC, 2015).

Local communities have been the beneficiary of PAs establishment but have also borne the social and economic costs. In many PAs, the costs have outweighed any such benefits. Allendorf and Gurung (2016) affirmed that finding win-win solutions for communities living adjacent to PAs is difficult. In Nepal too, there has been a continual search for such a solution. The fourth amendment (in 1992) of the 1973 National Park and Wildlife Conservation (NPWC) Act added 'Buffer Zones' as "a peripheral area of a national park or reserve to provide facilities to use forest resources on a regular and beneficial basis for the local people" (NPWCA, 1992). The goal of establishing such a buffer is to benefit local people, and provide opportunities for the local community to participate and assist the park authorities in PAs management activities (Bhusal, 2012). The revised act committed to 'plow-back' 30–50% of total park revenues into the communities involved with buffer zone management activities, mainly for community development (Bhatta et al., 2010; Thing & Poudel, 2017). However, this provision seems to have benefited only those communities in PAs with comparatively higher revenues derived from strong demand for nature-based tourism, e.g. in national parks such as Chitwan, Sagarmatha, and Langtang, together with Annapurna Conservation Area (Bhatta et al., 2010). PAs with lower revenues continue to face conflict and community resentment. Despite revisions to the relevant legislature, such as the NPWC Act (1973); Buffer Zone Regulations (1996); and Buffer Zone Guidelines (1999, 2015); to help the community economically and socially, actions to mitigate conflict did not yield better results, being overshadowed by conservation or by political agendas.

Figure 12.4 provides an overview of revenues earned by PAs from 2014 to 2019. Since 2016, there has been profound growth in tourism numbers thus revenue reaching 1767.36 million NPR 1.76 billion (USD 14.89 million; 1 USD = 118.772 NPR as of 27th August 2020) in 2019. A significant proportion of the revenues could be attributed to nature-based tourism. There have been other achievements too, particularly in the conservation areas, mainly in terms of community-managed conservation areas and provisions to prevent the relocation of the community from such areas (Bhandari et al., 2018). Similar cases of allowing extraction of thatch

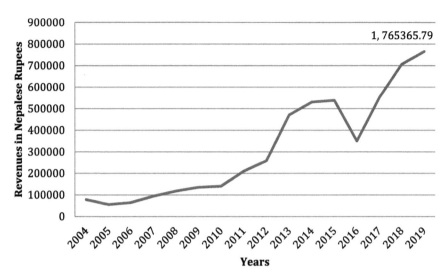

Fig. 12.4 Aggregate revenue from divisions and PAs under DNPWC jurisdiction. *Source* DNPWC (2020c) and MoCTCA (2018)

grasses on a season basis from PAs located in the Terai plains and the establishment of zoning regulations for extraction in conservation areas are some of the other initiatives implemented to minimize conflict between parks and people (DNPWC, 2015).

12.5 Tourism and PAs

12.5.1 Tourism in PAs

Tourism has played a pivotal role in the development of fragile, remote, and inaccessible countries like Nepal. Due to Nepal's geographical location, weather and climate, tranquility, and spirituality, the PA system has been attracting more international travelers. Just as PAs provide them with recreational opportunities, tourism provides opportunities for host communities with few other means to sustain their livelihoods, and the relationship has become symbiotic so that it is difficult to opt for development without tourism (Dahal, 2010). In 2019/2020, 0.42 million foreigners visited PAs in Nepal (refer Fig. 12.1). PAs are equally popular among domestic visitors for relaxation, recreation, religious and cultural pursuits. The reasons are the varied touristic resources these PA offers. There has been growing resentment over the PAs and the cost to maintain it at the disposal of the local's aspirations and needs. PAs offers socio-economic benefits to the community and the industry operating in and around them, but such has been inadequate. It is therefore quite natural to call for appropriate tourism development in the PAs. Table 12.2 provides details on tourists visiting the PAs of Nepal.

12.5.2 Tourism Dynamics (Activities and Tourism Types)

Nepal's PAs have an abundance of both natural and cultural manifestations. These, coupled with adventure activities, have positioned Nepal as a hotspot for nature-based tourism. Nepal's PAs offer arrays of touristic activities, such as mountaineering, trekking, rock climbing, canyoning, kayaking, canoeing, nature walking, jungle safari, jungle driving, paragliding, ultra-light aircraft, skydiving, mountain flight, trail running, yoga trekking, etc., which are dependent on nature. Also, the Shey Monastery of SheyPhoksundo NP, Tyangboche Monastery of Sagarmatha NP, Ne Me Pasal Cave, and the Kalinchowk Temple in Gaurishankar Conservation Area are some of the most revered cultural/religious sites of the PAs. Hundreds of thousands of tourists (including recreational) visit these PAs every year. Table 12.3 presents the factsheet on the place of visit by tourists visiting Nepal.

From Table 12.3, it is evident that the majority of visitors to Nepal (0.42 million) visit the PAs for touristic purposes. If the data of trekking to Manaslu, Mustang, Lower Dolpa, Upper Dolpa, and Kanchenjunga is considered, these areas, to a larger extent, fall under the jurisdiction of PAs. A safe estimate from Table 12.3 shows more than half of international visitors choosing destinations to PAs. Similarly, more than 60% of tourism activities in Nepal (including Nepalese) centers around four PAs of Nepal—Annapurna CA, Sagarmatha NP, Chitwan NP and the Langtang NP (MoCTCA, 2020). PAs other than these four receive considerably fewer visitors. The reason mainly centres around (i) lack of publicity and promotion of these PAs, and (ii) lack of tourism infrastructures and services.

12.5.3 Tourism Impacts

There has been significant improvement in the economic development of such areas; however, such benefits are skewed. Despite such, stakeholders of PAs of Nepal favor the increasing number of tourists and the positive impact of tourism. Some of the socio-economic benefits that tourism has brought to communities and stakeholders are increased participation in conservation and development activities (Bajracharya & Lama, 2008; Nepal, 2000; Paudel, 2016); revitalization and strengthening of cultural traditions and practices (Karanth & Nepal, 2012); increased income and employment opportunities for youths, women and marginalized groups (Cooper, 2001; Wells, 1994); revenue generation for government and parks (Aryal et al., 2019; DNPWC, 2020c); increased education and awareness level (DNPWC, 2020c); affordable access to housing and health-care (Wells, 1994); community's capacity building (Acharya & Halpenny, 2013; Paudel, 2016). On the other hand, negative impacts

Table 12.2 Tourist trends to Nepal's PAs

Protected areas	Number of visitors				
	17 July 2015–15 July 2016[a]	16 July 2016–15 July 2017[a]	16 July 2017–15 July 2018[a]	17 July 2018–16 July 2019[a]	17 July 2019–15 July 2020[a]
Chitwan National Park	178,257	87,391	139,125	118,621	142,486
Bardia National Park	13,548	10,638	1,7959	6773	8260
Langtang National Park	12,265	5016	8254	10,619	12,132
Sagarmatha National Park	34,412	27,794	45,112	56,303	57,289
Rara National Park	143	132	201	317	421
SheyPhoksundo National Park	383	431	535	469	578
Khaptad National Park	2	21	26	21	67
Makalu Barun National Park	1270	828	1537	1252	2057
ShivapuriNagarjun National Park	N/A	N/A	N/A	16,813	12,496
Banke National Park	0	0	4	1	12
Shukla Phata Wildlife Reserve	N/A	N/A	N/A	464	329
KoshiTappu Wildlife Reserve	N/A	N/A	N/A	436	388
Parsa Wildlife Reserve	263	235	417	95	15
Dhorpatan Hunting Reserve	0	91	163	119	424
Krishnasar Conservation Area	N/A	N/A	N/A	43	37
Api Nampa Conservation Area	3	29	19	36	38
Kanchenjunga Conservation Area	0	502	479	821	806
Annapurna Conservation Area	114,418	83,419	144,409	172,720	181,746
Manaslu Conservation Area	5658	2287	5745	7200	7655
Gaurishankar Conservation Area	2818	1840	2770	2668	2528

(continued)

Table 12.2 (continued)

Protected areas	Number of visitors				
	17 July 2015–15 July 2016[a]	16 July 2016–15 July 2017[a]	16 July 2017–15 July 2018[a]	17 July 2018–16 July 2019[a]	17 July 2019–15 July 2020[a]
Total	N/A	N/A	N/A	395,791	429,764

[a]Nepalese fiscal year is 1 Shrawan (4th month of Bikram calendar) to 31 Ashad (3rd month of Bikram calendar). Compared to the Gregorian calendar, the Nepali year starts and ends in mid-July
Source MoCTCA (2020)

Table 12.3 Place of visit by tourists 2018–2019

Places visited	2018	2019	% change
National Parks and Wildlife Reserve	395,791	429,764	8.6
Pashupati Area (excluding Indian)	163,311	171,937	5.3
Lumbini (excluding Indian Tourists)	169,180	173,083	2.3
Manaslu Trekking	7371	6070	−17.7
Mustang Trekking	4116	3739	−9.2
Humla Trekking	10,814	8670	−19.8
Lower Dolpa Trekking	1222	1360	11.3
Kanchanjunga Trekking	970	911	−6.1
Upper Dolpa Trekking	525	530	1.0

Source MoCTCA (2020)

from tourism have occurred because of economic reasons. The forced displacement because of the enactment of PAs has forced communities out of the area and has limited their employment opportunities. Even such opportunities are taken by outsiders. Some of the reported negative impacts of tourism are—the increased cost of living and deteriorating conditions (Yergeau, 2020), skewed economic benefits (Kandel et al., 2020; Yergeau, 2020), benefits controlled by outsiders (Aryal et al., 2019), acculturation leading to change in the local way of living (Kunwar, 2002), prostitution, drug and alcohol abuses (Lipton & Bhattarai, 2014).

In summary, there are various problems related to tourism in PAs of Nepal, including the need for education of personnel in the tourist industry on environmental issues, energy/fuel depots, litter control, medical assistance and on the demand for agricultural produce on major trekking routes (Cooper, 2001; Karanth & Nepal, 2012; Nepal, 1999; Paudel, 2016; Wells, 1994). Congestion on main tourism centers and routes is also perceived to be a problem, especially in the four main areas of Annapurna Conservation Area, Chitwan National Park, Sagarmatha National Park and the Langtang National Park. These negative environmental impacts increase when the movement and behavior of tourists are not well managed, thus threatening the sustainability of tourism and the PAs.

12.6 Sustainable Tourism in Protected Areas: Issues and Challenges

The major objective of PAs around the world has been balancing conservation needs for present and future generations by enriching visitors' experiences at such sites. Tourism in PAs of Nepal is centered on offering recreational activities to tourists, thereby maximizing revenues from such touristic activities. This seriously challenges the notion of resource sustainability. It is important to communicate the issue of tourism impacts and conservation goals to visitors, hotel/lodge operators and community alike. There remain unbalanced tourism growth and development in the PA system of Nepal. Some PAs, like the Annapurna CA, Chitwan NP, Sagarmatha NP, Langtang NP, receive a higher proportion of visitors, while others, like Kanchenjunga CA, Khaptad NP and Shuklaphanta WR, receive fewer visitors (see Table 12.3). Those receiving more tourists face problems of unmanaged tourism growth and development that poses challenges to the resources, while those receiving less have the challenges of economic sustainability. While discussing sustainable tourism in PAs of Nepal, it is important to discuss other factors that are not necessarily tourism bound but has a greater effect on the sustainability of tourism and the PAs itself. Those have been discussed as issues and challenges hereunder.

- Acculturation and culture change. The fast-changing culture and the conditions of the touristic places (mainly villages and settlements) will greatly affect the prospect of developing cultural and heritage tourism in the area. The change in the local way of life, custom and traditions, housing pattern, language, food habits, dresses, and ornaments are the greatest threats to develop and promote cultural and heritage tourism in the area.
- Geological condition. The volatile geological condition (tectonic and seismic movements) is the greatest threat to all types of tourism in the PAs.
- National policy favoring mass tourism. Nepal's PAs, blessed with natural and cultural resources, are best suited for soft ecotourism, ranging from adventure tourism, ecotourism, cultural and heritage tourism, and wellness tourism. However, the quest for most tourism will significantly affect the tourism resource base. Controlled tourism greatly assists in sustaining the resources and equally brings high yield to the area. However, the lack of education and awareness regarding the scale and size of tourism and the resultant impact of such have induced stakeholders towards volume-based tourism rather than locally owned and controlled type tourism.
- Climate change. The increasing climate change trend will surely affect the natural resources of the PAs. The increasing temperature and rise in weather changes will greatly affect the snow-based and water-based touristic activities in the region, particularly, mountaineering, rafting and kayaking. Similarly, the lack of snowfall and an increase in temperature will equally affect the serenity and spirituality of the area.
- Search-and-Rescue logistics. The natural vulnerability, coupled with a lack of logistic and rescue mechanisms, has led to tourist injuries and fatalities in PAs.

One such example is the effect of an avalanche in the Everest region in 2015. Similar incidents occurred in other PAs, albeit without attracting as much attention as the Everest case. An important yet underestimated challenge for PA managers is to have search and rescue capabilities ready along with a logistics system in case of emergencies.

Considering Nepal's PA system in terms of geographic coverage and context, it is difficult to holistically manage the issues outlined above. Lack of information about resources, norms/practices, hazards and conservation needs aggravates the problem of tourism in PAs. However, the key to these problems could be attributed to education and interpretation. Eagles et al. (2002) stated that tourism helps in protecting the natural and cultural heritage of PAs, as such "transmits conservation values, through education and interpretation" (p. 24). Providing access to PAs programs for interpretation and environmental education ultimately benefits locals. Additionally, adequate information to visitors, industry and the communities through interpretation (using displays, materials, and interpreters) will help mitigate negative impacts from tourism. In this sense, education and interpretation holds the key to sustainable tourism in PAs.

12.7 Tourism and Conservation at the World's Highest National Park: Sagarmatha National Park

The Sagarmatha National Park (SNP), established in 1976 and listed in UNESCO World Heritage Sites in 1979, is spread over an area of 1148 km^2, in the Himalayan ecological zone of Nepal (see Fig. 12.2) (SNP, 2018). The park elevation ranges from 2845 m at Monjo to 8848 m on top of Mt. Sagarmatha (also known as Mt. Everest or Mt. Qomolangma). Major tourist attractions of the park include: (a) Mountains (mainly, Lhotse, Cho Oyu, Nuptse, Thamserku, Amadablam); and (b) people and culture (famed Sherpa people and monasteries such as—Tengboche, Thame, Khumjung, and Pangboche). The park has diverse biodiversity. With 118 species of birds, the park is also home to the red panda, musk deer, Himalayan Tahr, marten, Himalayan mouse hare (pika).

Early tourism in the Sagarmatha National Park (SNP) could be attributed to mountaineering since the 1950s and trekking since the 1960s. With 20 tourists in 1963, the number of visitors to SNP has increased significantly (SNP, 2016). In 2018–2019, the park received 57,289 visitors (MoCTCA, 2020). The increased number of tourists and associated activities in the area has brought significant improvement in the livelihood of the community living along the main trekking trails/settlements. Communities on the trekking trails have benefited from tourism-induced economic activities, such as climbers, guides, porters, entrepreneurs, traders, etc.

The tourism growth in the 1960s evolved from camping tourism and has consolidated to tea-house/lodge-style tourism. Niche-adventure-centered tourism in the SNP has garnished attention globally. The reason could be attributed to natural heritage and the cultural richness of the area. The park offers unmatched mountaineering experiences to the top of the world—Mt. Sagarmatha.

Everest has been the prime attraction of the park. 280 people successfully climbed Everest in 2019 (MoCTCA, 2020). Besides Everest, other peaks are equally popular; however, Everest remains the most alluring for a range of reasons summarized simply by George Mallory in 1923 as—"because it's there" (Jacquemet, 2017). Tourism in the region could be classified into three: (1) Adventure sports tourism, comprises mountaineering expeditions, trekking, ice-climbing, sky-diving, etc.; (2) Cultural tourism—Monastery visits and stays, Mani Rimdu and Dumje festival, village tours to Thame, Thamo, etc., and; (3) Recreational tourism—Heli Tours, Photography, Golfing, Animal viewing, etc.

The growth of tourism over the years has significantly boosted the local economy and has greatly transformed the livelihood of the local people. The direct revenue from climbing Everest alone stood at USD 4.05 million in 2019 (MoCTCA, 2020). Table 12.4 presents a snapshot of mountaineering status and royalty fees to climb some of the peaks in the SNP. Tourism has equally helped in meeting the conservation goals of the SNP.

Despite the greater socio-economic benefits of tourism, there have been significant negative impacts too, such as seasonal employment; inflation; acculturation and culture change (Dahal, 2011; Rai, 2017); change in locals subsistence pattern (Nepal, 2016); littering and garbage due to climbing activities (SNP, 2016); trail degradation (Nepal, 2003, 2016); tourism impact on snow leopard and its habitat (SNP, 2016), etc.

Tourism in the PAs is anticipated to enrich the visitor's experience and contribute towards conservational needs (SNP, 2016). The sustainability of tourism and its role in heritage protection for future generations has been the prime focus of park managers. Here, interpretation has been pointed out as a key tool in meeting such objectives. Based on the author's own research (2010, 2011) and personal information from Sagarmatha National Park (Office, 2016) some of the major issues and challenges related to tourism and conservation can be outlined as follows:

(1) Seasonality and overcrowding—there is a surge in tourist number during peak seasons (September–November and March–May) which affects park experiences, visitors' satisfaction and also stresses the parks managers and the industry/operators with the cost of operation/management;

Table 12.4 SNP mountain climbers and revenue in 2019

Peak's name	Height (m)	Number of climbers	Amount of fees (royalty) (US$)
Mt. Everest	8848	397	4,058,000
Mt. Lhotse	8516	13	11,700
Mt. Nuptse	7855	27	16,200
Mt. Amadablam	6814	402	159,200

Source MoCTCA (2020)

(2) Skewed tourism benefits—people living on the main trekking trails are the greatest beneficiaries of tourism. The villages/settlements outside the Everest Base Camp trail loops are deprived of economic and other benefits;
(3) Sprawl, ribbon development and congestion—the issue of concentrated development along the trekking trails to Everest Base Camp has presented park managers with challenges to diversify tourism from the crowded trekking trails to the nearby villages and communities;
(4) Unplanned tourism development in and around the park—lack of tools for monitoring and controlling the tourism development and its impacts has posed challenges in terms of fragmented wildlife habitat
(5) Lack of policy and plans to guide the appropriate forms of tourism (such as ecotourism) within the park;
(6) Limited interpretation facilities in the park;
(7) Limited conservation awareness programs for the local communities and visitors; and
(8) Climate change (see above).

12.8 Conclusion

Nepal has been a frontrunner in biodiversity conservation through the enactment of conservation laws and PA systems that have benefited from the state's provision of ecological sensitivity and conservation policies. Such policies have evolved over time, from a strict top-down style to a more collaborative and inclusive approach. Nepal's PAs have achieved global recognition for conservation success with community participation at the core of all activities. However, PAs other than conservation areas, are strictly managed by authorities with less inclusion of the community. The Buffer Zone concept has benefited the PA and the community alike. However, there has been a growing resentment over the access and control of the resources.

The resource diversity at these PAs has enabled nature-based tourism to flourish and benefit the park economically. The Sagarmatha NP, Annapurna CA, Chitwan NP and Langtang NP are the most popular tourist destinations in Nepal. Their varied landscape, loftiest mountains, dense forest, and wild rivers offer tremendous opportunities for nature-based tourism, ranging from trekking and mountaineering tourism to canyoning, bungee jumping, kayaking and jungle safari, to name a few. More than 60% of total tourism in Nepal centers around the PA system in Nepal. Communities residing in the PAs that receive significant tourists have benefited too, but such benefits are confined to certain tourism centers/trails, such as Sauraha in Chitwan NP, a settlement along the main trekking trail to Sagarmatha (Everest) Base Camp, Annapurna Base Camp, Langtang Himal Base Camp, etc. However, residents outside of these particularly popular trails and tourist hubs are yet to share in the benefits, further fueling resentment amongst those deprived communities. The increasing pressure of economic development, migration, climate change and wildlife-human conflicts continues to ignite the debate over the status of many PAs in Nepal, mainly in terms of governance and control. The timely intervention of DNPWC to promote genuine ecotourism at PAs could help to minimize the negative impacts of mass tourism, but green-washing of the prevalent form of tourism will only aggravate the situation in the future. The long-term sustainability of these PAs subsequently depends upon community access, inclusion, and control; visitor management; mechanism of interpretation for visitors, community and industry alike; wildlife management to avoid wildlife-human conflicts, and attitudinal change in the mindset of the park officials when handling communities' grievances.

References

Acharya, B. P., & Halpenny, E. A. (2013). Homestays as an alternative tourism product for sustainable community development: A case study of women-managed tourism product in rural Nepal. *Tourism Planning and Development, 10*(4), 367–387.

Allendorf, T. D., & Gurung, B. (2016). Balancing conservation and development in Nepal's protected area buffer zones. *Parks, 22*(2), 69–82.

Allendorf, T. D., Radeloff, V. C., & Keuler, N. S. (2019). People's perceptions of protected areas across spatial scales. *Parks, 25,* 25–38.

Andrade, G. S., & Rhodes, J. R. (2012). Protected areas and local communities: An inevitable partnership toward successful conservation strategies? *Ecology and Society, 17*(4), art. 14.

Aryal, C., Ghimire, B., & Niraula, N. (2019). Tourism in protected areas and appraisal of ecotourism in Nepalese policies. *Journal of Tourism and Hospitality Education, 9,* 40–73.

Bajracharya, S., & Lama, A. K. (2008). Linking tourism to biodiversity conservation: A paradigm shift in protected area management. In S. B. Bajracharya & N. Dahal (Eds.), *Shifting paradigms in protected area management* (pp. 107–118). National Trust for Nature Conservation.

Bhandari, A. R., Rai, D. P., Wikramanayake, E., Thapa, G. J., Koirala, S., Gyawali, T. P., & Joshi, D. (2018). *Kangchenjunga conservation area: A retrospective Report, 2018.* WWF Nepal. https://www.wwfnepal.org/media_room/publications/?uNewsID=331530.

Bhatta, L. D., Koh, K. L., & Chun, J. (2010). Policies and legal frameworks of protected area management in Nepal. In L. Ling et al. (Eds.), *Sustainability matters: Environmental management in Asia* (pp. 157–188). World Scientific.

Bhattarai, S., Pant, B., & Timalsina, N. (2017). Conservation without participation: Detrimental effect of escaping people's participation in protected area management in Nepal. In S. A. Mukul & A. Z. M. M. Rashid (Eds.), *Protected areas: Policies, management and future directions* (pp. 83–104). Nova Science Publishers.

Bhusal, N. P. (2012). Buffer zone management system in protected areas of Nepal. *The Third Pole: Journal of Geography Education, 34–44.* http://dx.doi.org/10.3126/ttp.v11i0.11558.

Borrini-Feyerabend, G., & Hill, R. (2015). Governance for the conservation of nature. In G. L. Worboys, M. Lockwood, A. Kothari, S. Feary & I. Pulsford (Eds.), *Protected area governance and management* (pp. 169–206). ANU Press.

Braber, B. D., Evans, K. L., & Oldekop, J. A. (2018). Impact of protected areas on poverty, extreme poverty, and inequality in Nepal. *Conservation Letters, 11*(6),

Brockington, D., & Wilkie, D. (2015). Protected areas and poverty. *Philosophical Transactions of the Royal Society B: Biological Sciences, 370,* 1–6.

Budhathoki, P. (2004). Linking communities with conservation in developing countries: buffer zone management initiatives in Nepal. *Oryx, 38*(3), 334–341.

CBS. (2017). *Introduction to Central Bureau of Statistics, Nepal.* Presented as part of the First Training in Japan on Economic Census. Retrieved from https://www.stat.go.jp/info/meetings/nepal/pdf/pre01.pdf.

Central Bureau of Statistics (CBS). (2019). *Nepal in figures 2019.* https://cbs.gov.np/wp-content/uploads/2019/07/Nepal-in-Figures-2019.pdf.
Chauhan, Y. S. (2004). *Ecotourism in Nepal.* New Delhi: Kalinga Publication.
Cooper, M. (2001). Backpackers to Fraser Island: Why is ecotourism a neglected aspect of their experience? *Journal of Quality Assurance in Hospitality & Tourism, 1*(4), 45–59.
Dahal, R. (2010). *Development of an ecotourism model in the HinduKush Himalayan (HKH) Region for ensuring sustainable tourism: A case study of Sagarmatha (Mt. Everest) National Park (SNP).* The Setsutaro Kobayashi Memorial Fund.
Dahal, R. (2011). *Development of an ecotourism model for the Sagarmatha (Mt. Everest) National Park: Ensuring sustainable development through tourism* (Unpublished Doctoral Dissertation). Ritsumeikan Asia Pacific University.
Dahal, R. (2012). *Tourism and mountain protected areas of Nepal: Empirical study of motivations and preferences of tourist travelling to the Sagarmatha (Mt. Everest) National Park (SNP), Annapurna Conservation Area (ACA) and Langtang National Park (LNP).* The Setsutaro Kobayashi Memorial Fund.
Dearden, P., Bennett, M., & Johnston, J. (2005). Trends in global protected area governance, 1992–2002. *Environmental Management, 36*(1), 89–100.
Department of National Park and Wildlife Conservation (DNPWC). (2020a). *Welcome to Department of National Parks and Wildlife Conservation. DNPWC, Nepal.* http://www.dnpwc.gov.np/en/.
Department of National Park and Wildlife Conservation (DNPWC). (2020b). *Organisation structure.* http://www.dnpwc.gov.np/en/organizational-structure/.
Department of National Park and Wildlife Conservation (DNPWC). (2020c). *Annual Report FY 2075/76.* Kathmandu: DNPWC.
DNPWC. (2015). *DNPWC Law Book 2015.* DNPWC, Ministry of Forests and Soil Conservation (MOFSC). Kathmandu: DNPWC.
DNPWC. (2019). *Protected Areas of Nepal- 2076.* Kathmandu: DNPWC.
Eagles, P. F., McCool, S. F., Haynes, C. D., & Phillips, A. (2002). *Sustainable tourism in protected areas: Guidelines for planning and management* (Vol. 8). IUCN.
Ferraro, P. J., Hanauer, M. M., & Sims, K. R. E. (2011). Conditions associated with protected area success in conservation and poverty reduction. *Proceedings of the National Academy of Sciences of the United States of America, 108,* 13913–13918.
Fisher, B., & Christopher, T. (2007). Poverty and biodiversity: Measuring the overlap of human poverty and the biodiversity hotspots. *Ecological Economics, 62,* 93–101.
Garrod, B., & Fyall, A. (2000). Managing heritage tourism. *Annals of Tourism Research, 27*(3), 682–708.
Gautam, B. P. (2011). Tourism and economic growth in Nepal. *NRB Economic Review, 23*(2), 18–30.
Ghimire, K. B. (1994). Parks and people: Livelihoods issues in national parks management in Thailand and Madagascar. *Development and Change, 25,* 195–229.
Golder, B., & Gawler, M. (2005). Cross-cutting tool: Stakeholder analysis. *Resources for implementing the WWF standards.* WWF.
Grant, U., Kratli, S., Mahiba, Y., Magnussen, C., Saavedra, G. R., & Rodrigues, I. (1998). *Biodiversity and protected areas: The concept and case studies.* Institute of Development Studies (IDS).
Gurung, T. R. (2007). *Mountain tourism in Nepal.* Pratima Gurung.
Hashim, Z., Abdullah, S. A., & Nor, S. M. (2017, October). Stakeholders analysis on criteria for protected areas management categories in Peninsular Malaysia. In *IOP Conference. Series: Earth and Environmental Science* (Vol. 91, No. 1).
Ivanić, K. Z., Štefan, A., Porej, D., & Stolton, S. (2017). Using a participatory assessment of ecosystem services in the Dinaric Arc of Europe to support protected area management. *Parks, 23*(1), 61–74.

Jacquemet, E. (2017). Why do people come to see Mount Everest? Collective representations and tourism practices in the Khumbu Region. *Journal of Alpine Research\ Revue de géographie alpine* (105-3).

Joshi, N., Dhakal, K. S., & Saud, D. S. (2017). *Checklist of CITES listed flora of Nepal*. Department of Plant Resources (DPR).

Joshi, S., & Dahal, R. (2019). Relationship between social carrying capacity and tourism carrying capacity: A case of Annapurna conservation area, Nepal. *Journal of Tourism and Hospitality Education, 9*, 9–29.

Kandel, S., Harada, K., Adhikari, S., & Dahal, N. K. (2020). Ecotourism's impact on ethnic groups and households near Chitwan National Park, Nepal. *Journal of Sustainable Development, 13*(3), 113–127.

Karanth, K. K., & Nepal, S. K. (2012). Local residents perception of benefits and losses from protected areas in India and Nepal. *Environmental Management, 49*(2), 372–386.

Kothari, A. (2008). Protected areas and people: The future of the past. *Parks, 17*(2), 23–34.

Kunwar, R. R. (1999). *Fire of Himal: An anthropological study of the Sherpas of Nepal Himalayan region*. Nirala Publications.

Kunwar, R. R. (2002). *Anthropology of tourism: A case study of Chitwan-Sauraha*. Adroit Publishers.

Lipton, J. K., & Bhattarai, U. (2014). Park establishment, tourism, and livelihood changes: A case study of the establishment of Chitwan National Park and the Tharu People of Nepal. *American International Journal of Social Science, 3*(1), 12–24.

Lockwood, M. (2010). Good governance for terrestrial protected areas: A framework, principles and performance outcomes. *Journal of Environmental Management, 91*(3), 754–766.

Ministry of Culture, Tourism and Civil Aviation (MoCTCA). (2020). *Nepal Tourism Statistics 2019*. MoCTCA. https://www.tourism.gov.np/publications/1.

Ministry of Forests and Soil Conservation (MoFSC). (2016). *Conservation landscapes of Nepal*. MoFSC.

Ministry of Tourism and Civil Aviation (MoTCA). (2009). *Nepal Tourism Statistics 2008*. MOTCA.

Mishra, H. R. (1984). A delicate balance: Tigers, rhinoceros, tourists and park management vs. the needs of local people in Royal Chitwan National Park, Nepal. In J. McNeely & K. Miller (Eds.), *National parks, conservation and development: The role of protected areas in sustaining society* (pp. 197–205). Smithsonian Institution.

MoCTCA (2016). *National Tourism Strategy 2016–2025*. Kathmandu: MOCTCA.

MoCTCA. (2018). *Unveiling of main works to be performed within 100 days since the formation of this government*. https://old.tourism.gov.np/photonewsdetail.php?id=259.

National Park and Wildlife Conservation Act (NPWCA). (1992). *The fourth amendment (in 1992) of the 1973 National Park and Wildlife Conservation (NPWC) Act*. Government of Nepal.

National Trust for Nature Conservation (NTNC). (2020a). *Sustainable tourism enhancement in Nepal's Protected Areas: Environmental and social management framework*. NTNC.

National Trust for Nature Conservation (NTNC). (2020b). *Protected Areas: For species, habitats and people*. https://ntnc.org.np/thematic-area/protected-areas-and-ecosystems.

Nepal, S. K. (1999). *Tourism-induced environmental changes in the Nepalese Himalaya: A comparative analysis of the Everest, Annapurna, and Mustang regions* (Ph.D. dissertation). Faculty of Natural Sciences, University of Berne, Switzerland.

Nepal, S. K. (2000). Tourism in protected areas—The Nepalese Himalaya. *Annals of Tourism Research, 27*(3), 661–681.

Nepal, S. K. (2002). Tourism as a key to sustainable mountain development: The Nepalese Himalayas in retrospect. *UNASYLVA, 208*(53), 38–45.

Nepal, S. K. (2003). *Tourism and the environment: Perspectives from the Nepalese Himalaya*. Himal Books.

Nepal, S. K. (2016). 27 Tourism and change in Nepal's Mt Everest Region. *Mountain tourism: Experiences, communities, environments and sustainable futures, 270*.

Oldekop, J. A., Holmes, G., Harris, W. E., & Evans, K. L. (2016). A global assessment of the social and conservation outcomes of protected areas. *Conservation Biology, 30*(1), 133–141.

Paudel, R. P. (2016). Protected areas, people and tourism: Political ecology of conservation in Nepal. *Journal of Forest and Livelihood, 14*(1), 13–27.

Poudyal, B. H., Maraseni, T., & Cockfield, G. (2020). Scientific forest management practice in Nepal: Critical reflections from stakeholders' perspectives. *Forests, 11*(1), 27.

Rai, D. B. (2017). Tourism development and economic and socio-cultural consequences in Everest Region. *Geographical Journal of Nepal, 10*, 89–104.

Robinson, J. (2004). Squaring the circle? Some thoughts on the idea of sustainable development. *Ecological Economics, 48*, 369–384.

Sagarmatha National Park (SNP) (2018). *SNP: Home*. https://www.sagarmathanationalpark.gov.np/.

Sherbinin, A. D. (2008). Is poverty more acute near parks? An assessment of infant mortality rates around protected areas in developing countries. *Oryx, 42*, 26–35.

Shrestha, B. B., Shrestha, U. B., Sharma, K. P., Thapa-Parajuli, R. B., Devkota, A., & Siwakoti, M. (2019). Community perception and prioritization of invasive alien plants in Chitwan-Annapurna Landscape, Nepal. *Journal of Environmental Management, 229*, 38–47.

Shrestha, H. P. (2016). *Tourism management in Nepal*. Ultimate Marketing.

Shrestha, H. P., & Shrestha, P. (2012). Tourism in Nepal: A historical perspective and present trend of development. *Himalayan Journal of Sociology and Anthropology, 5*, 54–75.

Shrestha, T. B. (1995). *Mountain tourism and environment in Nepal*. International Centre of Integrated Mountain Development (ICIMOD).

SNP (2016). *Sagarmatha National Park and its Buffer Zone: Management Plan 2016–2020*. Namche Bazaar: SNP Office.

Stevens, S. (1988). Tourism and development in Nepal. *Kroeber Anthropological Society Papers, 67*(68), 67–80.

Struhsaker, T. T., Struhsaker, P. J., & Siex, K. S. (2005). Conserving Africa's rain forests: Problems in protected areas and possible solutions. *Biological Conservation, 123*(1), 45–54.

Thing, S. J., & Poudel, B. S. (2017). Buffer zone community forestry in Nepal: Examining tenure and management outcomes. *Journal of Forest and Livelihood, 15*(1), 57–70.

United Nations Economic and Social Commission for Asia and the Pacific (UNESCAP). (1991). *Environmental management of mountain tourism in Nepal*. UNESCAP.

Varvasovszky, Z., & Brugha, R. (2000). A stakeholder analysis. *Health Policy and Planning, 15*(3), 338–345.

Wells, M. P. (1994). Parks tourism in Nepal: Reconciling the social and economic opportunities with the ecological and cultural threats. In M. Munasinghe & J. McNeely (Eds.), *Protected area economics and policy: Linking conservation and sustainable development* (pp. 319–391). The World Bank.

West, P. C., & Brechin, S. R. (1991). *Resident peoples and national parks: Social dilemmas and strategies in international conservation*. University of Arizona Press.

Wight, P. A. (1993). Sustainable ecotourism: Balancing economic, environmental and social goals within an ethical framework. *Journal of tourism studies, 4*(2), 54–66.

World Bank. (2020). *Nepal*. https://data.worldbank.org/country/NP.

World Commission on Protected Areas (WCPA). (2003). *Durban action plan*. IUCN.

World Travel and Tourism Council (WTTC). (2019). *Travel & tourism: Economic impact 2019 Nepal*. World Travel & Tourism Council.

World Travel and Tourism Council (WTTC). (2020). *Nepal 2020 annual research: Key highlights*. https://wttc.org/Research/Economic-Impact.

Yergeau, M. E. (2020). Tourism and local welfare: A multilevel analysis in Nepal's protected areas. *World Development, 127*,.

Zafra-Calvo, N., & Geldmann, J. (2020). Protected areas to deliver biodiversity need effectiveness and equity. *Global Ecology and Conservation*, e01026.

Zurick, D. N. (1992). Adventure travel and sustainable tourism in the peripheral economy of Nepal. *Annals of the Association of American Geographers, 82*(4), 608–628.

Rajiv Dahal is a Visiting Faculty (Associate Professor) in the Department of Travel and Tourism Management, Faculty of Humanities and Social Sciences, Lumbini Buddhist University (LBU) in Lumbini, Nepal. His research interests include Ecotourism, Adventure-Tourism, Protected Areas Planning, and Management. Rajiv completed his Ph.D. at the Ritsumeikan Asia Pacific University and has done extensive research in the mountain protected areas of Nepal, particularly, Sagarmatha (Everest) National Park, Annapurna Conservation Area, Langtang National Park, and Gaurishankar Conservation Area.

Chapter 13
Mountainous Protected Areas in Sri Lanka: The Way Forward from Tea to Tourism?

Renata Rettinger, Dinesha Senarathna, and Ruwan Ranasinghe

13.1 Introduction

Sri Lanka is a tropical country with a history of protection of natural resources, specifically forests and wildlife protection, that dates back to the third century BC where an area was declared as a reserve for wildlife by the then King Devanampiyatissa. The territory of Sri Lanka, formerly known as Ceylon, is an island in the Indian Ocean and located approximately 80 km south of India. Tectonically, Ceylon is the southernmost fragment of the old Precambrian platform known as the Deccan (Maciejowski, 2003) platform.

Existing vegetation diversity around the island is based on climatic and ecological diversity, and they in turn are a result of the geological formation of Sri Lanka. This formation consists of three peneplains (Dahanayake, 1982): the highest (central highland), the middle (intermediate peneplain) and the lowest (coastal peneplain). The

---

R. Rettinger (✉)
Department of Tourism and Regional Studies, Institute of Geography, Pedagogical University of Krakow, Krakow, Poland
e-mail: renata.rettinger@up.krakow.pl

D. Senarathna
Department of Geography, University of Kelaniya, Kelaniya, Sri Lanka

New Zealand Tourism Research Institute, Auckland, New Zealand

D. Senarathna
e-mail: dsenarat@aut.ac.nz

R. Ranasinghe
Department of Tourism Studies, Faculty of Management, Uva Wellassa University of Sri Lanka, Badulla, Sri Lanka
e-mail: ruwan@uwu.ac.lk

© The Author(s), under exclusive license to Springer Nature Switzerland AG 2021
T. E. Jones et al. (eds.), *Nature-Based Tourism in Asia's Mountainous Protected Areas*, Geographies of Tourism and Global Change,
https://doi.org/10.1007/978-3-030-76833-1_13

central highland acts as a barrier and unevenly distributes the north-eastern and south-western monsoons, which are the prominent method of receiving the country's rainfall. This geographical feature has caused the creation of four main climatic zones: wet, intermediate, dry and semi-arid and varying the forest cover around the island. It has also made Sri Lanka a global biodiversity hotspot (Gunasinghe, 2011).

Sri Lanka's central highlands occupy a unique position among the main geographical zones of the country. It is an area elevated 300 m above the mean sea level and occupies about 17% of the country's land area. Highland forest areas are extremely wet with an average annual rainfall of 5000 mm while lowland areas are much drier with less than 2500 mm average annual rainfall (Mathanraj & Kaleel, 2017). Some areas have ground frost during January to March. The area is vulnerable to strong winds. The central highlands are also the watershed for 103 main rivers and more than 1000 feeder streams. Picturesque water cascades and waterfalls are numerous in the central plateau, with the highest ones—Bambarakanda (263 m—one of the highest in Asia), Diyaluma (220 m) and Kurundu Oya (206 m). Most of them were established on tectonic rapids created as a result of vertical movements of the Earth's crust in the late Tertiary (Maciejowski, 2003).

The central highland range provides shelter to 128 bird species, 20 amphibian species, 60 butterfly species, 17 mollusk species, 31 mammal species, 53 reptile species and 15 fish species. Among the total vertebrate animals recorded in Sri Lanka, five endemic species consisting of three freshwater fish species (Phillips garra, blotched filamented barb, and Martenstyn barb), one amphibian (Kirthisingha's rock frog) and one lizard (leaf nose lizard) are confined solely to the central highlands. Unfortunately, commercial goals and the human desire for profit were much more important than the protection of this unique environment. Humans realised that agroclimatic conditions of the island were perfect to grow tea and by the 19th century, the plantation era had begun.

The country's biodiversity is significant and worth protection amidst the agricultural, industrial developments and urbanisation. Backed by the Buddhist cultural influence, the people of this country tend to extend their kind-hearted protection not only to animals but also towards trees. The Sinharaja rainforest and Udawattakale, belonging to the Tooth temple of Kandy, are such historically protected areas, which still exist today (Niven et al., 1999). Thus, the protection of natural reserves, animals and water bodies was a part of the socio-cultural system of the people of Sri Lanka. At present, the island boasts nearly 27% of its area under protection which is technically one of the highest in Asia as well as in the world. However, due to plantation development (mainly tea), huge swathes of the island were transformed. Consequently, numerous plants and animals disappeared forever, and the natural landscape has been significantly modified, which is why it is extremely important to protect the remaining natural heritage of Sri Lanka.

In this chapter, we will focus on the mountainous Protected Areas (PAs). Those parts that have survived in their natural state and have not been turned into plantations are now under increasing tourist pressure. Furthermore, we make proposals for the future of mountain tourism in Sri Lankan PAs.

13.2 Sri Lankan National System of Protected Areas

13.2.1 History and Development of Sri Lanka's Protected Areas

The evolution of Sri Lankan PAs began with the establishment of the world's ancient wildlife sanctuary called 'Mihinthale' in the third century BC (Breuste & Jayathunga, 2010) and some of the early reserves such as Udawattakale and Sinharaja still exist to this day (Cummings, 2006; Ranwala & Thushari, 2012; Wijesinghe, 2003). Nature conservation strategies and approaches date back many centuries to the appointment of forest officers (*Kelekorala*) and the establishment of state organisations to protect the nature in the country under the British colonisation (Breuste & Jayathunga, 2010). The modern laws and regulations of PAs were introduced in the country after the arrival of British colonists, due to the intense hunting that took place after 1796. William Henry Gregory, the then Governor of the Island, introduced the historical Wildlife Preservation Act in 1872. Further, fulfilling a rising need against the intense hunting of British colonists the Ceylon Game Protection Society, today known as Wildlife and Nature Protection Society of Sri Lanka, was formed in 1894. The Ceylon Game Protection Society pushed the government to increase the country's PAs as well as to appoint responsible personnel to take care of the wildlife (Uragoda, 1994). In 1889, upon the recommendation of the Game Protection Society, the first sanctuary in Asia in modern times was declared as Yala, and after 1938 it was upgraded to national park status. Due to this significant achievement, the Game Protection Society continued to declare sanctuaries throughout the island and later many of them were designated as national parks (Uragoda, 1994).

The two central bodies currently responsible for wildlife protection in Sri Lanka are the Department of Forest Conservation and the Department of Wildlife Conservation (Department of Wildlife Conservation, 2020; Ekanayake & Theodore, 2017). The National Heritage Wilderness Act enacted in 1988 serves as the foundation for the operations of the department of Forest Conservation. Managing the island's forest reserves towards sustainable development goals is the key objective of this body. Regional offices are established in each reserve and the officers/rangers are equipped with necessary training and weapons as well as legal protection in order to serve the mission. For example, the Sinharaja rainforest was declared as a UNESCO World Heritage site in 1988 and it is managed by the Department of Forest Conservation as per the guidelines of UNESCO natural heritage sites.

The PAs of Sri Lanka are organised into several categories namely Strict Nature Reserves; National Parks; Nature Reserves; Jungle Corridors; and Sanctuaries (Fig. 13.1). The greatest extent of strict nature reserve remains Yala National Park which was declared in 1938 and spans 978.9 km^2. Hakkgala, the first declared strict nature reserve, has an extent of 1142 ha and it was designated in 1938. Ritigala, a strict nature reserve of 1528 ha was designated in 1941 (Senarathna, 2005).

The national parks are administered by the Department of Wildlife Conservation based on the Fauna and Flora Protection ordinance of 1937. In a national park

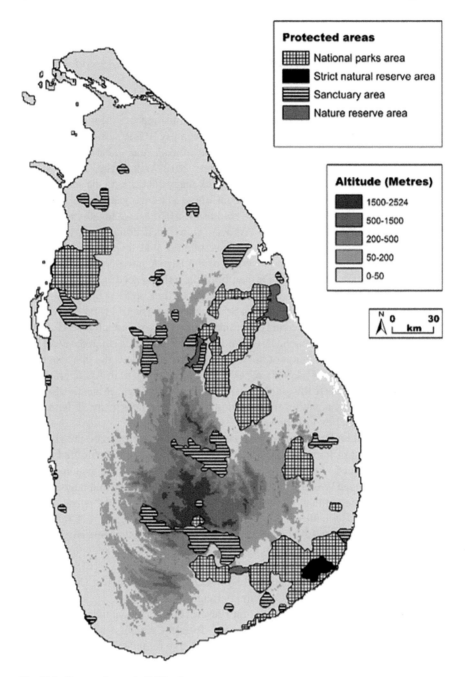

Fig. 13.1 Protected areas in Sri Lanka

Table 13.1 List of national parks in Sri Lanka

National park	Year established	Area (km^2)
Adam's Bridge	June, 2015	190
Angamadilla	June, 2006	75
Bundala	January, 1993	62
Chundikkulam	June, 2015	1976
Delft	June, 2015	18
Flood Plains	August, 1984	174
Galoya	February, 1954	259
Galway's Land	May, 2006	0.5
Hikkaduwa	October, 2002	1
Horagolla	July, 2004	0.3
Horowpothana	December, 2011	26
Horton Plains	March, 1988	32
Kaudulla	April, 2002	69
Kumana	January, 1979	181
Lahugala Kithulana	October, 1989	16
Lunugamwehera	December, 1995	235
Madhu Roard	June, 2015	164
Maduruoya	November, 1983	588
Minneriya	August, 1997	89
Pigeon's Island	June, 2003	5
Somawathiya	September, 1986	376
Udawalawa	June, 1972	308
Ussangoda	May, 2010	3
Wasgomuwa	August, 1984	371
Wipattu	February, 1938	1317
Yala	February, 1938	979
Total extent		5734

Source Compiled by author based on URL data collected in 2020

the entire wildlife is protected and activities such as hunting, removing of animals, destroying eggs, nests and habitats, carrying of fire and related activities are prohibited. The total area protected under national parks in the island is 5734 km^2 as at June 2020 (Table 13.1), that accounts for 8.74% of the landmass.

Nature reserves are different from strict nature reserves since traditional human activities are allowed inside the boundary. Wildlife studies and research initiatives are authorized under the supervision of the Department of Wildlife Conservation. The total area protected of nature reserves is 1016 ha on the island and Vidataltive, Triconamady remains the largest nature reserve. Jungle corridors were declared under the accelerated Mahaweli Development project in the 1970s and 1980s to protect the forests along the water bodies and the catchment areas of reservoirs such as Victoria, Randenigala and Rantambe. An area of 2780 km² is protected under the category of sanctuaries. Sanctuaries are least restricted in terms of regulation and are the lands outside the state claim since wildlife protection as well as human activities take place concurrently. Seruwila, Pallekele and Bar Reef are among the largest sanctuaries in the island.

13.2.2 Sri Lankan Mountain Protected Areas

Sri Lankan PAs have been classified by considering the forests' ecological value and services. Their administration was delegated to the Forest Department or the Department of Wildlife Conservation by considering the diversity of flora and fauna. Specifically, the open forests in dry and intermediate zones that contain more wildlife are administered by the Department of Wildlife Conservation. In contrast, the Forest Department governs the areas that have a higher density of tree cover. However, there are some overlaps in terms of services and conservation efforts provided by both institutions (Ekanayake & Theodore, 2017).

Sri Lanka's PAs belong to the state and not the private sector or community. Among the PAs that belong to the Forest Department or the Department of Wildlife Conservation, the forests lie under the National Heritage Wilderness Area, and Strict Natural Reserves categories which have high ecological value and place strict limits on any human activities. Access is only allowed to researchers with special permission obtained by the Directors General of these departments. Next to these categories, Conservation Forests, National Parks and Sanctuaries are given the highest priority, but any of the land that belongs to these government bodies are legally prohibited for community consumption. Typically, national parks and sanctuaries serve for tourism. Plantation Forests which consist of mahogany, teak, eucalyptus, pine and other local species provide support to agriculture, timber demand and safeguard the wildlife habitat to a certain extent (Forest Department, 2009).

Most of the mountainous PAs are clustered within the central highlands (Fig. 13.1) and were recognised as a UNESCO World Heritage site in 2010 under category IX and X, concerning, "the significant ongoing ecological and biological process", and "threatened species with outstanding universal value" (UNESCO 2010). This UNESCO heritage site covers 56,844 ha of property and 72,645 ha of the buffer zone in the central highlands including three significant sites: Peak Wilderness Conservation Forest; Horton Plains National Park; and Knuckles Conservation Forest. Table 13.2 shows ten mountain PAs which have a significant value for the tourism industry. Out of these, six are located in the central highlands, and all of them are higher than 1200 m. Four are situated in the wet zone while the other two mountain ranges, Namunukula and Knuckles, are located in the intermediate zone. Due to their magnificent peaks, these two are famous attractions for hiking and camping for both foreign and local tourists alike.

Table 13.2 Selected mountainous protected areas in Sri Lanka

	Mountain	Height (m)	Climatic zone	Topographic feature	PA name	Administration	PA established year	Size of the PA (ha)	Value of this mountain	Tourism activities	Type of tourists (local/foreign)
1	Piduruthalagala	2524	Wet	The tallest mountain in the island	Piduruthalagala Conservation Forest	FD	2007	7625	Ultra-prominent peak	Strictly not allowed	–
2	Adam's peak	2243	Wet	Conical mountain	Peak Wilderness Conservation Forest	DWC	1940	22,400	Religious/Spiritual	Pilgrims	Mostly local
3	Hakgala	2170	Wet	Dome mountain	Hakgala Strict Nature Reserve	DWC	1938	1142	Ecological/Legendary	Strictly not allowed	–
4	Namunukula	2036	Intermediate	Nine Peaks mountain range	Namunukula Conservation Forest	FD	2009	248	Scenic/Agricultural (Turpentine)	Hiking, Camping	Both
5	Knuckles	1863	Intermediate	Series of recumbent folds and peaks	Knuckles Conservation Forest	FD	2000	17,835	Ecological	Hiking	Both
6	World's End	1200	Wet	Sheer precipice	Horton Plains National Park	DWC	1988	3160	Ecological/Prehistoric	Hiking	Both
7	Sinhagala[a]	743	Wet	Residual mountain	Sinharaja National Heritage Wilderness Area	FD	1988	11,428	Ecological	Hiking (limited)	Both
8	Ritigala[a]	572	Dry	Residual mountain	Ritigala Strict Nature Reserve	DWC	1941	1528	Ecological/Archaeological	Strictly not allowed	–

(continued)

Table 13.2 (continued)

Mountain	Height (m)	Climatic zone	Topographic feature	PA name	Administration	PA established year	Size of the PA (ha)	Value of this mountain	Tourism activities	Type of tourists (local/foreign)
9 Dimbulagala[a]	534	Dry	Isolated residual mountain	Dimbulagala Forest	Dimbulagala Buddhist Temple	1950	1250	Religious/Archaeological	Religious/Spiritual	Mostly local
10 Wedasitikanda[a]	424	Semi-arid	Residual mountain	Wedasitikanda Reserved Forest	FD	1978	966	Legendary	Spiritual	Mostly local

Explanation: [a]Outside the central highland; PA—Protected Area; FD—Forest Department; DWC—Department of Wildlife Conservation

Source Created by author based on data collected from various websites, 2020

The highest mountain is Piduruthalagala (2524 m) which is covered by a conservation forest administrated by the Forest Department (Werner, 1995). Being an ultra-prominent peak, because of its high altitude, a special permit must be obtained to visit the summit. Adam's peak is a sacred conical mountain which has been popular among local people from ancient times. Haggala was designated as a Strict Nature Reserve in 1938, and has value not only ecologically but also in a religious and cultural sense. Being a Strict Nature Reserve, this mountain is not open for visitors. However, the country's second largest botanical garden, located on its north-eastern slope is an appealing tourist attraction. The Horton Plains National Park records the highest visitor numbers (foreign and local) of the mountain PAs, and its symbolic feature is the sheer precipice called the World's End, which is 1200 m in height. This plain is believed to have existed as an agricultural land in 15,000 BC, which has been proven by fossil evidence that adds a prehistoric value on top of its ecological and biological significance (Premathilake, 2006).

All four mountains located outside the central highlands are residual and are less than 1000 m (Fig. 13.1 and Table 13.2). Of these, only Sinhagala is located in the wet zone, while Ritigala and Dimbulagala are in the dry zone. Wedasitikanda is situated in the semi-arid zone which is in the south-east of Sri Lanka. Two out of the three strict nature reserves, which are administrated by the Department of Wildlife Conservation, are mountainous PAs. One of those is Ritigala, and visitors are not allowed into the strict nature reserve zone. Sinharaja is the only National Heritage Wilderness Area in the country owing to its high biodiversity, where the Sinhagala is located and offers limited visits (Forest Department, 2020). Dimbulagala and Wedasitikanda are popular mostly among locals for religious and spiritual tourism. The Dimbulagala Forest belongs to the Dimbulagala Buddhist Temple and is not an officially reserved forest. However, this is protected as one of many forest hermitages that have spread around the country and which existed since ancient times. Not limited to scenic and ecological significance, the Sri Lankan mountain protected areas have different and distinctive values such as archaeological, prehistoric, religious, spiritual, legendary and topographic significance which have great potential yet are rarely promoted in the tourism industry at present.

13.2.3 The Tea Factors

The natural landscape of Sri Lanka's natural protected area has undergone a major transformation, mainly due to the development of agriculture and the cultivation of the tea bush (*Camellia sinensis var assamica*). Tea requires special topographic conditions and this variable determines specific climatic conditions. Tea cultivation is highly dependent on agro-climatic conditions such as solar radiation, temperature and rainfall (Jayasinghe et al., 2020; Shalleck, 1972). Due to the need for year-round rainfall, tea is grown in the central highlands of Sri Lanka and on the wetter western slopes of the mountains. Although the crop does not require irrigation, drought significantly reduces the yield of tea. In turn, heavy rains cause soil erosion through surface washout (Wijeratne, 1996). These specific topographic and climatic conditions determine the spatial distribution of tea plantations and their species diversity depending on the height above sea level and the exposure of the slopes.

In the second half of the twentieth century there was a significant expansion of the tea-growing acreage, but after 1992 there was a significant decline due to state policy in the framework of the privatisation program and the new registration of plantations. During this period, there was a significant increase in the tea area of the low floor. In 2010, 60% of total tea production was produced from low-level tea fields (Palihakkara et al., 2015).

Currently, tea is grown on an area covering about 221,969 ha at various altitudes, whose criterion form the basis for distinguishing the following regions: low (<600 m above sea level), medium (600–1200 m above sea level) and high (>1200 m above sea level). There are seven main tea growing areas: Ruhuna and Sambaragamuwa on the low floor, Kandy on the middle floor in the central part of the country, and Nuwara Eliya, Dimbula, Uva and Uda Pussallawa in highly cultivated areas on the high floor (Jayasinghe et al., 2020). The aroma of the teas is influenced by specific conditions resulting from the unique combination of climatic conditions at different altitudes and with different exposures.

The tea industry is the main source of currency earnings in Sri Lanka accounting for 15% of net foreign trade income (Jayasinghe et al., 2020), generates 1.2% of gross domestic product and is the source of income for 2.2 million farm workers (Wijeratne, 1996). Tea fields, along with tea factories, are very important tourist destinations (Joliffe, 2004), having specific economic impacts. Landscape elements (mountain areas), visual elements of tea cultivation, observation of the tea harvesting work and learning about the tea production process are the most important tourist attractions. Visiting tea-producing regions is very popular among international tourists visiting Sri Lanka.

13.3 Tourism in PAs

13.3.1 Visitor Profile

The number of international tourist arrivals was 1,913,702 in 2019 with an average stay-length of 10.4 nights and an overall contribution of US$3607 million in foreign exchange to the Sri Lankan economy (SLTDA, 2020). By mid-2018, tourism was reckoned to have become the country's second-largest foreign exchange earner (Xinhua, 2018). In particular, Sri Lanka's PAs collectively form the second-largest foreign visitor drawcard and income generator after the Cultural Tringle (Anuradhapura, Polonnaruwa and Kandy) (SLTDA, 2020).

Tourists' motivations differ according to the experience they wish to gain while they are travelling through the mountainous PAs, an experience comprised of two types. The first type is through what tourists can see, feel and do *en route* to and from

PAs such as sight-seeing and mountain scenery. The second type is the experience gained from hiking, camping, fire nights (e.g., Knuckles), adventure, viewing wild animals and flora. The breath-taking view that can be enjoyed on the summit is the key motivation, for example, in the case of the sunrise view from the summit of Adam's Peak (Wickramasinghe, 2005).

Even though there are a number of Sri Lankan mountains that have become tourist attractions, the number of tourist arrivals and the income they generate are not reported in the literature. Table 13.3 provides this information only for five main mountain PAs, which was published by the Sri Lanka Tourist Development Authority (SLTDA, 2020). Although the Horton Plains received the highest number of domestic and international tourists, the highest income was recorded by Singharaja (for international tourism) and the Knuckles (for domestic tourism). All the destinations were dominated by domestic visitors, but generate more income through international tourist arrivals (mainly from entrance fees). Consequently, promoting Sri Lankan mountain PAs as an international tourist attraction can bring more economic benefits into the country as a whole and the proximate communities in particular.

Figures 13.2 depicts the number of domestic and foreign tourist arrivals between 2008 and 2019 in Horton Plains National Park. Horton Plains is Sri Lanka's most

Table 13.3 Mountain protected area visitor numbers and income in 2019

Attraction	Visitors			Income (Rs)		
	Local	Foreign	Total	Domestic	International	Total
Horton Plains	248,864	80,928	329,792	13,990,150	214,817,851	228,808,001
Galway's Land	5024	112	5136	144,750	196,200	340,950
Ritigala	52,788	9207	61,995	845,500	3,239,342	4,084,842
Sinharaja	42,867	13,877	56,744	802,562,500	7,580,938,000	8,383,500,500
Knuckles	78,263	1964	80,227	1,625,896,460	1,039,342,250	2,665,238,710

(USD 1 = SLR 185)
Source SLTDA (2020)

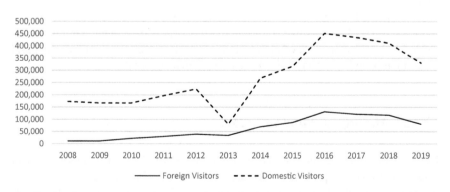

Fig. 13.2 Horton Plains' international and domestic visitation (2008–2019). *Source* SLTDA (2020)

Table 13.4 Longitudinal trends in mountain protected area visitor numbers

	2016		2017		2018		2019	
Galways land	198	3345	280	6085	277	5767	112	5024
Ritigala	3380	49,432	10,972	45,647	10,140	57,128	9207	52,788
Sinharaja	–	–	15,466	42,792	15,549	42,149	13,877	42,867
Knuckles	–	–	2450	66,090	2604	61,501	1964	78,263

Source SLTDA (2020)

popular destination for mountain PA tourism among both types of tourists. The visitation data of other attractions like Galway's land, Ritigala, Sinharaja and Knuckles are available only for recent years (Table 13.4). The domestic market dominates all destinations, e.g. 319,999 local visitors were recorded at Horton Plains while foreign visitors numbered 131,670 in 2016, the year that recorded the highest number of visitors for both types of tourists. After the civil war ended in 2009, both types of tourist arrivals showed an increment, even fluctuating, and then they have been decreasing. The sharp drop in 2019 could have been caused by the unfortunate situation of the Easter terrorist attack in Sri Lanka. According to the statistics in recent years, Ritigala and Sinharaja are popular among foreign tourists while Knuckles is for locals.

In order to maintain the infrastructure facilities in the mountainous PAs to a satisfactory level, garbage management and information availability need to be upgraded. Tourists such as backpackers should be able to reach any mountain PA within a day using the road network and regular public transport system whereas tour groups guided by international tour operators arrive at the destinations within a few hours by luxury bus. Most facilities, like accommodation, supermarkets, and communication, are sufficient in the nearest town of any mountain PA. However, inadequate or non-existent facilities such as information centres, souvenir shops, canteens and toilets in some mountain PAs need to be addressed (Rathnayake, 2016). Online ticketing, creating an official website for mountain PAs, giving safety instructions at the entrance, and placing enough signboards will be useful especially for self-guided tourists (Senevirathna & Perera, 2013). Visitor congestion at certain attractions within the PAs may lead to a reduction in the quality of their experience (Buultjens et al., 2005). In places like Horton Plains it is strictly prohibited to bring plastics inside the park, but at most of the other sites, mismanagement of garbage disposal is commonplace (e.g., Adam's Peak) (Karunarathna et al., 2011).

13.3.2 Managing Tourism in PAs

The mission of Department of Wildlife Conservation (2020) is "conservation of wildlife heritage for present and future generations." The central environmental authority is closely linked to the operations of the department in serving the above mission. Despite the strong legal structure and the regulations, the continuous deforestation and extensive development threaten the island's PAs. This can clearly be seen by the reduction of 70% of natural forest in the twentieth century to a mere 20% by 2010 (Mangala De Zoysa & Inoue, 2015; Subasinghe, 2013). Sri Lanka remains in the list of most rapidly deforesting nations in recent years and the forest coverage is now down to around 18%.

The present management profile is sound enough to continue the expected service. However, a discrepancy is observed between the core goals and commercial money-making activities that do not reflect the conservation priorities. For example, wildlife offices operate three reserves which have become financially lucrative for the country, yet the priority conservation mission mentioned above is left behind. Particularly, in the case of tourism, issuing tickets and guiding are sensitive issues to be closely monitored and redesigned. Secondly, the re-gazetting of PAs mainly to release land for industrial development, agricultural activities, community settlement and so forth are frequently observed and have been strongly backed by the political authorities in recent years. Such economic and community development projects are important but conservation should be a priority as well in order to protect the island's remaining wildlife resources. Chena cultivation is one of the key drivers of deforestation and community awareness of such unsustainable economic activities is much needed.

According to the Sri Lanka Environmental Journalist Forum (SLEJF), there are 7000 non-governmental organisations (NGOs) with an interest in environmentally-friendly activities. Government institutions directly contributing to environmental protection are the Wildlife Environment and Forest Conservation Department, the Central Environment Authority, the Geological Survey and Mines Bureau, the Wildlife Trust of Sri Lanka and the Mahaweli Authority; they protect the highland environmental resources and values. Other government authorities and 700 other NGOs provide additional services related to influencing, encouraging and promoting sustainable development. They communicate environmental issues and build awareness among the public, not only in Sri Lanka's central highlands but across the whole country. Government agencies and NGOs are trying to provide informal environmental education, especially for farmers and households. To reach the global environmental targets, Sri Lanka faces big challenges but also huge opportunities. Managing Sri Lanka's central highlands ecologically will be an important contribution to both global and national environmental targets.

Nature-based tourism (NBT) is one of the fastest growing tourism segments in the world (Higgins, 1996; OECD, 2009). This trend is confirmed in Sri Lanka's case, where tourists are interested in seeing wild animals in their natural habitat, especially elephants. Therefore, PAs play a key role in the development of Sri Lankan NBT. Extensive research indicates that tourists tend to focus more on the observation of fauna (especially wild animals) than flora (Higginbottom et al., 2001; Rodger et al., 2007). Sri Lanka typifies this lop-sided segmentation of the NBT market, since PAs where elephants can be observed are the most popular. This overwhelming focus on interaction with iconic wildlife puts the relevant PAs under great pressure, but could conversely save mountainous national parks with less popular fauna from over-tourism and negative impacts on delicate mountain ecosystems.

However, the park authorities are facing a new set of challenges: the first basic one is nature conservation, which is a key function of national parks. The protection of areas of exceptional importance due to biodiversity and the presence of endemic species especially concerns the mountainous areas of Sri Lanka. Another key issue is the relationship between PAs and the economic activity of the local population, mainly in agriculture. As previously mentioned, agroclimatic conditions (especially the amount of precipitation) are changing in connection with climate change. These new conditions will result in the shifting of plantation cultivation areas to higher-lying areas, which will be a real threat to natural plant communities. A significant part of forest areas, especially those in the buffer zone of national parks, will be threatened with deforestation. Based on the analysis of the literature, it can be concluded that it is currently one of the greatest problems for mountain areas (Seo et al., 2005).

A third challenge facing national parks is the management capacity to deal with increasing volumes of tourism. It is especially difficult to limit visitation to national parks because the revenues from admission tickets are an important revenue stream for PA budgets, especially in developing countries (Sumanapala, 2018). Therefore, an important task will be to reconcile high-quality tourist services with the protection of the area of national parks. Wildlife observation is one of the most important motivations for making a trip to national parks. Contemporary tourists set specific requirements related to the handling of tourist traffic, and the scope and level of service results from the imperative to see as many sights in as short a time as possible. Therefore, the tourist is interested in visiting national parks using a means of transport, which are very often off-road vehicles. By default, we are dealing with the emergence of a dense network of unplanned road connections and increased traffic by cars or other means of transport. This, unfortunately, causes roadkill and an increase in noise pollution to the environment (Karunarathna et al., 2017). Therefore, it is necessary to take actions aimed at the unification of the design and management policy of routes aimed at reducing the pressure of tourist traffic by rationalising the capacity and flow of tourist traffic. This is less of a problem in backcountry mountain areas, where hikers are more frequently encountered. In sum, it is necessary to create a management plan regulating the movement and stays of tourists in the national park. Sri Lanka's mountainous PAs require investment in the delineation and proper marking of mountain hiking trails and training of mountain guides.

Currently, the most important strategic activity of national park authorities is the development of a scheme of actions aimed at achieving common goals within three main groups of stakeholders (park authorities, local population and tourists). Unfortunately, a substantial number of legal enactments to dispel all undesirable aspects due to political interference, implementation of imposed legal enactments have been delayed or abandoned. If the legal enactments were enforced, the present problems related to environmental, social, cultural, and economic aspects could have been avoided (Breuste & Dissanayake, 2013).

Therefore, in developing a code of conduct in PAs for both tourists and service providers, it should be remembered that the priority activity must be the conservation of nature, while the remaining functions must be subordinated to the main one. All other activities must be aimed at adapting the conditions of the natural environment to the increasing tourist traffic by building tourist infrastructure in national parks in an organised and thoughtful manner. The planning of tourist development in mountain national parks should allow for the monitoring of tourist traffic and should diversify tourist traffic spatially without leading to the creation of hotspots in which the tourist capacity will be exceeded. Spatial diversification should be based on unique varieties of products linked to mountain areas; offering various forms of tourism will increase the tourist attractiveness of these regions.

13.4 Case Study: Tourism in the Knuckles Massif

The Knuckles massif is situated north of the country's central highlands, an isolated mountain range which is separated from the main Sri Lankan highlands by the Dumbara valley. Dothalugala MAB Reserve extends over the southern and southeastern parts of the Knuckles Conservation Region. Located in the central part of Sri Lanka in the Kandy and Matale districts, it covers an area of 18,500 ha and includes 34 peaks ranging from 900 to 2000 m. Knuckles features a great diversity in its forest cover: dry evergreen forests, montane forests, submontane forests, dry and wet pathana, savanna, etc. It has been found that 20% of the plants are endemic to Sri Lanka (Gunatilleke et al., 2017). A wide range of hardwood as well as herbal plants are also found there. In recognition of these biological and hydrological values, 17,500 ha of forest was designated as Knuckles conservation area in 1985 (Hemachandra et al., 2014). The entire KFR was declared a National Man and the Biosphere (MAB) Reserve in 2000 and later nominated as an international MAB site (Bambaradeniya & Ekanayake, 2003). The Knuckles Mountain Range was subsequently declared a UNESCO World Heritage Natural Site in 2009 (along with two other sites) and is one of Sri Lanka's major eco-tourism destinations. KFR was also included in Sri Lanka's central highlands in 2010 (Hemachandra et al., 2014). It is considered one of the most important biodiversity hotspots of national and global importance.

As a PA, this region struggles with two fundamental problems, the growth in tourism traffic (both domestic and foreign), and the direction of socio-economic development. The Knuckles massif, apart from its protective functions, plays a very important role as a place to live and work for the local community, with most residents from agriculture and tourist sectors. Unfortunately, there are often environmental conflicts between these two sectors and nature conservation. The range is inhabited by indigenous communities that live in 37 ancient villages whose existence depends upon traditional cultivation of rice, Chena and cardamom. Until recently, it was one of Sri Lanka's least accessible regions.

In relation to the aforementioned issues related to the cultivation of tea plantations, a similar set of problems can be identified in this region concerning the cultivation of cardamom. Cardamom is a perennial spice plant which was introduced to Sri Lanka in 1805 (Gunawardana, 2003). The government granted permission to cultivate cardamom on leased forested lands and consequently the understory of montane forests was cleared to various extents (from 3 ha to over 20 ha plots) in order to cultivate cardamom (Gunawardana, 2003). In the Knuckles region, cardamom has been identified as a major driver of deforestation, leading to biodiversity loss and increased soil erosion (Gunawardana, 2003) and reduction in the richness of taxonomies, especially endemic plant species (Adikaram & Perera, 2005). Official data show that around 3000 ha of forest were grown with cardamom inside the KFR (Bambaradeniya & Ekanayake, 2003; Gunawardana, 2003). Cardamom cultivation and management involves the selective removal of canopy trees and regular maintenance by removal of competing plants in the forest undergrowth (Bandaratillake, 2005). This system of cultivation may hinder the capacity of forests to regenerate and undermine their conservation value (Ashton et al., 2001). Since 1985, cardamom has been banned in this area on conservation grounds. This ban caused the deterioration of the living conditions of the local people who cultivated cardamom.

As the cultivation of cardamom in this region was the basis for the livelihood of many families, the ban on growing this plant forced a change in the development model of this region. Tourism has been trying to fill this gap for over 35 years, and mountain tourism can offer an alternative to decline in such areas of traditional agriculture that offers positive directions for sustainable development (Apollo & Andreychouk, 2020). The additional income from tourism has affected economic structures and contributed to a shift of agricultural activities towards part-time farming (Uhlig & Kreutzmann, 1995). Overall, the region has optimal conditions for the development of ecotourism because it is characterised by rich biodiversity and is relatively under-developed for tourism.

The most important tourist attractions of this region are 34 tourist trails (paved routes), leading through the most interesting natural places and through the surrounding villages. One of the means of socio-economic activation of the local population is the Small Grants Program, financed by the Global Environmental Facility (GEF). Friends of Dumbara, or Dumbara Mithuro, a voluntary organisation, whose main goal is to spread environmental awareness and fight for the sustainability and biodiversity of these places, has been operating since 1987. These areas are characterised by exceptional tourist attractiveness, the most visited places are: hiking trails from Mini World's End from Deanston; trail to Dothalugala from Deanston; trail to Nitro Caves from Corbtt's Gap; trail to Augallena cave via Thangappuwa from Corbett's Gap; trail to Kalupahana from 'Meemure' village (Table 13.5).

Since the Knuckles Mountain Range is a protected reserve, the construction of hotels and restaurants is not permitted. This means there are few food and beverage options within the range, other than occasional convenience stores found in the villages. Due to the growing pressure of NBT on the sensitive mountain system, it is now necessary to take measures to monitor the volume of tourism and diversify the product offerings towards sustainable tourism.

Table 13.5 Characteristics of the most popular tourism destinations in the Knuckles massif

Name	Characteristic of the spot
Bambarella hike of 5 peaks	This evergreen forest in the central highlands was known as *Dumbara anduvetiya* meaning "mist-laden mountains" in Sinhalese. The mountain range in the middle of the 21,000 hectare area resembles the shape of five knuckles in a clenched fist, hence the English name. It takes around 5-6 hrs to complete all five peaks, weather permitting, but camp sites are also available
Corbet's Gap	A landmark in the Knuckles mountain range. This is a deep valley between the Knuckles mountain range and also a rain shadow area that has a mixture of vegetation types such as wet, dry and montane type. Strong winds blow at certain times of the year and the forest trees grown in that area have a special feature, that is stunted and gnarled. It is possible to see here the panoramic view of Aliyawetunaela and Kinihirigala mountains and the Dumbanagala mountain
Dothalugala nature trail	The entire trail is 5.8 km and starts at the Knuckles Conservation Centre in Deanston. A permit, which can be obtained at the centre, is needed to enter the area. Dothalugala has the landscape values of the mountain trail
Nitro Caves	Popular choice is the nature trail that leads to Nitro Caves, a massive cave inhabited by hundreds of bats. The 11-km path starts at Corbett's Gap and will take about five hours to complete. The caves do make for a great, albeit slightly unnerving, sight
Mini World's End	Located towards the south of the mountains is Mini World's End, a 1192-m cliff that offers sweeping views of the mountains. The trail to Mini World's End is one of the two trails in the mountain range. It begins at the Knuckles Conservation Centre (KCC) and is approximately 1.5 km in length. Again, a permit is required
Duwili Ella trail	While all the trails are impressive, the crown goes to the Duwili Ella trail, a 40-m waterfall hidden deep within the forest. What makes Duwili Ella special is the cave inside the waterfall, allowing hikers to get behind the waterfall, akin to the entrance of the Bat Cave

Source Created by author based on data collected from various websites, 2020

13.5 Conclusion

Nature-based tourism is one of Sri Lanka's main tourism products. With their iconic fauna such as elephants, national parks already own optimal conditions for the development of NBT. In the future, it is imperative to promote Sri Lanka's mountainous NBTs as one of the most interesting destinations on the island. The unique character of mountain areas, in particular their fauna and flora biodiversity and the endemic nature of plant communities, constitutes a strong basis for actions aimed at conservation. However, this fundamental objective stands in stark contrast to the other two functions, tourism and agriculture. Mountain national parks include areas modified by humans for agricultural activities and this economic function may influence environmental changes concerning the species composition of forests, their range and hydrological relations. In the coming years, the most important issue will be to define the relationship between PAs and the local socio-economic system. This issue focuses primarily on issues related to the economic use (agriculture) of national parks and conflict situations arising from this fact. Accordingly, the enlargement of PAs in mountain areas may lead to functional changes consisting of replacing the agricultural sector with nature-based tourism. In the near future, the scale of pressure from local communities on mountain PAs should be determined. Climate change and global warming is another significant threat, as it not only affects the range and character of mountain forests, but also modifies the growing range of plantation plants. Climate change will alter the spatial extent of primary crops with a consequent increase in pressure on the mountainous areas of Sri Lanka.

The second extremely important issue is the organisation and service of NBT in mountain areas. Recent years have seen an upward trend in the volume of tourism to national parks, generating new problems both in terms of tourist services and satisfaction with their stay in the region. Above all, tourism management is aimed at reducing the pressure of NBT on sensitive mountain ecosystems. Reconciling these two elements is difficult but necessary because it could lead to the sustainable development of mountain areas. Tourist activity in national parks generates specific financial revenues, which contribute to the funding basis for the statutory activities of national parks and the maintenance of the local population. That is, increasing tourism improves the financial condition of these two groups, but the capacity should be determined to predict and prevent major environmental consequences. Mountain national parks also require improvement of the services offered to visitors, this applies to accommodation, catering and transport services. These activities will form the basis for a better competitive position in relation to other lowland parks and PAs that offer the viewing of elephants in their natural environment. Better promotion of mountain national parks will thus lead to product and spatial diversification of NBT across Sri Lanka. However, development of tourism requires more extensive analysis, for example the share and ratio of domestic and foreign tourism in the volume of tourism and generated revenues. Determining this situation will facilitate the creation of a clear development strategy for mountainous national parks.

References

Adikaram, N. W. A. M. M. D., & Perera, G. A. D. (2005). *Impacts of Cardamom cultivation on the floristic diversity of the montane forests of the Knuckles Range.* Proceedings of the University of Peradeniya Annual Research Sessions 2005, University of Peradeniya, Sri Lanka.

Apollo, M., & Andreychouk, V. (2020). Mountaineering and the natural environment in developing countries: An insight to a comprehensive approach. *International Journal of Environmental Studies*, 1–12. https://doi.org/10.1080/00207233.2019.1704047.

Ashton, M. S., Gunatilleke, C. V. S., Singhakumara, B. M. P., & Gunatilleke, I. A. U. N. (2001). Restoration pathways for rain forest in southwest Sri Lanka: A review of concepts and models. *Forest Ecology and Management, 154*(3), 409–430.

Bambaradeniya, C. N., & Ekanayake, S. P. (2003). *A guide to the biodiversity of Knuckles Forest Region*. IUCN Sri Lanka.

Bandaratillake, H. M. (2005). The Knuckles Range: Protecting livelihoods, protecting forests. In P. B. Durst, C. Brown, H. D. Tacio & M. Ishikawa (Eds.), *In search of excellence: Exemplary forest management in Asia and the Pacific* (pp. 167–174). FAO.

Breuste, J., & Dissanayake, L. (2013). Socio-economic and environmental change of Sri Lanka's Central Highlands. In A. Borsdorf (Ed.), *IGF research report 5* (pp. 11–31). Steigerdruck Ges.m.b.H.

Breuste, J., & Jayathunga, S. (2010). Representatives of nature conservation and ecotourism in different biomes of Sri Lanka. *Hercynia-Ökologie und Umwelt in Mitteleuropa, 43*(2), 257–276.

Buultjens, J., Ratnayake, I., Gnanapala, A., & Aslam, M. (2005). Tourism and its implications for management in Ruhuna National Park (Yala), Sri Lanka. *Tourism Management, 26*(5), 733–742.

Cummings, J. (2006). *Sri Lanka*. Lonely Planet Publications Ltd.

Dahanayake, K. (1982). Laterites of Sri Lanka—A reconnaissance study. *Mineralium Deposita, 17*(2), 245–256.

Department of Wildlife Conservation. (2020). *Protected areas*. http://www.dwc.gov.lk/Aoldsite/index.php/en/component/content/category/97-protected-areas.

Ekanayake, E. M. B. P., & Theodore, M. (2017). Forest policy for sustainability of Sri Lanka's forest. *International Journal of Sciences, 6*(1), 28–33.

Forest Department. (2009). Sri Lanka Forestry Outlook Study, Working Paper No. APFSOS II/WP/2009/29, Asia-Pacific Forestry Sector Outlook Study II. Food and Agriculture Organization of the United Nations Regional Office for Asia and The Pacific, Bangkok.

Forest Department. (2020). *Introduction*. http://www.forestdept.gov.lk/index.php/en/.

Gunasinghe, K. G. S. D. (2011). Conservation of biodiversity and sustainability of the tourism industry of Sri Lanka. *Economic Review, 7*, 27–32.

Gunatilleke, N., Pethiyagoda, R., & Gunatilleke, S. (2017). Biodiversity of Sri Lanka. *Journal of the National Science Foundation of Sri Lanka, 36*, 25–62.

Gunawardana, H. G. (2003). Ecological implication of cardamom cultivation in the high altitudes of Knuckles Forest Reserve, Sri Lanka. *The Sri Lanka Forester, 26*, 1–9.

Hemachandra, I. I., Edirisinghe, J. P., Karunaratne, W. I. P., Gunatilleke, C. S., & Fernando, R. S. (2014). Diversity and distribution of termite assemblages in montane forests in the Knuckles Region, Sri Lanka. *International Journal of Tropical Insect Science, 34*(1), 41–52.

Higginbottom, K., Northrope, C., & Green, R. (2001). *Positive effects of wildlife tourism on wildlife*. CRC for Sustainable Tourism.

Higgins, B. R. (1996). The global structure of the nature tourism industry: Ecotourists, tour operators, and local businesses. *Journal of Travel Research, 35*(2), 11–18.

Jayasinghe, S. L., Kumar, L., & Hasan, M. K. (2020). Relationship between environmental covariates and Ceylon Tea cultivation in Sri Lanka. *Agronomy, 10*(4), 476.

Joliffe, L. (2004). The lure of tea: History, traditions and attractions. In C. M. Hall, L. Sharples, R. Mitchell, N. Macionis & B. Cambourne (Eds.), *Food tourism around the world* (pp. 133–148). Butterworth-Heinemann.

Karunarathna, D. M. S. S., Amarasinghe, A. T., & Bandara, I. N. (2011). A survey of the avifaunal diversity of Samanala Nature Reserve, Sri Lanka, by the Young Zoologists' Association of Sri Lanka. *Birding Asia, 15*, 84–91.

Karunarathna, S., Ranwala, S., Surasinghe, T., & Madawala, M. (2017). Impact of vehicular traffic on vertebrate fauna in Horton plains and Yala national parks of Sri Lanka: Some implications for conservation and management. *Journal of Threatened Taxa, 9*(3), 9928–9939.

Maciejowski, W. (2003). Sri Lanka: w kraju Syngalezów i Tamilów [Eng. Sri Lanka: in the country of Sinhalese and Tamils]. In Z. Górka & J. Więcław-Michniewska (Eds.), *Badania i podróże naukowe krakowskich geografów* [Research and scientific journeys of Krakow's geographers] (pp. 112–125). Polskie Towarzystwo Geograficzne Oddział w Krakowie.

Mangala De Zoysa, W. D. & Inoue, M. (2015). Sri Lanka: Forest governance of community-based forest management. In M. Inoue & G. P. Shivakoti (Eds.), *Multi-level forest governance in Asia: Concepts, challenges and the way forward* (pp. 81–101). SAGE Publications.

Mathanraj, S., & Kaleel, M. I. M. (2017). Rainfall variability in the wet-dry seasons: An analysis in Batticaloa District, Sri Lanka. *World News of Natural Sciences, 9,* 71–78.

Niven, C., Noble, J., Forsyth, S., & Wheeler, T. (1999). *Sri Lanka: Cool Coasts, high teas, hot curries* (7th ed.). Lonely Planet.

OECD. (2009). *Wildlife and nature-based tourism for pro-poor growth, in natural resources and pro-poor growth. The economics and politics.* https://doi.org/10.1787/9789264060258-en.

Palihakkara, I. R., Mohammed, A. J., & Inoue, M. (2015). Current Livelihood condition of and futurity of Tea farming for Marginal Small Tea Farm Holders (MSTH) of Sri Lanka: Case study from Badulla and Matara District. *Environment and Natural Resources Research, 5*(1), 11.

Premathilake, R. (2006). Relationship of environmental changes in central Sri Lanka to possible prehistoric land-use and climate changes. *Palaeogeography, 240*(3–4), 468–496.

Ranwala, S. M. W., & Thushari, P. G. I. (2012). Current status and management options for plant invaders at Mihintale. *Journal of the National Science Foundation of Sri Lanka, 40*(1), 67–76.

Rathnayake, R. M. W. (2016). Economic values for recreational planning at Horton Plains National Park, Sri Lanka. *Tourism Geographies, 18*(2), 213–232.

Rodger, K., Moore, S. A., & Newsome, D. (2007). Wildlife tours in Australia: Characteristics, the place of science and sustainable futures. *Journal of Sustainable Tourism, 15*(2), 160–179.

Senarathna, P. M. (2005). *Sri Lankawe Wananthara (in Sinhala).* Sarasavi Publishers.

Senevirathna, H. M. M. C., & Perera, P. K. P. (2013). Wildlife viewing preferences of visitors to Sri Lanka's national parks: Implications for visitor management and sustainable tourism planning. *Journal of Tropical Forestry and Environment, 3*(2), 1–10.

Seo, S. N. N., Mendelsohn, R., & Munasinghe, M. (2005). Climate change and agriculture in Sri Lanka: A Ricardian valuation. *Environment and Development Economics, 10*(5), 581–596.

Shalleck, J. (1972). *Tea.* Viking Press.

SLTDA. (2020). *Annual statistical report of Sri Lanka tourism—2016.* https://www.sltda.gov.lk/download.

Subasinghe, S. M. C. U. P. (2013). *Plantation forestry in Sri Lanka challenges and constraints.* AGM of Institute of Biology.

Sumanapala, D. (2018). A review: National parks in Sri Lanka and impending development and research. *Asian Journal of Tourism Research, 3,* 121–147.

Uhlig, H., & Kreutzmann, H. (1995). Persistence and change in high mountain agricultural systems. *Mountain Research and Development, 15*(3), 199–212.

UNESCO. (2010). *Central highlands of Sri Lanka* (Report of the decisions adopted by the World Heritage Committee at its 34th session). https://whc.unesco.org/en/list/1203/.

Uragoda, C.G., (1994). *Wildlife conservation in Sri Lanka. A history of Wildlife and nature Protection Society of Sri Lanka 1894–1994.* Centenary Publication.

Werner, W. L. (1995). Biogeography and ecology of the Upper Montane Rainforest of Sri Lanka (Ceylon). In L. S., Hamilton, J. O., Juvik & F. N. Scatena (Eds.), *Tropical montane cloud forests: Ecological studies (analysis and synthesis)* (pp 343–351). Springer-Verlag.

Wickramasinghe, A. (2005). Adam's Peak sacred Mountain Forest. In *The importance of sacred natural sites for biodiversity conservation* (pp. 109–118). Proceedings of the International Workshop held in Kunmingand Xishuangbanna Biosphere Reserve, Kunming and Xishuangbanna Biosphere Reserve, People's Republic of China, People's Republic of China, 17–20 February 2003.

Wijeratne, M. A. (1996). Vulnerability of Sri Lanka tea production to global climate change. *Water, Air, and Soil pollution, 92*(1–2), 87–94.

Wijesinghe, L. C. A. (2003). Forestry in Sri Lanka a voyage through time. *Tropical Agricultural Research and Extension, 2,* 14–21.

Xinhua. (2018). *Tourism emerges as second largest forex earner in Sri Lanka.* http://www.xinhuanet.com/english/2018-06/19/c_137265529.htm.

Renata Rettinger is Assistant Professor at the Pedagogical University of Krakow, Institute of Geography, Department of Tourism and Regional Studies. Renata's academic interests focus on issues related to determinants and consequences—as well as the process itself—of tourism development in selected regions of the world, the Caribbean in particular.

Dinesha Senarathna is a geography lecturer at the University of Kelaniya, Sri Lanka. Her current PhD study at the Auckland University of Technology, New Zealand lies in the field of 'protected area tourism, community development and conservation'. Her Masters' thesis focused on 'the sustainability of canopy adventure tours in New Zealand'.

Ruwan Ranasinghe is senior lecturer at the Department of Tourism Studies of Faculty of Management of Uva Wellassa University of Sri Lanka. He is also the Pioneering Head of the Department of Tourism Studies, Director for World Bank funded projects of the University as well as functioning as the pioneering Chairman of Uva Tourism Promotion Bureau.

Part V
Conclusion

Chapter 14
Reflections for Trans-Regional Mountain Tourism

Huong T. Bui, Thomas E. Jones, and Michal Apollo

14.1 Introduction

Asia is both a generating source and a destination region for tourism, with a large share of intra-regional visitors. Interest in nature-based tourism (NBT) in Asia has proliferated with increasing disposable income resulting from economic development and expansion of population. Economic, social and cultural changes within the region affect both demand and supply of NBT. Frost et al. (2014) identified major trends of the future of NBT in Asia including (1) shifting demographics, (2) increasing urbanization, (3) climate change, and (4) development. This edited volume centres on analysing the extent to which mountainous Protected Areas (PAs) have been utilized for tourism activities across Northeast, Southeast and South Asia.

Papers in this edited volume focus on the mountainous PAs across Asia. Mountains are not only distinctive for biological characteristics, and landscape values, but also are sacred places for religion and culture. For example, Mt. Fuji in Japan has been recognized by UNESCO as a Cultural Heritage owing to its legacy of art and worship-ascent. Mountains in Southeast Asia such as Mt. Bromo (Indonesia) or Mt. Ramelau (Timor-Leste), and numerous Himalayan peaks are sacred in indigenous beliefs. Owing to cultural and religious symbolic meaning of the peaks in addition to surrounding forests and landscapes, many mountains are designated as PAs.

H. T. Bui (✉) · T. E. Jones
College of Asia Pacific Studies, Ritsumeikan Asia Pacific University (APU), Beppu, Japan
e-mail: huongbui@apu.ac.jp

T. E. Jones
e-mail: 110054tj@apu.ac.jp

M. Apollo
Department of Tourism and Regional Studies, Institute of Geography, Pedagogical University of Krakow, Krakow, Poland
e-mail: michal.apollo@up.krakow.pl

© The Author(s), under exclusive license to Springer Nature Switzerland AG 2021
T. E. Jones et al. (eds.), *Nature-Based Tourism in Asia's Mountainous Protected Areas*, Geographies of Tourism and Global Change,
https://doi.org/10.1007/978-3-030-76833-1_14

Fig. 14.1 Mountain tourists and prayers flags in Poon Hill with Dhaulagiri Peaks in the background, Annapurna range, Western Nepal. *Source* Author

The national designation for conservation purposes to protect natural, cultural and religious outstanding values of the mountain, however, generates a 'placebo' effect (Yang et al., 2019) that triggers diverse demand for mountainous NBT across the region (Fig. 14.1).

Although climbers' motivations and activities might align with natural and cultural outstanding values of the mountains, the profiles, characteristics and activities of modern mountain climbers in Asia are quite different from historical pilgrimages for sacred belief. Peak-experience hunting of professional climbers, or learning and experience nature for both hard and soft adventurous eco-tourists have been frequently discussed in the tourism literature (Ewert, 1994). For example, Jones and Bui (2018) analysis of climber motivations at Mt. Fuji and Mt. Kinabalu in Malaysia found the 'peak-hunting' pull of seeking to climb a world famous icon to be most important, following by social motivation to accompany friends. In South and Southeast Asia, domestic visitors are outnumbered by pilgrims to sacred places on the mountains, for example in Mt Bromo and to a lesser extent, climbing mountain for recreational purposes (Choe & Hitchcock, 2018). The same phenomenon is recorded each year in the Himalayas, which are visited by hundreds of thousands of pilgrims each year (Apollo, 2017a). Under the supply-demand rule, climbers' motivations and preferences of mountain-related activities shape the way infrastructures and services are developed and expanded.

Change in mountain landscapes is also attributed to state government policies to encourage or discourage developmental initiatives (Apollo, 2015; Wengel, 2020).

Policy-led and demand-led development fundamentally alters the human-nature nexus in fragile montane areas (Apollo, 2015; Musa et al., 2015). Generally, a geographic (mountain) ecosystem is complex, and includes elements of different natures (abiotic, biotic, anthropic, etc.) that are interconnected with each other in many different ways, thus forming the entire ecosystem (Andreychouk, 2015). External influences on this relatively stable, but fragile mountain geosystem provoke a series of reactions (internal interactions). The main target of human activity is the high mountain zone, however, effects (positive or negative) of hiking, trekking, and climbing can be seen at a much lower altitude. This leads to various changes that can be perceived as being positive, negative, or neutral (Apollo & Andreychouk, 2020). The balance of profit and loss depends on a number of factors and circumstances that are clearly presented in these chapters.

The unique relationship between humans and nature, in particular mountains, is deeply rooted in Asian cultural and religious characteristics, that draw our attention to the other side of the mountain, detaching from a conventional approach to study with focus on biological values for conservation purposes. Intertwining socio-cultural and religious meanings of the mountain with its ecological values has challenged the conventional position of mountains in eco-centric Western philosophy. Evidently, analysis and observations of Asian mountain visitation presented in this edited volume reflect Frost et al. (2014) contention "there appears to be widespread respect for and interest in nature among Asia-Pacific societies, which has strong cultural and religious roots, contrary to the popular view that this is mainly a Western concern" (p. 721). Romanticised nature has historically been characterized as a Western trait, in contrast, it is often argued that Asian societies have a more pragmatic and less romanticized view (Kellert, 1995). Common philosophies and traditions, for instance Confucian and Buddhist thinking tends to "regard all elements of the universe as one entity" (Lee et al., 2013, p. 523). Human elements such as cultural attractions in natural surroundings are not seen as problematic (Weaver et al., 2020). Nature and human beings are interdependent and one can shape or even improve the other. This contrasts with Western notions of stewardship and the importance of nature 'unsullied' by human intervention (Buckley et al., 2008). Arguably, NBT in developing Asia is economically driven, based primarily on leisure, and characterized by resorts utilizing nature leading to overdevelopment (Bui & Dolezal, 2020). Case study evidence documents how humans can contribute to 'improve' nature for their consumption and comfort, towards enlightened mass tourism in the mountains (Weaver et al., 2020).

The fourteen chapters covered in this edited volume tackle diverse aspects of mountain PAs and NBT, from single-site case studies, to national scale PA management and governance, and toward more tightly integrated regional networks. For the characteristics and operations of mountain tourism in each country are highly dependent on regional geographic, cultural, and political conditions. In the sections below, we review intra-regional issues pertinent to Northeast, Southeast and South Asia before synthesising inter-regional issues to highlight our concluding remarks.

14.2 Regional Characteristics

14.2.1 Northeast Asia

Northeast Asia covers 11.5 million km^2 and 1.7 billion people, equivalent to around 22% of the world's population. The region examined in this book covers four countries (*China, Japan, South Korea* and *Taiwan*) but excluded Hong Kong, Macau, Mongolia and North Korea due to data deficiencies. The region contains over 8,600 protected areas (PAs), and over 20% of all UNESCO World Heritage sites (MacKinnon & Yan, 2008). Yet the rich biodiversity is under threat in the most highly-densely populated part of the planet, that includes the world's most populous country (China, home to about 1.4 billion people) and city (Tokyo, with 38 million residents when defined according to its metropolitan catchment area). However, even as Tokyo continues to grow, the overall population of Japan has been declining since 2010 due to falling birth rates. Although shrinking more slowly, the fundamental demographic outlook for South Korea and Taiwan is also similar and in China, the population is projected to decline by 31.4 million, or 2.2%, between 2019–2050 (UN, 2020).

The widespread urbanization that has enabled the respective 'economic miracles' of East Asia has exacerbated this demographic decline by funnelling younger generations toward urban hubs for education and employment (Matanle & Rausch, 2011). Japan leads the way, with 92% of the population living in urban areas followed by South Korea (82%) and Taiwan (79%) (Worldometers, 2020). Although urbanization has boosted economic efficiency and improved economies of scale (Quigley, 2009) the obverse effects include depopulation, aging and decline of rural regions. Urban economic indicators of the four East Asian powerhouses are positive, but national-level benchmarks such as GDP belie the regional realities of countryside towns and villages that have been 'hollowed-out' or abandoned due to the rural exodus (Liu et al., 2013).

The 'ghost villages' of East Asia's increasingly deserted countryside raise issues beyond landscape conservation. It is estimated that by 2050, over 60% of rural Japanese populations will have declined by half compared to the 2015 benchmark, and 19% of villages will be abandoned altogether (Tsunoda & Enari, 2020). As the mountain populaces shrink, populations of boar, deer and monkeys are on the rise, triggering an increase in human-wildlife conflict. Village shops and post offices close or consolidate, steepening the vicious circle of rural decline by reducing employment opportunities. The outflow of human resources from peripheral regions thus reflects and further exacerbates the decline in primary industries, such as farming and forestry, once seen as viable alternatives to tourism in upland areas (Knight, 2000).

The dichotomy of urban growth and rural decline has more significant ramifications for less accessible areas. Mountains are of vital importance to species distributions, biodiversity, and fundamental ecosystem services including clean water and air (Peters et al., 2019). Mountains' steep gradient and corresponding climatic variations compress 'life zones' into vertical strips that create countless niches to support different biota (Körner & Spehn, 2019). As mountainous environments increasingly come under threat, there is a greater need for the tangible benefits they offer. NBT in PAs can provide otherwise isolated, inaccessible and inhospitable areas a range of economic benefits including "job-creation, inward investment, enhanced tourism, and a stronger identity for the marketing of areas and their products" (Warren, 2002, p. 215).

14.2.2 Southeast Asia

Southeast Asia has a population of 655 million people, about 8.5% of the world's population. It combines continental countries (Cambodia, Laos, *Myanmar*, Thailand and *Vietnam*) with island nations (Brunei, *Indonesia*, Malaysia, *the Philippines*, Singapore and *Timor-Leste*—countries covered by this book are in italics). Founded in 1967 as a non-communist block of East Asian countries, ASEAN today plays a major role in the region's economic, social and political development. Since environmental issues were first inscribed on the ASEAN agenda in 1977, the member states have developed an increasingly complex web of soft-law declarations, resolutions, plans of action, issue-specific programs and two binding multilateral agreements (Elliott, 2011).

In 1976, the organization created a Sub-Committee on Tourism for the development of coordinated tourism projects and their enhanced marketing and, four years later, the ASEAN Tourism Forum was established as an annual event. Later on, in 2015, the ten members of ASEAN signed a declaration on the formal establishment of the ASEAN community, a broad framework of regional integration made up of three pillars: the ASEAN Economic Community (AEC), which focuses on economic integration; the Political-Security Community, which aims to link up regional foreign affairs and security interests, and the Socio-Cultural Community, which seeks to build people-to-people connections (Hall & Page, 2017).

Southeast Asia covers about 4.5 million km^2, some 3% of the world's land area that still supports between 20–25% of the world's inventory of plant and animal species (Corlett, 2005). Most are endemic to the region, or are shared only with South or East Asia. In practice, the long-term survival of the great majority of Southeast Asia's native biota, and thus a large fraction of the global biota, will depend on the protection of their natural habitats within the region (Corlett, 2005). In the past two centuries, the physical environment was significantly degraded, first by the establishment of plantation and extraction economies of the colonial times, and then, since the second half of the twentieth century, by widespread deforestation, agricultural expansion, resettlement, and urbanization (Gupta, 2005).

Mountains dominate Southeast Asian landscapes, creating geographic and political boundaries, separate the region from India and China, forming natural barriers between and within mainland Southeast Asian countries; the Arakan Yoma Range in western Myanmar; the Bilauktaung Range, which runs along the border between Myanmar and Thailand; and the Annamite Cordillera, the mountain range that separates Vietnam from Laos and Cambodia. Mountains on Southeast Asia's islands form part of the Ring of Fire, including many active volcanoes (Hutchison, 2005). Indigenous peoples who call the mountains of Southeast Asia their home account for almost 20% of the global indigenous population (Ariza et al., 2013). They are the custodians of a vast diversity of cultures, languages and traditional knowledge, but often marginalized, poor and underserviced by their respective states (Ariza et al., 2013) and thus face major challenges in carrying on traditional lifestyles at the edge of modernization.

14.2.3 South Asia

South Asia covers about 5.2 million km^2, or 3.5% of the world's land surface area and is home to almost 1.9 billion people, about one-fourth of the world's population. The region consists of Afghanistan, Bangladesh, Bhutan, *India*, Iran, the Maldives, *Nepal*, Pakistan, and *Sri Lanka* (this volume focused on the three italicized). The countries (excluding Iran) are affiliated with the South Asian Association for Regional Cooperation (SAARC), the regional intergovernmental organization and geopolitical union of states in South Asia. The SAARC comprises 3% of the world's area, 21% of the world's population and 4.21% (US$3.67 trillion) of the global economy. The estimated average GDP per capita (PPP) of all of the SAARC countries was $6,150 (IMF, 2020) and the highest value was $21,760 (Maldives), while the lowest was $2,017 (Afghanistan), making South Asia the poorest region in Asia. Furthermore, the mountains, being the home of the poorest people on earth, face political and economic marginalization by both national and local administrations (Ives, 2006; Messerli & Ives, 1997).

The dramatic growth and low levels of development can also pose a huge problem for the natural environment, such as where residents do not pay enough attention to waste management. For example, over 70% of people living in Nepal's Kathmandu Valley dump their garbage on the streets or along the river banks (Zurick & Pacheco, 2006). Bhatt and Pathak (1992) studied river pollution in Lesser Himalaya, finding that increased human influx near the banks damages the river basin to the extent that proper sewage treatment facilities become necessary (Apollo, 2017a). Multiple factors have led to environmental degradation in South Asia such as habitat conversion; agricultural conversion; fuelwood and fodder extraction; logging; grazing; flooding and wildfire (both natural and anthropogenic) and tourism (Mondal & Nagendra, 2011). As one of the global population growth epicentres, environmental management is paramount, but improved compliance to mitigate habitat loss within PAs is likely to prove difficult in South Asia, where rural population densities are high and many livelihoods rely on small-scale agriculture (Singh, 1985). Moreover, a quarter of the land *inside* South Asia's PAs is now classified as human modified (Clark et al., 2013). Consequently, in recent decades, South Asia countries have delineated a variety of areas with PA status, even though the conservation model is often highly pressured by commercial demands, especially in countries at the lower end of the development spectrum.

14.3 Nature-Based Tourism in Mountainous Areas

14.3.1 Northeast Asia: Back to the Future?

Despite their de facto image as urban economic powerhouses, the nature and culture of Japan, Korea and Taiwan has been shaped by the mountainous topography of each. The national parks share a combined colonial legacy that stems from 'sacred peaks.' In Japan, the first batch of national parks designated in 1934 centred on highland destinations such as Daisetsuzan, Aso-Kuju and the Japan Alps. The criteria in colonial Korea and Taiwan closely resembled that of the mainland. On the Korean peninsula, a national park plan was devised but never delivered at Mt Kongo (Mizuuchi et al., 2016).

Japanese park planners such as Tamura Tsuyoshi (1890–1979) borrowed from the British colonial myth of *terra nullius* to legitimize their site selection. Large areas of unoccupied, primeval wilderness like Kamikochi in the Japan Alps were prioritized due to the lack of permanent population. Although it had been selectively logged, the ridgelines which overlooked this remote Alpine valley had not been explored or systematically surveyed, so along with aesthetic landscape-appreciation, such "high-altitude zones were a treasure trove for scientific observation" (Wigen, 2005 cited in Jones, 2016). This drive for science-based exploration coincided closely with the patriotic fervour fuelling Japan's expansion overseas, underpinned by newly-imported academic disciplines such as geography and civil engineering (Jones, 2016).

Nonetheless, Japan's national park philosophy still tended to reflect the 'Yellowstone model' whereby scenery and sightseeing was prioritized over science-based criteria, unlike the Swiss National Park, that was designated in 1914. Today, the Swiss park represents the most strictly protected IUCN category 'Ia' area that covers 174 km^2 of a mountainous valley lacking in manifestly spectacular geoheritage such as glaciers or geysers. Although visitors are allowed to access the Swiss National Park, tourism is not the main purpose but rather it is managed as a sanctuary to conserve the natural environment for scientific research whilst removing evidence of extractive human activity (Kupper, 2009). Japan's parks followed a more pragmatic approach that would today be termed a mixture of IUCN category II 'national parks' with category V 'protected areas.' This compromise left room for sacred but immensely popular pilgrimage peaks such as Mt Fuji and Tateyama to be included in the national parks network alongside significant swathes of privately-owned forest and farmland, and shrine-temple complexes. Japan's 'multi-purpose' park system outlined in Chap. 3 seems equally relevant to Korea's parks (Chap. 4), that are mostly IUCN category 'V' containing substantial sections of privately-owned land.

However, category 'V' PAs face ambiguous management standards when it comes to implementing environmental legislation. Japan's Nature Parks Act, for example, does not permit the construction and operation of mountain huts inside core zones, unless specified in park plans. But in reality, many private mountain hut businesses pre-date park designation and in some cases own the land on which the hut is built (either outright, or on long-term leases). Hence for-profit operation of the huts continues with little legislative oversight and without the need for special approval as a park enterprise (Kato, 2008). This can result in a lack of strict environmental compliance within the core zone, as in the case study example of Mt Fuji where climbers' trash was buried around the huts and toilet waste periodically dumped down the mountainside at the end of each season (Jones et al., 2018). After the 2012 listing as a UNESCO world heritage site, management has been much improved thanks to a combination of financial subsidies from the state and heightened involvement of NGOs such as the Fujisan Club, but the fundamental lack of legislative firepower to deal with exceptional circumstances and private landowners remains unchanged. In a comprehensive historical review of Japanese parks, Murakushi (2005) criticizes this 'low-cost' style of administration that has undermined the effectiveness of the PA network.

Fragmented governance is a related issue, although South Korea's National Parks Service was introduced as a 'one-stop-shop' with jurisdiction over all parks except for Hallasan. In Taiwan, restructuring efforts are underway to set-up a new state organ to coordinate the roles of different agencies (Chap. 5). Likewise, in China, the potential is being explored for new—or re-configured—national parks in mountainous PAs such as Pudacuo. China's novel national park network symbolizes the renewed efforts to tackle the barriers to effective conservation that plague the forest park system as outlined in Chap. 2. However, Weaver et al. (2020) propose an 'endogenized' sustainable tourism model that cautions against simple criticism of the rampant commercialization undergone by many mountainous PAs as symbolized by the glass elevators of Zhangjiajie National Forest Park.

It is difficult to deny that the sustainable tourism literature, and indeed the IUCN classification of PAs used by this book, follows predominantly Anglo-Western centric philosophies that often ignore or underestimate "destination-specific culture" (ibid.). It is thus hoped this book can contribute in some small way to offer a counter-proposal for a more regionally relevant set of indicators based on 'Asian' cultural values. To that end, more research is needed to foster the mutual transfer of management tools and ideologies between park authorities and scientists across Asia and promote best-practise initiatives. PAs in China could represent a bridge between Northeast and Southeast Asia. As described above, China has followed a similar model of dynamic economic development with concurrent urbanization. Rural, mountain regions are increasingly sparsely populated with ageing permanent residents, but must meet the needs of seasonal influxes of increasing volumes of nature-based tourists. It is anticipated that lessons from effectively-managed mountain destinations in Northeast and Southeast Asia could thus cross-pollinate toward a more efficient and resilient model of Asian PA administration.

Clues as to the content of such a model can be gleaned from case studies such as Mt. Jade in Taiwan that operates a lottery system to distribute climber permits based on the overnight bed capacity of the mountain lodge. At Mt. Fuji, a cost recovery mechanism has already been introduced in the form of a donation collected from climbers, with discussions underway to convert the scheme into a fully-fledged entrance fee. In South Korea, the nominal admission fees were conversely abolished in 2007, but mountainous parks such as Hallasan use persuasive communication to spatially diffuse some 15 million annual visitors to Jeju, while incorporating environmental education goals. In Huangshan's 'Yellow Mountains', an instant carrying capacity approach has been tested, all tools with potential applications for mountain honeypots in Northeast and the Southeast Asian region discussed next.

14.3.2 Southeast Asia: 'Eco'- or 'Ethno'-Tourist?

Efforts in Southeast Asian countries to foster NBT or ecotourism have faced problems associated with the lack of infrastructure, adequacy of personnel training, absence of—or delays in—plan implementation, and political instability. The focus shifted to community-based ecotourism because of the industry's heavy reliance on national parks and other PAs (Razal et al., 2012). The operation of ecotourism in Southeast Asian mountains, however, centers around two fundamental questions: What is ecotourism? and Who are the ecotourists?

Dating back to the era when ecotourism was first introduced to Southeast Asia, Hitchcock and Jay's (1998) review of the situation in Malaysia and Indonesia highlighted several major issues: (1) Ecotourism remains problematic because of a lack of clear definition; and (2) the relationship between local people, environmental reserves and ecotourism is also problematic. However, Hitchcock and Jay's definition of "ecotourists" to Southeast Asian NBT destinations as international, independent travellers from Global North to South reflect a narrow Western-centric model of ecotourism. The shift in market viewpoint concerning ecotourism exemplified in this volume demonstrates that the visitation to NBT destination is dominated by the domestic market. Although domestic demand does not necessarily seek unique environmental features, it merges with the relaxation, rejuvenation purpose of conventional travel, and pilgrimage to sacred mountain sites. Therefore, the question of what is ecotourism and who are ecotourists be redefined in the context of increasing expansion of domestic Southeast Asian demand, and burgeoning interregional visitors to the Southeast Asia's mountainous PAs.

Some of the original colonial legacies behind the designation of national parks and wildlife reserves in Southeast Asia were readily adaptable to early ecotourism models, with the underlying assumption that small groups of privileged elites were more likely to engage in non-destructive and non-consumptive activities than other kinds of tourists. Funds from ecotourism could cover the parks' overheads in order to stave off demands for other kinds of commercial exploitation that may be environmental detrimental such as mining or deforestation. Evidence from research in this volume show that mass volumes of domestic visitors and their leisure-oriented behaviour in NBT destinations might challenge every aspect of this ecotourism definition and criteria for mitigation of negative impacts. In fact, modern ecotourists from Asia tend to visit mountain sites for relaxation, seeking cool climate to escape from the heat of summer, or for pilgrimage. Eco- and ethno-tourism are closely related in such cases.

The major contribution of our volume is to open up an intra-regional research forum for NBT in the mountainous PAs, and to include countries which are relatively under-researched in prior literature. For example, research on NBT in the mountains of Myanmar and Timor-Leste is at an early stage, with limited publications to date. Addressing this gap, Chap. 10 provides a comprehensive review of Myanmar's PAs, its management and the potential for NBT in mountainous areas. However, due to the current economic and socio-political challenges, the capacity to implement sustainable NBT in mountain PAs remains in doubt. In Timor-Leste, Chap. 7 presents an in-depth discussion on community NBT at Mount Ramelau, where issues concerning stakeholder's collaboration are highlighted.

Our book also shines a light on PAs in emerging tourist destinations in Southeast Asian countries, such as Vietnam, Indonesia, and the Philippines. In Vietnam (Chap. 9), the host-guest tensions inherent in mountainous PAs are discussed in the context of Hoang Lien Mountain Range against the backdrop of NBT development. From the host perspective of ethnic groups in the mountains, the challenges are linked to their lifestyles, capacity for and perceptions of NBT. From the guest perspective, NBT challenges are rooted in market trends, product preferences and accessibility to the mountain PAs supported by a radical modernization approach. Chapter 6 discusses the challenges faced by Indonesia's PAs, such as land tenure conflicts, poaching and deforestation, and the potential for NBT to offer a viable alternative. However, when NBT criteria are evaluated according to the contemporary realities of tourism to the Bromo Tengger Semeru National Park, it can be seen that the conceptual standards are not yet realized. Furthermore, a mismatch between Mount Bromo's mass tourism with the notion of NBT can be detected from the motivational make-up of tourists who tend to be casual pleasure-seekers. Whereas most contributions have a meta-analytical method, Chapter 8 takes a different approach based on primary data collected at the Mount Apo Nature Park in the Philippines. The economic potential for watershed protection, biodiversity conservation, and mountain climbing is explored via schemes to Pay for Environmental Service that could address the revenue shortfall of PA management and generate significant supplemental funds for management and conservation.

14.3.3 South Asia—Tourism Impacts in Highlight

South Asia tourism has experienced exponential growth in the last few years. In 2018, over 32.8 million foreign tourists (UNTWO, 2019) visited the region. Tourism development has mostly been driven by the strong performance of the dominant destination, India, with increased demand from western source markets due to simplified visa procedures. However, it is noteworthy that regional domestic tourism markets are much larger than the international tourist segments. For example, in 2011, the Amarnath Holy Cave (3,888 m) received 634,788 pilgrims, while the Badrinath Temple (3,133 m) had 980,667 visitors, and the Jamunotri Temple (3,291 m) welcomed

287,688 visitors (BKTC, 2016). Most of the crowded paths lead to Hindu temples and are visited by hundreds of thousands of pilgrims each year (Apollo, 2017a).

Overall, the tourism sector is dominated by regional domestic trips, often performed by pilgrims practicing Buddhist or Hindu religious or cultural rites. For example, surveys carried out by the Nepalese (NTS, 2013) and Bhutanese (BTM, 2013) government tourism institutions show that around 10% of the total foreign tourists in Himalayan regions visit their mountain range. Two main tourism segments can be classified based on their motivation: (1) mountain climbers and trekkers for adventure; and (2) pilgrims. These two groups can be distinguished by different motivations (secular vs. sacred). Adventurous mountain tourism involving hiking, trekking and mountain climbing, broadly defined as mountaineering, is becoming more popular (Apollo, 2017b). In the new inventory of mountain NBT, adventure tourism has already become a 'mass' sport, climbing activities and climber's profiles have changed completely, and fewer mountain users understand or care about the impacts of their activities on the fragile environment. On the other hand, one of the oldest forms of travel is pilgrimage, "a journey resulting from religious causes, externally to a holy site, and internally for spiritual purposes and internal understanding" (Barber, 1993). This is another staple as mountains of the world are known for their serene landscape and sanctity, and they are major centers of recreation and spirituality (Sati, 2015). Across every religion's sacred book, the creator and the prophets all agree that the environment is sacred (Taylor, 2007; Watling, 2009). However, not all pilgrims take their god's will to be sacrosanct. For example, Hindu sacred places are the most polluted around the world (Apollo, Andreychouk, et al., 2020; Timothy & Nyaupane, 2009). In highland pilgrimages to the Himalayas, open dumping of waste leads to air and water pollution (Sati, 2014, 2015).

Mountain NBT has the potential to contribute to sustainable development and conservation of PAs (Apollo, 2017b; Apollo, Wengel, et al., 2020; Musa et al., 2015). Apollo and Andreychouk (2020) note that mountain NBT has positive impacts on environment in populated areas (inhabited and agricultural exploited), but at higher altitude above populated areas, these impacts become negative. Mountain NBT could be an alternative livelihood for areas of declining traditional agriculture, as discussed above in the context of Northeast Asia's depopulated rural regions. Additional income from tourism can facilitate a shift from agricultural activities to part-time farming (Uhlig & Kreutzmann, 1995). Researchers highlight the urgent need for revision of PA management, and identification of factors attributed to the success or failure of PAs requires more detailed studies (Hanson et al., 2020; Joppa & Pfaff, 2011; Yang et al., 2020). With regards to South Asia, particular attention should be paid to those regions of high priority for PA establishment, namely the Western Ghats, Sri Lanka and the Himalayas (Clark et al., 2013; Rodrigues et al., 2004). The chapters on South Asia mountain PAs respond to this call. By using a comprehensive and holistic approach to PAs of the Indian Himalaya (Chap. 11), Nepal (Chap. 12) and Sri Lanka (Chap. 13), authors were able to draw the basic conceptual framework of the management of mountain tourism in PAs. There are ongoing efforts to conserve

PAs characterised by rare flora and fauna, where nature protection is on a par with commercial goals, that is tourism. All chapters mention the needs of creating or improving development and protected plans.

Chapter 11 describes the challenges and opportunities for mountain tourism in the Indian Himalayan PAs. The authors concluded that there is still a lack of clarity in policies and guidelines that could assist tourism planners when NBT is introduced in the conservation areas. It was also demonstrated in the example of Stok Kangri in Hemis National Park. As this mountain is now closed to tourists until 2022 due to environmental degradation, it can be a useful case study for future ecological research into resilience and the speed of recovery.

In Chap. 12, Nepal has been globally recognized for its conservation success with community participation at the core of all activities. However due to the increasing pressure of economic development, migration, climate change, wildlife-human conflicts and unequal distribution of profit from tourism will surely fuel the debate over the status of many PAs in Nepal. In addition, attitudinal change is necessary for the mindset of the park officials when handling communities' grievances and outstanding issues.

The final chapter of this volume deals with Ceylon (Sri Lanka). The unique character of the island, in particular their fauna and flora biodiversity and the endemic nature of plant communities, constitutes a strong basis for actions aimed at conservation and NBT. Unfortunately, even mountain national parks contain considerable human economic activity, including agriculture (mainly tea plantations) that has influenced environmental changes. Chapter 13 calls for authorities to defend the relationship between PAs and the local socio-economic system. This is exemplified by the Knuckles Mountain Range, where increasing number of tourists and very limited NBT possibilities in Sri Lanka mountains requires planning resolutions adapted to the new realities.

14.4 Toward Co-management of Mountainous PAs in Asia

14.4.1 Colonial Legacies

All across Asia, natural resources have been appropriated from local communities by colonial governments and their state-run equivalents. During the nineteenth and early twentieth centuries large swathes of South and Southeast Asia fell under European rule, while Northeast Asia was later colonized by Japan. World War II eventually signalled a phasing out of colonialism. From 1946 to 1957 almost all Asian countries gained their sovereign independence. Many colonial governments, however had long-established forestry departments and Euro-centric principles of 'science-based' resource management, so planning and administration proved harder to reorganize (Boomgaard, 2007). Likewise, European patterns of seasonal trips to upland areas proved surprisingly durable, as typified by the concept of holidays to 'hill stations' that date back two centuries or more. Development of the hill stations—towns founded by European colonial rulers as a refuge from the summer heat, mostly located in the hills, where the climate is much cooler (Kennedy, 1996)—also contributed to improving the accessibility of mountain regions (Apollo, 2017a). In the past, emphasis was placed on the accessibility of hill stations (e.g., Bà Nà Hills, Mussoorie, Nagarkot, Nuwara Eliya), and the majority of these settlements could be reached by railway within a range of less than one hundred kilometres (Kreutzmann, 2000). The travel time has also dropped significantly. Many of the same expatriate destinations still survive and are prospering in a new era of domestic-dominated NBT (Inagaki, 2008). The same search for temperate respite from the tropical climate that once propelled the development of NBT thus remains relevant today (DeWald, 2008). On a less positive note, mistakes made by colonial regimes that sought to unilaterally impose their will on mountainous regions can be recurrent in the current crop of PAs. In some cases, such as the example of Indonesia, this is due to lingering legislative issues linked to the 'domain declaration' (*Domein verklaring*), a legacy of the Dutch colonial era that has continued to blight land ownership into the twenty-first century with ongoing struggles over land tenure in and around the national parks (Kano, 2008). Yet a tendency to 'blame history' diverts responsibility away from contemporary ruling elites, whose decisions determine the potential of PAs to revitalize mountainous regions via domestic interpretations of innovative NBT solutions.

14.4.2 PA Governance: From Colonial Roots Toward Co-management?

In spite of the rhetoric, many of the PAs networks detailed in this volume continue to be administered in a top-down manner by state institutions whose planning emphasis on conservation and visitor regulation is exacerbated by a lack of NBT projects and programs. In most of the national studies explored herein, a central government agency is legally responsible for the management of national parks and PAs. This often results in an overtly 'official' focus on conservation, at the expense of more feasible compromises to actively involve private sector and local government partners and create employment opportunities via NBT. The problem of '*paper parks*' arises when PA planning is confined to central government agencies without due recognition of the range of stakeholders, including local governments and communities who rely on the mountainous resources. For example, Mallari et al. (2016) refer to PAs in the Philippines that are delineated on maps but exist in a sub-optimal state that belies the goals of Aichi Target 11.

In addition, many of the iconic peaks and PAs in this volume have also been concurrently listed by international organizations, such as UNESCO, whose World Heritage Program requires the creation of a management plan and periodic monitoring. However, UNESCO management models also espouse Euro-centric ideals of holistic participation between comprehensive round-tables of public and private organizations, NGOs, religious institutions, and local residents. Yet in the post-millennium era, a loosely-knit transnational movement has emerged, comprising a range of actors from government and NGOs, along with environmental organizations and citizens groups. This co-management style, otherwise referred to as 'collaborative,' 'joint' or 'partnership' management, was widely heralded as a new form of 'governance' in response to the shortcomings of centralized, state-run approaches (Persoon et al., 2003). However, the limitations of 'co-management' are all too apparent based on the evidence from sites examined in this volume. Progressive modes of governance remain elusive in PAs whose legal framework often has no requirement for citizen involvement; co-management seeks to actively include members of civil society and the tourism sector in its decision-making, achieving the kind of collaborative strategy necessary to achieve a significant uptick in sustainable NBT.

At a regional level, collaborative PA management is loosely linked to regionalism and inter-governmental diplomacy, as typified by Southeast Asia. Instead of being constructed through collaborative, bottom-up and organic processes, ASEAN environmental regionalism has been implemented by member states in a top-down style (Elliott, 2009, 2011). Consequently, regional collaborative efforts remain at the planning stage. Cultural similarity of South Asia's countries directly affects the approach to environmental protection. Overall, establishment of PAs in South Asia has been broadly accepted on the grounds of biodiversity conservation, but South Asia's mountains are marginalised politically and economically. Due to the low level of economic development in the Himalayas, there is currently no possibility of introducing a comprehensive, rational and balanced approach to the natural environment in the region (Sachs, 2015). Despite the historical legacy and parallel challenges faced by mountainous NBT destinations, few enduring efforts could be found for meaningful trans-regional partnerships between Asian countries. In an era of renewed border tensions and geo-political chauvinism, peaceful solutions such as transboundary PAs are needed more urgently than ever to protect vulnerable biodiversity according to ecological catchment areas rather than arbitrarily-drawn political boundaries of more than one country or sub-national entity. Transboundary PAs are more commonly known as 'peace parks,' and various examples exist in Asia such as Taxkorgan on the border of Afghanistan, China, and Pakistan; Turtle Islands Wildlife Sanctuary between Malaysia and the Philippines; and the 'Heart of Borneo' which includes mountainous PAs of Brunei, Indonesia and Malaysia. The latter is especially relevant for its potential to connect existing PAs into a 220,000 km^2 network on Borneo island that could protect some of Asia's last surviving rainforests along with iconic species such as the Bornean rhinoceros and peaks such as Mt. Kinabalu (Persoon & Osseweijer, 2008).

14.5 Conclusion

14.5.1 A Transboundary Collection of Case Studies from Asia's Peaks and Parks

Having first compared the national networks of PAs, the destination-level analysis now enables us to draw several conclusions on the current issues facing mountainous NBT across Asia's PAs. These corroborate the regional synthesis and shape directions for future development presented in the earlier sections of this chapter as follows.

First, mountains are a shared space for natural, religious, socio-cultural and recreational values. Distinctive geological features include volcanic activity (e.g. Bromo) or lava tubes (e.g. Hallasan) that have contributed to religious value in local beliefs, turning many mountains into pilgrimage sites. Many of the peaks in Table 14.1 also represent the country's highest point (e.g. Fansipan, Hallasan, Yushan etc.). Adopted as patriotic symbols, most are located inside national parks or other PAs, designated at both national and international level, as in the UNESCO World Heritage list. Yet despite the 'universal' socio-cultural values of UNESCO listing, distinct ethnographic features of indigenous groups emerged that earn their livelihood and form unique cultural associations with the iconic peaks, that are increasingly used for marketing NBT. Local tourism operators play a key role in the NBT economy, for example as guides and porters (e.g. Sagarmatha, Fansipan, and Knuckles). But disputes over models of land use, cultivation and economic opportunities are also inherent in these marginal communities, not least because many still have permanent residents including indigenous communities living in and around the parks. Many are threatened by NBT development, including mega-projects such as roads (e.g. Natma Taung), cable cars (Fansipan, Fujisan), or a staircase (in the planning stage at Mt. Apo). Yet without stable access infrastructure, an unplanned increase in visitation is inevitable via four-wheel-drive SUVs and motorbikes that transport casual and inexperienced climbers up into mountain destinations inspired by Instagram views and non-technical ascents. All of our PA case studies thus face serious NBT management issues related to traffic, trail management, toilets, trash etc. Although most have some form of a cost recovery mechanism, such as an entry fee or donation system, the recreational value of these mountains is often underestimated as demonstrated quantitatively in Chap. 8.

We end this edited volume with a preliminary, four-stage life-cycle analysis of NBT development in different iconic mountain destinations across Asia's PAs and their associated dimensions:

Early-stage NBT development: emerging destination countries such as Myanmar, Sri Lanka and Timor-Leste consider mountains as potential NBT resources. However, due to socio-political unrest, less attention and resources have been invested in NBT development. Consequently, mountain tourism potential remains untapped due to a lack of facilities, incoherent policies and insufficient human and financial resources.

Table 14.1 Intra-chapter synopsis of case study peaks' current issues and counter strategies

Case study peak	Findings on the park's uniqueness, current issues and counter strategies
Huangshan (1,864 m) in Huangshan Scenic Area, China (**C2**) ⓥ	> 3 million annual visits, mainly domestic to the 'loveliest mountain of China'; cultural acclaim for art and literature (e.g. the Shanshui 'mountain and water' school of landscape painting); partnership model for NBT development; cable car system has seasonal pricing strategy to mitigate congestion
Fujisan (3,776 m) in Fuji-Hakone-Izu National Park, Japan (**C3**) ⓥ	300,000 summer climbers and 3–4 million annual visitors to Fuji-Yoshida Fifth station; iconic shape of solitary stratovolcano; modernized pilgrimage epitomized by 'bullet climbs'; all climbers requested to pay conservation donation (¥1000) with mandatory entry fee system also under negotiation
Hallasan (1,947 m) in Hallasan National Park, South Korea (**C4**) ⓥ	3 million annual visitors to WHS, listed as Jeju Volcanic Island and Lava Tubes; includes Geomunoreum system of caves; park managed by Province; island developed for leisure and recreational tourism attracting domestic and international (especially Chinese) visitors; moving toward mass tourism model
Yushan National Park inc. Mt. Jade (*Yushan*, 3,952 m) in Taiwan (**C5**)	32,000 climbers staying at Paiyun lodge in 2019; highest peak in north-east Asia and Taiwan's largest terrestrial national park; well-established regulation for carrying capacity control of trails; visitor management measure as hikers pre-register online for a permit allocated using a lottery system
Mt. Bromo (2,329 m) in Bromo Tengger Semeru National Park, Indonesia (**C6**) ⓥ	127,000 visits in 2016; popular destination for leisure and pilgrimage for domestic tourists; dual national park entrance fee system (internationals pay ten-fold more than domestic visitors); all fees raised in 2014; rapid development of facilities and unregulated tourism activities; active volcano (>50 eruptions since 1804) but claims of poor risk and disaster management
Mt. Ramelau (2,986 m) in Timor Leste (**C7**)	31,000 visits in 2018; sacred mountain in local belief & popular for domestic pilgrims; visitation spike during the pilgrimage season; early stage of NBT development with poor infrastructure, visitor management and safety measures
Mt. Apo (2,954 m) in Mount Apo Natural Park, the Philippines (**C8**)	around 3,500 climbers per year; nominated for WHS listing in 2009 but removed from the tentative list (2015); recommended to ease pressures from logging and poaching etc.; > 120,000 people live within park boundaries; findings suggest entry fee should be increased and Payment for Environmental Services proposed

(continued)

Table 14.1 (continued)

Case study peak	Findings on the park's uniqueness, current issues and counter strategies
Fansipan (3,147 m) in Hoàng Liên National Park, Vietnam (**C9**)	67,414 visitors to national park in 2018; cable car to the summit since 2016 (after official opening); opened Vietnam's highest peak for mass tourism development; cable car fee raised used to mitigate the visitor numbers and generate higher income for operators; reduced livelihood for local ethnic tourism operators
Mt. Natma Taung (3,053 m) in Natma Taung National Park, Myanmar (**C10**)	> 30,000 domestic and 3,762 international visitors to Chin State, of which 8,029 domestic and 769 international tourists summited Mt Natma Taung in 2019. On the WHS tentative list, but still in an early state of tourism development owing to lack of transportation and supporting facilities; provision of livelihood for the local Chin community; national park Zone Fee Entry 10,000 MMK (~10 USD)
Stok Kangri (6,153 m) in Hemis National Park, India (**C11**)	150,000 visitors to Ladakh ('Lesser Tibet') in 2011; this is the highest peak in the Stok Range of Zazkar Mountains; overcrowding, poor infrastructure, water pollution and trash results in trail closure to trekkers (2020–2023); checkpoints set up to restrict motorized access by tourists, who must pay an entrance fee
Sagarmatha (8,848 m) in Sagarmatha National Park, Nepal (**C12**) ⊙	57,289 visitors to the national park (2018–19); around 400 Everest climbers in 2019 input 4 million USD via climbing fees; niche adventure tourism embedded in unique local cultural and religious tradition with trekking trails but seasonality and overcrowding, skewed tourism benefits, sprawling development, lack of interpretation and conservation, and severe impacts of climate change
Knuckles Masif (1,906 m) in Knuckles Conservation Forest, Sri Lanka (**C13**) ⊙	80,000 visitors in 2019, mostly domestic; montane forest with large indigenous communities living in 37 villages; conflicts in sharing resources for tourism development and livelihood for indigenous communities; changes in land use and conservation owing to introduction of cash-crop planation and cultivation
	⊙ *denotes a UNESCO World Heritage Site*

Early-transitional-stage NBT development: a less breakneck pace of development is evident in PAs in India, Indonesia, Nepal and the Philippines, but the increase in visitation nonetheless outstrips efforts to institutionalise policies or measures to mitigate negative impacts, leading to criticism for pollution and other impacts on fragile mountain environment.

Mid-transitional-stage NBT development: rapid reduction in access time via distance demolishing technology (China, Vietnam) such as giant elevators and cable cars. The intervention of technology presents a deliberate ideology for development-at-all-cost, transforming NBT into mass tourism to maximise economic or political gain.

Established NBT development: where the PAs have institutionalised regulations, systems and infrastructure, such as Taiwan, Japan and South Korea. However, despite carrying capacity and climber permit systems pioneered at Yushan and other destinations, sudden spikes still undermine efforts to steer visitor management toward sustainability.

14.5.2 COVID-19 Regional Impacts and Recovery

In tourism, warnings of pandemics have been sounded for years (Gössling, 2002; Hall et al., 2020; Page & Yeoman, 2007). However, no one was prepared for the scale and speed of the novel coronavirus pandemic that began in Wuhan, China in 2019. The COVID-19 pandemic has caused an unprecedented disruption to both domestic and international travel, bringing destination and source markets worldwide to a standstill. Northeast Asia, the region that first recorded the out-break of the pandemic, has implemented strong and sweeping policies to address the pandemic, including closing borders, restricting international travel, closure of public facilities and popular tourist destinations. To date, these procedures appear to have slowed the spread of the virus. Unfortunately, the situation may be worse in developing countries struggling with other problems related to health care, such as India (Apollo, Wengel, et al., 2020; Sarkar et al., 2020). Wu et al. (2017) mentioned that, among other factors, the high-risk areas for the emergence and spread of infectious disease are cultural practices that increase contact between humans. Thus, pollution and a lack of hygiene can increase the risk of diseases (Niu & Xu, 2019) and this is a common problem in many parts of South Asia, with mountainous NBT destinations no exception. As of October 2020, countries in South Asia remain more or less closed (Lancet, 2020; Paul et al., 2020) while actions to restart tourism are primarily adopted in Asia and the Pacific. Direct fiscal and monetary support for tourism ranges from economic relief for tourism businesses, especially SMEs, to the introduction of financial instruments such as special lines of credit, new loan schemes, and investment programs (UNWTO, 2020a).

Most recent measures announced are moving forwards with initiatives to restart tourism and promote domestic demand. With domestic tourism as the priority, marketing and promotional campaigns, product development initiatives and special discounts have begun to emerge in a few countries. Among the world's biggest domestic tourism markets that had over 100 million domestic trips in 2018, seven of the 16 countries are from Asia with India, China and Japan near the top of the list (see Table 14.1) (UNWTO, 2020b). Domestic tourism has proven its resilience to recover faster and become the main source of growth for tourism industry while international borders remain close. The current crisis offers an opportunity for countries to re-evaluate their tourism policies to encourage domestic travel. Although international tourism often receives more attention due to its capacity to generate valuable export revenues, domestic tourism represents a much larger share of travellers and spending in many countries (UNWTO, 2020b) (Table 14.2).

Table 14.2 World's major domestic tourism markets 2018

	Tourist trip (arrivals)		Domestic visitor trip (million)	Domestic visitor		Population	Domestic tourism trips per capita
	Domestic (million)	Inbound (million)		Guests (million)	Nights (million)		
India	1,855	17.4	.	.	.	1,334	1.4
China	.	62.9	5,539	.	.	1,395	.
United States	1,659	79.7	2,291	.	.	327	5.1
Japan	291	31.2	561.8	317	406	126	2.3
Brazil	191	5.4	.	.	.	197	1
France	190	89.4	268	82	135	65	2.9
Spain	170	82.8	455	51	117	46	3.7
Russian Federation	.	24.6	.	48	136	146	.
South Korea	163	15.3	311	24	26	52	3.2
Germany	159	38.9	.	114	235	83	1.9
Indonesia	.	13.4	303	73	.	264	.
Thailand	131	38.2	228	107	.	68	1.9
United Kingdom	119	36.3	1,822	42	91	66	1.8
Australia	106	9.2	312	37	101	25	4.2
Malaysia	102	25.8	302	52	.	32	3.1
Mexico	100	41.3	.	62	109	125	0.8

Source Adapted from UNWTO (2020b)

Throughout this volume, authors address the importance of domestic markets to mountain NBT in Asia. Due to remote locations, limited facilities, and symbolic natural, social and religious meaning of mountains, whether domestic travellers are seeking short weekend breaks, relaxation or pilgrimage. With large volume, high propensity to pay, increasing health concerns, safety issues and social distance during the pandemic, along with the international travel restrictions, NBT in remote mountains has the potential to recover quickly and sustain growth momentum. Many scholars are hopeful that the current time provides an opportunity to reshape tourism into models that are more sustainable, inclusive, and caring of the many stakeholders that rely on it (Cheer, 2020; Lapointe, 2020; Nepal, 2020). Post-pandemic NBT, however, runs the risk of repeating the patterns of unsustainable, *en masse*, and volume-oriented approaches described in some of the case studies of this book. Therefore, the fundamental question remains unchanged: How to balance conservation and NBT development in the fragile mountainous PAs across Asia?

14.5.3 Limitations and Future Research

Our book spans multiple levels of analysis ranging from single-site case studies, to country-wide issues, towards intra-regional and inter-regional synthesis. However, there are certain limitations that open opportunities for future research. Due to the macro scale and trans-regional approach, the Northeast Asia section could not include some mountainous areas such as Mongolia and North Korea, and has barely scratched the surface when it comes to comparative research on mountainous PAs. Countries such as Japan, South Korea and Taiwan have much in common when it comes to resource management issues, and more co-research is required to break down barriers and pioneer innovative NBT products tailored to meet the needs of Asia's unique PA destinations. The chapters of the Southeast Asia region, moreover, do not include all countries that have been well-documented in tourism research. Mt. Kinabalu in Malaysian Borneo is one such example that has attracted considerable prior research interest (Jones & Bui, 2018) but was excluded from this research for logistical reasons. Thailand is also relevant for mountainous NBT research site to complement hill-tribe tourism focused on socio-cultural perspectives. Three other countries with less extensive research are Cambodia, Laos and Brunei, were also outside the scope of this book. Such gaps serve to highlight future research areas for mountainous NBT in Southeast Asia. The chapters on South Asia PAs have also had to exclude certain countries due to data deficiencies. Mountainous PAs of Pakistan, Bhutan and Iran should certainly be taken into account in future studies. All of them have a significant, largely untapped NBT potential, and on the other hand the unique character of their mountains, in particular their biodiversity including habitats for endemic species, constitutes a strong basis for action aimed at conservation.

References

Andreychouk, V. (2015). Cultural landscape functions. In M. Luc, U. Somorowska, & J. B. Szmańda (Eds.), *In landscape analysis and planning* (pp. 3–19). Springer.

Apollo, M. (2015). The clash–social, environmental and economical changes in tourism destination areas caused by tourism. The case of Himalayan villages (India and Nepal). *Current Issues of Tourism Research, 5*(1), 6–19.

Apollo, M. (2017a). The population of Himalayan regions—By the numbers: Past, present, and future. In R. Efe & M. Ozturk (Eds.), *Contemporary studies in environment and tourism* (pp. 145–160). Cambridge Scholars Publishing.

Apollo, M. (2017b). The true accessibility of mountaineering: The case of the high Himalaya. *Journal of Outdoor Recreation and Tourism, 17*, 29–43.

Apollo, M., & Andreychouk, V. (2020). Mountaineering and the natural environment in developing countries: An Insight to a comprehensive approach. *International Journal of Environmental Studies, 77*(6), 942–953.

Apollo, M., Andreychouk, V., Moolio, P., Wengel, Y., & Myga-Piątek, U. (2020). Does the altitude of habitat influence residents' attitudes to guests? A new dimension in the residents' attitudes to tourism. *Journal of Outdoor Recreation and Tourism, 31*, 100312.

Apollo, M., Wengel, Y., Schänzel, H., & Musa, G. (2020). Hinduism, ecological conservation, and public health: What are the health hazards for religious tourists at Hindu temples? *Religions, 11*(8), 416.

Ariza, C., Maselli, D., & Kohler, T. (2013). *Mountains: Our life, our future. progress and perspectives on sustainable mountain development from Rio 1992 to Rio 2012 and beyond.* Retrieved from Bern, Switzerland:

Barber, R. (1993). *Pilgrimages*. The Boydell Press.

Bhatt, S., & Pathak, J. (1992). Assessment of water quality and aspects of pollution in a stretch of River Gomti (Kumaun: Lesser Himalaya). *Journal of Environmental Biology, 13*(2), 113–126.

BKTC (Producer). (2016). Shri Badarinath—Shri Kedarnath Temples Committee, Pilgrims Statistics. Retrieved from www.badarikedar.org

Boomgaard, P. (2007). *Southeast Asia: An environmental history*. ABC Clio.

BTM. (2013). *Bhutan tourism monitor, annual report 2012*. The Tourism Council of Bhutan.

Buckley, R., Cater, C., Linsheng, Z., & Chen, T. (2008). Shengtai luyou: Cross-cultural comparison in ecotourism. *Annals of Tourism Research, 35*(4), 945–968.

Bui, H. T., & Dolezal, C. (2020). The tourism-development nexus in Southeast Asia: History and current issues. In C. Dolezal, A. Trupp, & H. T. Bui (Eds.), *Tourism and development in Southeast Asia* (pp. 23–40). Routledge.

Cheer, J. M. (2020). Human flourishing, tourism transformation and COVID-19: A conceptual touchstone. *Tourism Geographies, 22*(3), 1–11.

Choe, J. Y., & Hitchcock, M. (2018). Pilgrimage to Mount Bromo, Indonesia. In D. H. Olsen & A. Trono (Eds.), *Religious pilgrimage routes and trails: Sustainable development and management* (pp. 180–195). CABI.

Clark, N. E., Boakes, E. H., McGowan, P. J., Mace, G. M., & Fuller, R. A. (2013). Protected areas in South Asia have not prevented habitat loss: A study using historical models of land-use change. *PLoS ONE, 8*(5), e65298.

Corlett, R. (2005). Vegetation. In A. Gupta (Ed.), *The physical geography of Southeast Asia* (pp. 105–119). Oxford University Press.

DeWald, E. (2008). The development of tourism in French colonial Vietnam, 1918–1940. In J. Cochrane (Ed.), *Asian tourism: Growth and change* (pp. 221–231). Elsevier.

Elliott, L. (2009). Environmental challenges, policy failure and regional dynamics in Southeast Asia. In M. Beeson (Ed.), *Contemporary Southeast Asia* (pp. 248–265). Palgrave MacMillian.

Elliott, L. (2011). ASEAN and environmental governance: Rethinking networked regionalism in Southeast Asia. *Procedia-Social and Behavioral Sciences, 14*, 61–64.

Ewert, A. W. (1994). Playing the edge: Motivation and risk taking in a high-altitude wilderness like environment. *Environment and Behavior, 26*(1), 3–24.

Frost, W., Laing, J., & Beeton, S. (2014). The future of nature-based tourism in the Asia-Pacific region. *Journal of Travel Research, 53*(6), 721–732.

Gössling, S. (2002). Global environmental consequences of tourism. *Global Environmental Change, 12*(4), 283–302.

Gupta, A. (2005). The landforms of Southeast Asia. In A. Gupta (Ed.), *The physical geography of Southeast Asia* (pp. 38–64). Oxford University Press.

Hall, C. M., & Page, S. (Eds.). (2017). *The Routledge handbook of tourism in Asia*. Routledge.

Hall, C. M., Scott, D., & Gössling, S. (2020). Pandemics, transformations and tourism: Be careful what you wish for. *Tourism Geographies, 22*(3), 1–22.

Hanson, J. O., Rhodes, J. R., Butchart, S. H., Buchanan, G. M., Rondinini, C., Ficetola, G. F., & Fuller, R. A. (2020). Global conservation of species' niches. *Nature, 580*(7802), 232–234.

Hitchcock, M., & Jay, S. (1998). Eco-tourism and environmnetal change in Indonesia, Malaysia and Thailand. In V. King (Ed.), *Environmental challanges in Southeast Asia* (pp. 305–316). Cuzon.

Hutchison, C. S. (2005). The geologial framework. In A. Gupta (Ed.), *The physical geography of Southeast Asia* (pp. 3–23). Oxford University Press.

IMF. (2020). *Report for selected countries and subjects*. Retrieved September. 29 2020 https://www.imf.org/external/pubs/ft/weo/2015/01/weodata/weorept.aspx?pr.x=88&pr.y=15&sy=2015&ey=

2015&scsm=1&ssd=1&sort=country&ds=.&br=1&c=512,556,513,514,558,564,524,534&s= NGDPD,PPPGDP&grp=0&a=.

Inagaki, T. (2008). Hill stations in Asia: A discovery of scenery and environmental change. *Global Environmental Research, 12*, 93–99.

Ives, J. D. (2006). *Himalayan perceptions. Environmental change and the Well-being of mountain peoples*. Routledge.

Jones, T. E. (2016). The role of the Shin Nihon Hakkei in redrawing Japanese attitudes to landscape. In *Environment, modernization and development in East Asia* (pp. 139–156). Palgrave Macmillan.

Jones, T. E., Beeton, S., & Cooper, M. (2018). World heritage listing as a catalyst for collaboration: Can Mount Fuji's trail signs point the way for Japan's multi-purpose national parks? *Journal of Ecotourism, 17*(3), 220–238.

Jones, T. E., & Bui, H. T. (2018). Comparing international and domestic climber profiles, motivation and the influence of World Heritage Site Status at Mount Fuji and Mount Kinabalu. *Journal of Environmental Information Science, 2018*(1), 67–72.

Joppa, L. N., & Pfaff, A. (2011). Global protected area impacts. *Proceedings of the Royal Society B: Biological Sciences, 278*(1712), 1633–1638.

Kano, H. (2008). *Indonesian exports, peasant agriculture and the world economy, 1850-2000: Economic structures in a Southeast Asian State* (p. 282). NUS Press.

Kato, M. (2008). *National Park system of Japan, National Park & protected area management series (Vol. III)*. Kokon Shoin.

Kellert, S. R. (1995). Concepts of nature east and west. In M. Soule & G. Lease (Eds.), *Reinventing Nature? Responses to postmodern deconstrcution* (pp. 103–212). Island Press.

Kennedy, D. (1996). *Magic mountains. Hill stations and the British Raj*. University of California Press.

Knight, J. (2000). From timber to tourism: Recommoditizing the Japanese forest. *Development and Change, 31*(1), 341–359.

Körner, C., & Spehn, E. (2019). *A Humboldtian view of mountains*. American Association for the Advancement of Science.

Kreutzmann, H. (2000). *Improving accessibility for mountain development. Role of transport networks and urban settlements*. Paper presented at the Growth, Poverty Alleviation and Sustainable Resource Management in the Mountain Areas of South Asia, Kathmandu.

Kupper, P. (2009). Science and the national parks: A transatlantic perspective on the interwar years. *Environmental History, 14*(1), 58–81.

Lancet, E. (2020). *COVID-19 in India: The dangers of false optimism*. Elsevier.

Lapointe, D. (2020). Reconnecting tourism after COVID-19: The paradox of alterity in tourism areas. *Tourism Geographies, 22*(3), 1–6.

Lee, Y.-S., Lawton, L. J., & Weaver, D. B. (2013). Evidence for a South Korean model of ecotourism. *Journal of Travel Research, 52*(4), 520–533.

Liu, Y., Yang, R., & Li, Y. (2013). Potential of land consolidation of hollowed villages under different urbanization scenarios in China. *Journal of Geographical Sciences, 23*(3), 503–512.

MacKinnon, J., & Yan, X. (2008). *Regional action plan for the protected areas of East Asia* (p. 82). IUCN.

Mallari, N. A. D., Collar, N. J., McGowan, P. J. K., & Marsden, S. J. (2016). Philippine protected areas are not meeting the biodiversity coverage and management effectiveness requirements of Aichi Target 11. *Ambio, 45*, 313–322.

Matanle, P., & Rausch, A. S. (Eds.). (2011). *Japan's shrinking regions in the 21st century: Contemporary responses to depopulation and socioeconomic decline*. Cambria Press.

Messerli, B., & Ives, J. D. (1997). *Mountains of the world: A global priority*. Parthenon Publishing Group.

Mizuuchi, Y., Awano, T., & Furuya, K. (2016). A study on planning idea of Tsuyoshi Tamura and Keiji Uehara based on Mt. Kongo National Park Plan. *Journal of The Japanese Institute of Landscape Architecture, 79*(5), 431–436.

Mondal, P., & Nagendra, H. (2011). Trends of forest dynamics in tiger landscapes across Asia. *Environmental Management, 48*(4), 781.
Murakushi, N. (2005). *The history of the establishment of the National Parks: The feud between development and nature conservation*. Hosei University.
Musa, G., Higham, J., & Thompson-Carr, A. (2015). Mountaineering tourism: Looking to the Horizon. In G. Musa, J. Higham, & A. Thompson-Carr (Eds.), *Mountaineering tourism* (pp. 328–348). Routledge.
Nepal, S. K. (2020). Travel and tourism after COVID-19–business as usual or opportunity to reset? *Tourism Geographies, 22*(3), 1–5.
Niu, S., & Xu, M. (2019). Impact of Hajj on global health security. *Journal of Religion and Health, 58*(1), 289–302.
NTS. (2013). *Nepal tourism statistics 2012. Ministry of culture, tourism & civil aviation, planning & evaluation division, statistical section*. Singha Durbar.
Page, S., & Yeoman, I. (2007). How VisitScotland prepared for a flu pandemic. *Journal of Business Continuity & Emergency Planning, 1*(2), 167–182.
Paul, A., Chatterjee, S., & Bairagi, N. (2020). Prediction on Covid-19 epidemic for different countries: Focusing on South Asia under various precautionary measures. *medRxiv*.
Persoon, G., Van Est, D. M., & Sajise, P. E. (Eds.). (2003). Co-management of natural resources in Asia: A comparative perspective (No. 7). Nias Press.
Persoon, G. A., & Osseweijer, M. (Eds.). (2008). Reflections on the Heart of Borneo (Vol. 24). Tropenbos International.
Peters, M. K., Hemp, A., Appelhans, T., Becker, J. N., Behler, C., Classen, A., & Frederiksen, S. B. (2019). Climate–land-use interactions shape tropical mountain biodiversity and ecosystem functions. *Nature, 568*(7750), 88–92.
Quigley, J. M. (2009). *Urbanization, agglomeration, and economic development*: Worldbank Group.
Razal, R. A., Karki, M., Sanchez, B., Aksha, S., & Mahat, T. J. (2012). *Sustainable mountain development 1992-2012 and beyond*. Retrieved from
Rodrigues, A. S. L., Akcakaya, H. R., Andelman, S. J., Bakarr, M. I., Boitani, L., Brooks, T. M., Chanson, J. S., Fishpool, L. D. C., da Fonseca, G. A. B., Gaston, K. J., Hoffmann, M., Marquet, P. A., Pilgrim, J. D., Pressey, R. L., Schipper, J., Sechrest, W., Stuart, S. N., Underhill, L. G., Waller, R. W., Watts, M. E. J., & Yan, X. (2004). Global gap analysis: priority regions for expanding the global protected-area network. BioScience, 54(12), 1092–1100.
Sachs, J. (2015). *The age of sustainable development*. Columbia University Press.
Sarkar, K., Khajanchi, S., & Nieto, J. J. (2020). Modeling and forecasting the COVID-19 pandemic in India. *Chaos, Solitons & Fractals, 139*, 110049.
Sati, V. P. (2014). *Towards Sustainable livelihoods and ecosystems in mountain regions*. Springer.
Sati, V. P. (2015). Pilgrimage tourism in mountain regions: Socio–economic and environmental implications in the Garhwal Himalaya. *South Asian Journal of Tourism and Heritage, 8*(2), 164–182.
Singh, S. (1985). An overview of the conservation status of the national parks and protected areas of the Indomalayan realm. In J. W. Thorsell (Ed.), *Conserving Asia's natural heritage: The planning and management of protected areas in the Indomalayan realm* (pp. 1–5). IUCN.
Taylor, S. M. (2007). What if religions had ecologies? The case for reinhabiting religious studies. *Journal for the Study of Religion, Nature & Culture, 1*(1).
Timothy, D. J., & Nyaupane, G. P. (Eds.). (2009). *Cultural heritage and tourism in the developing world: A regional perspective*. Routledge.
Tsunoda, H., & Enari, H. (2020). *A strategy for wildlife management in depopulating rural areas of Japan*. In press.
Uhlig, H., & Kreutzmann, H. (1995). Persistence and change in high mountain agricultural systems. *Mountain research and development, 15*(3), 199–212.
UN. (2020). *Department of economic and social affair*. Retrieved 23 September 2020. https://www.un.org/development/desa/en/news/population/world-population-prospects-2019.html.
UNTWO. (2019). *International tourism highlights* (2019th ed.). World Tourism Organisation.

UNWTO. (2020a). *Briefing note—Tourism and COVID-19, Issue 1—How are countries supporting tourism recovery*. UNWTO (World Tourism Organization).

UNWTO. (2020b). *UNWTO Briefing note—Tourism and COVID-19, Issue 3. Understanding domestic tourism and seizing its opportunities.* UNWTO (World Tourism Organisation).

Warren, C. (2002). *Managing Scotland's environment.* Edinburgh University Press.

Watling, T. (2009). *Ecological imaginations in the world religions: An ethnographic analysis.* A&C Black.

Weaver, D., Tang, C., & Zhao, Y. (2020). Facilitating sustainable tourism by endogenization: China as exemplar. *Annals of Tourism Research, 81*, 102890.

Wengel, Y. (2020). The micro-trends of emerging adventure tourism activities in Nepal. *Journal of Tourism Futures.*

Worldometers. (2020). East Asia Trends. Retrieved September 23, 2020. https://www.worldometers.info/population/asia/eastern-asia/.

Wu, T., Perrings, C., Kinzig, A., Collins, J. P., Minteer, B. A., & Daszak, P. (2017). Economic growth, urbanization, globalization, and the risks of emerging infectious diseases in China: a review. *Ambio, 46*(1), 18–29.

Yang, R., Cao, Y., Hou, S., Peng, Q., Wang, X., Wang, F., Convery, I., Zhao, Z., Shen, X., Li, S., Zheng, Y., Liu, H., Gong, P., Ma, K. (2020). Cost-effective priorities for the expansion of global terrestrial protected areas: Setting post-2020 global and national targets. *Science Advances, 6*(37), eabc3436.

Yang, Y., Xue, L., & Jones, T. E. (2019). Tourism-enhancing effect of World Heritage Sites: Panacea or placebo? A meta-analysis. *Annals of Tourism Research, 75*, 29–41.

Zurick, D., & Pacheco, J. (2006). *Illustrated atlas of the Himalaya.* The University Press of Kentucky.

Huong T. Bui is Professor of Tourism and Hospitality cluster, the College of Asia Pacific Studies, Ritsumeikan Asia Pacific University (APU), Japan. Her research interests are Heritage Conservation; War and Disaster-related Tourism, Sustainability and Resilience of the Tourism Sector.

Thomas E. Jones is Associate Professor in the Environment & Development Cluster at Ritsumeikan APU in Kyushu, Japan. His research interests include Nature-Based Tourism, Protected Area Management and Sustainability. Tom completed his PhD at the University of Tokyo and has conducted visitor surveys on Mount Fuji and in the Japan Alps.

Michal Apollo is an Assistant Professor at the Pedagogical University of Krakow, Institute of Geography, Department of Tourism and Regional Studies, and a Fellow of Yale University's Global Justice Program, New Haven, USA. Michal's areas of expertise are tourism management, consumer behaviours as well as environmental and socio-economical issues. Currently he is working on a concept of sustainable use of environmental and human resources. He is also an enthusiastic, traveller, diver, mountaineer, ultra-runner, photographer, and science populariser.

Printed by Printforce, the Netherlands